Francisco Javier Frías Orrieta

Graduate Texts in Mathematics

11

Graduate Texts in Mathematics

continued after Index

John B. Conway

Functions of
One Complex Variable

Second Edition

With 30 Illustrations

Springer-Verlag
New York   Berlin   Heidelberg   Tokyo

John B. Conway
Department of Mathematics
Indiana University
Bloomington, IN 47405
USA

AMS Subject Classification: 30–01

Library of Congress Cataloging in Publication Data

Conway, John B
    Functions of one complex variable.

    (Graduate texts in mathematics ; 11)
    Bibliography: p.
    Includes index.
    1. Functions of complex variables.   I.   Title.
II.   Series.
QA331.C659   1978   515′.93   78-18836

Printed and bound by R. R. Donnelley & Sons, Harrisonburg, Virginia.
Printed in the United States of America.

9 8 7 6 5 4 3

Third printing, 1984

ISBN 0-387-90328-3   Springer-Verlag   New York   Berlin   Heidelberg   Tokyo
ISBN 3-540-90328-3   Springer-Verlag   Berlin   Heidelberg   New York   Tokyo

To
Ann

PREFACE

This book is intended as a textbook for a first course in the theory of functions of one complex variable for students who are mathematically mature enough to understand and execute $\epsilon - \delta$ arguments. The actual prerequisites for reading this book are quite minimal; not much more than a stiff course in basic calculus and a few facts about partial derivatives. The topics from advanced calculus that are used (e.g., Leibniz's rule for differentiating under the integral sign) are proved in detail.

Complex Variables is a subject which has something for all mathematicians. In addition to having applications to other parts of analysis, it can rightly claim to be an ancestor of many areas of mathematics (e.g., homotopy theory, manifolds). This view of Complex Analysis as "An Introduction to Mathematics" has influenced the writing and selection of subject matter for this book. The other guiding principle followed is that all definitions, theorems, etc. should be clearly and precisely stated. Proofs are given with the student in mind. Most are presented in detail and when this is not the case the reader is told precisely what is missing and asked to fill in the gap as an exercise. The exercises are varied in their degree of difficulty. Some are meant to fix the ideas of the section in the reader's mind and some extend the theory or give applications to other parts of mathematics. (Occasionally, terminology is used in an exercise which is not defined—e.g., group, integral domain.)

Chapters I through V and Sections VI.1 and VI.2 are basic. It is possible to cover this material in a single semester only if a number of proofs are omitted. Except for the material at the beginning of Section VI.3 on convex functions, the rest of the book is independent of VI.3 and VI.4.

Chapter VII initiates the student in the consideration of functions as points in a metric space. The results of the first three sections of this chapter are used repeatedly in the remainder of the book. Sections four and five need no defense; moreover, the Weierstrass Factorization Theorem is necessary for Chapter XI. Section six is an application of the factorization theorem. The last two sections of Chapter VII are not needed in the rest of the book although they are a part of classical mathematics which no one should completely disregard.

The remaining chapters are independent topics and may be covered in any order desired.

Runge's Theorem is the inspiration for much of the theory of Function Algebras. The proof presented in section VIII.1 is, however, the classical one involving "pole pushing". Section two applies Runge's Theorem to obtain a more general form of Cauchy's Theorem. The main results of sections three and four should be read by everyone, even if the proofs are not.

Chapter IX studies analytic continuation and introduces the reader to analytic manifolds and covering spaces. Sections one through three can be considered as a unit and will give the reader a knowledge of analytic continuation without necessitating his going through all of Chapter IX.

Chapter X studies harmonic functions including a solution of the Dirichlet Problem and the introduction of Green's Function. If this can be called applied mathematics it is part of applied mathematics that everyone should know.

Although they are independent, the last two chapters could have been combined into one entitled "Entire Functions". However, it is felt that Hadamard's Factorization Theorem and the Great Theorem of Picard are sufficiently different that each merits its own chapter. Also, neither result depends upon the other.

With regard to Picard's Theorem it should be mentioned that another proof is available. The proof presented here uses only elementary arguments while the proof found in most other books uses the modular function.

There are other topics that could have been covered. Some consideration was given to including chapters on some or all of the following: conformal mapping, functions on the disk, elliptic functions, applications of Hilbert space methods to complex functions. But the line had to be drawn somewhere and these topics were the victims. For those readers who would like to explore this material or to further investigate the topics covered in this book, the bibliography contains a number of appropriate entries.

Most of the notation used is standard. The word "iff" is used in place of the phrase "if and only if", and the symbol ■ is used to indicate the end of a proof. When a function (other than a path) is being discussed, Latin letters are used for the domain and Greek letters are used for the range.

This book evolved from classes taught at Indiana University. I would like to thank the Department of Mathematics for making its resources available to me during its preparation. I would especially like to thank the students in my classes; it was actually their reaction to my course in Complex Variables that made me decide to take the plunge and write a book. Particular thanks should go to Marsha Meredith for pointing out several mistakes in an early draft, to Stephen Berman for gathering the material for several exercises on algebra, and to Larry Curnutt for assisting me with the final corrections of the manuscript. I must also thank Ceil Sheehan for typing the final draft of the manuscript under unusual circumstances.

Finally, I must thank my wife to whom this book is dedicated. Her encouragement was the most valuable assistance I received.

John B. Conway

PREFACE FOR THE SECOND EDITION

I have been very pleased with the success of my book. When it was apparent that the second printing was nearly sold out, Springer-Verlag asked me to prepare a list of corrections for a third printing. When I mentioned that I had some ideas for more substantial revisions, they reacted with characteristic enthusiasm.

There are four major differences between the present edition and its predecessor. First, John Dixon's treatment of Cauchy's Theorem has been included. This has the advantage of providing a quick proof of the theorem in its full generality. Nevertheless, I have a strong attachment to the homotopic version that appeared in the first edition and have proved this form of Cauchy's Theorem as it was done there. This version is very geometric and quite easy to apply. Moreover, the notion of homotopy is needed for the later treatment of the monodromy theorem; hence, inclusion of this version yields benefits far in excess of the time needed to discuss it.

Second, the proof of Runge's Theorem is new. The present proof is due to Sandy Grabiner and does not use "pole pushing". In a sense the "pole pushing" is buried in the concept of uniform approximation and some ideas from Banach algebras. Nevertheless, it should be emphasized that the proof is entirely elementary in that it relies only on the material presented in this text.

Next, an Appendix B has been added. This appendix contains some bibliographical material and a guide for further reading.

Finally, several additional exercises have been added.

There are also minor changes that have been made. Several colleagues in the mathematical community have helped me greatly by providing constructive criticism and pointing out typographical errors. I wish to thank publicly Earl Berkson, Louis Brickman, James Deddens, Gerard Keough, G. K. Kristiansen, Andrew Lenard, John Mairhuber, Donald C. Meyers, Jeffrey Nunemacher, Robert Olin, Donald Perlis, John Plaster, Hans Sagan, Glenn Schober, David Stegenga, Richard Varga, James P. Williams, and Max Zorn.

Finally, I wish to thank the staff at Springer-Verlag New York not only for their treatment of my book, but also for the publication of so many fine books on mathematics. In the present time of shrinking graduate enrollments and the consequent reluctance of so many publishers to print advanced texts and monographs, Springer-Verlag is making a contribution to our discipline by increasing its efforts to disseminate the recent developments in mathematics.

John B. Conway

TABLE OF CONTENTS

XII.   The Range of an Analytic Function

Chapter I

The Complex Number System

§1. The real numbers

We denote the set of all real numbers by $\mathbb{R}$. It is assumed that each reader is acquainted with the real number system and all its properties. In particular we assume a knowledge of the ordering of $\mathbb{R}$, the definitions and properties of the supremum and infimum (sup and inf), and the completeness of $\mathbb{R}$ (every set in $\mathbb{R}$ which is bounded above has a supremum). It is also assumed that every reader is familiar with sequential convergence in $\mathbb{R}$ and with infinite series. Finally, no one should undertake a study of Complex Variables unless he has a thorough grounding in functions of one real variable. Although it has been traditional to study functions of several real variables before studying analytic function theory, this is not an essential prerequisite for this book. There will not be any occasion when the deep results of this area are needed.

§2. The field of complex numbers

We define $\mathbb{C}$, the complex numbers, to be the set of all ordered pairs (a, b) where a and b are real numbers and where addition and multiplication are defined by:

$$(a, b)+(c, d) = (a+c, b+d)$$

$$(a, b)\,(c, d) = (ac-bd, bc+ad)$$

It is easily checked that with these definitions $\mathbb{C}$ satisfies all the axioms for a field. That is, $\mathbb{C}$ satisfies the associative, commutative and distributive laws for addition and multiplication; $(0,0)$ and $(1,0)$ are identities for addition and multiplication respectively, and there are additive and multiplicative inverses for each nonzero element in $\mathbb{C}$.

We will write a for the complex number $(a, 0)$. This mapping $a \to (a, 0)$ defines a field isomorphism of $\mathbb{R}$ into $\mathbb{C}$ so we may consider $\mathbb{R}$ as a subset of $\mathbb{C}$. If we put $i = (0, 1)$ then $(a, b) = a+bi$. From this point on we abandon the ordered pair notation for complex numbers.

Note that $i^2 = -1$, so that the equation $z^2+1 = 0$ has a root in $\mathbb{C}$. In fact, for each z in $\mathbb{C}$, $z^2+1 = (z+i)\,(z-i)$. More generally, if z and w are complex numbers we obtain

$$z^2+w^2 = (z+iw)\,(z-iw)$$

1

By letting z and w be real numbers a and b we can obtain (with both a and $b \neq 0$)

$$\frac{1}{a+ib} = \frac{a-ib}{a^2+b^2} = \frac{a}{a^2+b^2} - i\left(\frac{b}{a^2+b^2}\right)$$

so that we have a formula for the reciprocal of a complex number.

When we write $z = a+ib$ $(a, b \in \mathbb{R})$ we call a and b the *real* and *imaginary parts* of z and denote this by $a = \operatorname{Re} z$, $b = \operatorname{Im} z$.

We conclude this section by introducing two operations on $\mathbb{C}$ which are not field operations. If $z = x+iy(x, y \in \mathbb{R})$ then we define $|z| = (x^2+y^2)^{\frac{1}{2}}$ to be the *absolute value* of z and $\bar{z} = x-iy$ is the *conjugate* of z. Note that

2.1 $$|z|^2 = z\bar{z}$$

In particular, if $z \neq 0$ then

$$\frac{1}{z} = \frac{\bar{z}}{|z|^2}$$

The following are basic properties of absolute values and conjugates whose verifications are left to the reader.

2.2 $$\operatorname{Re} z = \tfrac{1}{2}(z+\bar{z}) \quad \text{and} \quad \operatorname{Im} z = \frac{1}{2i}(z-\bar{z}).$$

2.3 $$(\overline{z+w}) = \bar{z}+\bar{w} \quad \text{and} \quad \overline{zw} = \bar{z}\bar{w}.$$

2.4 $$|zw| = |z|\,|w|.$$

2.5 $$|z/w| = |z|/|w|.$$

2.6 $$|\bar{z}| = |z|.$$

The reader should try to avoid expanding z and w into their real and imaginary parts when he tries to prove these last three. Rather, use (2.1), (2.2), and (2.3).

Exercises

1. Find the real and imaginary parts of each of the following:

$$\frac{1}{z}; \; \frac{z-a}{z+a} \,(a \in \mathbb{R}); \; z^3; \; \frac{3+5i}{7i+1}; \; \left(\frac{-1+i\sqrt{3}}{2}\right)^3;$$

$$\left(\frac{-1-i\sqrt{3}}{2}\right)^6; \; i^n; \; \left(\frac{1+i}{\sqrt{2}}\right)^n \quad \text{for} \quad 2 \leq n \leq 8.$$

2. Find the absolute value and conjugate of each of the following:

$$-2+i; \; -3; \; (2+i)(4+3i); \; \frac{3-i}{\sqrt{2}+3i}; \; \frac{i}{i+3};$$

$$(1+i)^6; \; i^{17}.$$

3. Show that z is a real number if and only if $z = \bar{z}$.

4. If z and w are complex numbers, prove the following equations:

$$|z+w|^2 = |z|^2 + 2\operatorname{Re} z\bar{w} + |w|^2.$$

$$|z-w|^2 = |z|^2 - 2\operatorname{Re} z\bar{w} + |w|^2.$$

$$|z+w|^2 + |z-w|^2 = 2(|z|^2 + |w|^2).$$

5. Use induction to prove that for $z = z_1 + \ldots + z_n$; $w = w_1 w_2 \ldots w_n$: $|w| = |w_1| \ldots |w_n|$; $\bar{z} = \bar{z}_1 + \ldots + \bar{z}_n$; $\bar{w} = \bar{w}_1 \ldots \bar{w}_n$.

6. Let $R(z)$ be a rational function of z. Show that $\overline{R(z)} = R(\bar{z})$ if all the coefficients in $R(z)$ are real.

§3. The complex plane

From the definition of complex numbers it is clear that each z in $\mathbb{C}$ can be identified with the unique point $(\operatorname{Re} z, \operatorname{Im} z)$ in the plane $\mathbb{R}^2$. The addition of complex numbers is exactly the addition law of the vector space $\mathbb{R}^2$. If z and w are in $\mathbb{C}$ then draw the straight lines from z and w to $0 \ (=(0, 0))$. These form two sides of a parallelogram with 0, z and w as three vertices. The fourth vertex turns out to be $z+w$.

Note also that $|z-w|$ is exactly the distance between z and w. With this in mind the last equation of Exercise 4 in the preceding section states the *parallelogram law*: The sum of the squares of the lengths of the sides of a parallelogram equals the sum of the squares of the lengths of its diagonals.

A fundamental property of a distance function is that it satisfies the triangle inequality (see the next chapter). In this case this inequality becomes

$$|z_1 - z_2| \le |z_1 - z_3| + |z_3 - z_2|$$

for complex numbers z_1, z_2, z_3. By using $z_1 - z_2 = (z_1 - z_3) + (z_3 - z_2)$, it is easy to see that we need only show

3.1 $$|z+w| \le |z| + |w| \ (z, w \in \mathbb{C}).$$

To show this first observe that for any z in $\mathbb{C}$,

3.2 $$-|z| \le \operatorname{Re} z \le |z|$$

$$-|z| \le \operatorname{Im} z \le |z|$$

Hence, $\operatorname{Re}(z\bar{w}) \le |z\bar{w}| = |z| \, |w|$. Thus,

$$|z+w|^2 = |z|^2 + 2\operatorname{Re}(z\bar{w}) + |w|^2$$

$$\le |z|^2 + 2|z| \, |w| + |w|^2$$

$$= (|z| + |w|)^2,$$

from which (3.1) follows. (This is called the *triangle inequality* because, if we represent z and w in the plane, (3.1) says that the length of one side of the triangle $[0, z, z+w]$ is less than the sum of the lengths of the other two sides. Or, the shortest distance between two points is a straight line.) On encounter-

ing an inequality one should always ask for necessary and sufficient conditions
that equality obtains. From looking at a triangle and considering the geometrical significance of (3.1) we are led to consider the condition $z = tw$
for some $t \in \mathbb{R}$, $t \geq 0$. (or $w = tz$ if $w = 0$). It is clear that equality will
occur when the two points are colinear with the origin. In fact, if we look
at the proof of (3.1) we see that a necessary and sufficient condition for
$|z+w| = |z|+|w|$ is that $|z\bar{w}| = \text{Re}\,(z\bar{w})$. Equivalently, this is $z\bar{w} \geq 0$ (i.e., $z\bar{w}$
is a real number and is non negative). Multiplying this by w/w we get
$|w|^2(z/w) \geq 0$ if $w \neq 0$. If

$$t = z/w = \left(\frac{1}{|w|^2}\right) |w|^2(z/w)$$

then $t \geq 0$ and $z = tw$.

By induction we also get

3.3 $$|z_1+z_2+\ldots+z_n| \leq |z_1|+|z_2|+\ldots+|z_n|$$

Also useful is the inequality

3.4 $$\Big| |z|-|w| \Big| \leq |z-w|$$

Now that we have given a geometric interpretation of the absolute value
let us see what taking a complex conjugate does to a point in the plane.
This is also easy; in fact, $\bar{z}$ is the point obtained by reflecting z across the
x-axis (i.e., the real axis).

Exercises

1. Prove (3.4) and give necessary and sufficient conditions for equality.
2. Show that equality occurs in (3.3) if and only if $z_k/z_l \geq 0$ for any integers
k and l, $1 \leq k, l \leq n$, for which $z_l \neq 0$.
3. Let $a \in \mathbb{R}$ and $c > 0$ be fixed. Describe the set of points z satisfying

$$|z-a|-|z+a| = 2c$$

for every possible choice of a and c. Now let a be any complex number
and, using a rotation of the plane, describe the locus of points satisfying the
above equation.

§4. Polar representation and roots of complex numbers

Consider the point $z = x+iy$ in the complex plane $\mathbb{C}$. This point has
polar coordinates (r, θ): $x = r \cos \theta$, $y = r \sin \theta$. Clearly $r = |z|$ and θ is
the angle between the positive real axis and the line segment from 0 to z.
Notice that θ plus any multiple of 2π can be substituted for θ in the above
equations. The angle θ is called the *argument of z* and is denoted by $\theta = \arg z$.
Because of the ambiguity of θ, "arg" is not a function. We introduce the
notation

4.1 $$\text{cis } \theta = \cos \theta + i \sin \theta.$$

Let $z_1 = r_1 \text{ cis } \theta_1$, $z_2 = r_2 \text{ cis } \theta_2$. Then $z_1 z_2 = r_1 r_2 \text{ cis } \theta_1 \text{ cis } \theta_2 = r_1 r_2$ $[(\cos \theta_1 \cos \theta_2 - \sin \theta_1 \sin \theta_2) + i (\sin \theta_1 \cos \theta_2 + \sin \theta_2 \cos \theta_1)]$. By the formulas for the sine and cosine of the sum of two angles we get

4.2 $$z_1 z_2 = r_1 r_2 \text{ cis } (\theta_1 + \theta_2)$$

Alternately, $\arg (z_1 z_2) = \arg z_1 + \arg z_2$. (What function of a real variable takes products into sums?) By induction we get for $z_k = r_k \text{ cis } \theta_k$, $1 \le k \le n$.

4.3 $$z_1 z_2 \ldots z_n = r_1 r_2 \ldots r_n \text{ cis } (\theta_1 + \ldots + \theta_n)$$

In particular,

4.4 $$z^n = r^n \text{ cis } (n\theta),$$

for every integer $n \ge 0$. Moreover if $z \ne 0$, $z \cdot [r^{-1} \text{ cis } (-\theta)] = 1$; so that (4.4) also holds for all integers n, positive, negative, and zero, if $z \ne 0$. As a special case of (4.4) we get *de Moivre's formula*:

$$(\cos \theta + i \sin \theta)^n = \cos n\theta + i \sin n\theta.$$

We are now in a position to consider the following problem: For a given complex number $a \ne 0$ and an integer $n \ge 2$, can you find a number z satisfying $z^n = a$? How many such z can you find? In light of (4.4) the solution is easy. Let $a = |a| \text{ cis } \alpha$; by (4.4), $z = |a|^{1/n} \text{ cis } (\alpha/n)$ fills the bill. However this is not the only solution because $z' = |a|^{1/n} \text{ cis } \dfrac{1}{n} (\alpha + 2\pi)$ also satisfies $(z')^n = a$. In fact each of the numbers

4.5 $$|a|^{1/n} \text{ cis } \frac{1}{n} (\alpha + 2\pi k), \; 0 \le k \le n-1,$$

in an nth root of a. By means of (4.4) we arrive at the following: for each non zero number a in $\mathbb{C}$ there are n *distinct* nth roots of a; they are given by formula (4.5).

Example

Calculate the nth roots of unity. Since $1 = \text{ cis } 0$, (4.5) gives these roots as

$$1, \text{ cis } \frac{2\pi}{n}, \text{ cis } \frac{4\pi}{n}, \ldots, \text{ cis } \frac{2\pi}{n} (n-1).$$

In particular, the cube roots of unity are

$$1, \frac{1}{\sqrt{2}} (1 + i\sqrt{3}), \frac{1}{\sqrt{2}} (1 - i\sqrt{3}).$$

Exercises

1. Find the sixth roots of unity.

2. Calculate the following:
   - (a) the square roots of i
   - (b) the cube roots of i
   - (c) the square roots of $\sqrt{3} + 3i$

3. A *primitive* nth *root of unity* is a complex number a such that $1, a, a^2, \ldots, a^{n-1}$ are distinct nth roots of unity. Show that if a and b are primitive nth and mth roots of unity, respectively, then ab is a kth root of unity for some integer k. What is the smallest value of k? What can be said if a and b are nonprimitive roots of unity?

4. Use the binomial equation

$$(a+b)^n = \sum_{k=0}^{n} \binom{n}{k} a^{n-k}b^k,$$

where

$$\binom{n}{k} = \frac{n!}{k!(n-k)!},$$

and compare the real and imaginary parts of each side of de Moivre's formula to obtain the formulas:

$$\cos n\theta = \cos^n \theta - \binom{n}{2} \cos^{n-2} \theta \sin^2 \theta + \binom{n}{4} \cos^{n-4} \theta \sin^4 \theta - \ldots$$

$$\sin n\theta = \binom{n}{1} \cos^{n-1} \theta \sin \theta - \binom{n}{3} \cos^{n-3} \theta \sin^3 \theta + \ldots$$

5. Let $z = \text{cis} \dfrac{2\pi}{n}$ for an integer $n \geq 2$. Show that $1 + z + \ldots + z^{n-1} = 0$.

6. Show that $\varphi(t) = \text{cis } t$ is a group homomorphism of the additive group $\mathbb{R}$ onto the multiplicative group $T = \{z : |z| = 1\}$.

7. If $z \in \mathbb{C}$ and $\text{Re}(z^n) \geq 0$ for every positive integer n, show that z is a non-negative real number.

§5. Lines and half planes in the complex plane

Let L denote a straight line in $\mathbb{C}$. From elementary analytic geometry, L is determined by a point in L and a direction vector. Thus if a is any point in L and b is its direction vector then

$$L = \{z = a + tb : -\infty < t < \infty\}.$$

Since $b \neq 0$ this gives, for z in L,

$$\text{Im}\left(\frac{z-a}{b}\right) = 0.$$

In fact if z is such that

$$0 = \text{Im}\left(\frac{z-a}{b}\right)$$

then

$$t = \left(\frac{z-a}{b}\right)$$

implies that $z = a + tb$, $-\infty < t < \infty$. That is

5.1
$$L = \left\{ z \colon \operatorname{Im}\left(\frac{z-a}{b}\right) = 0 \right\}.$$

What is the locus of each of the sets

$$\left\{ z \colon \operatorname{Im}\left(\frac{z-a}{b}\right) > 0 \right\},$$

$$\left\{ z \colon \operatorname{Im}\left(\frac{z-a}{b}\right) < 0 \right\}?$$

As a first step in answering this question, observe that since b is a direction we may assume $|b| = 1$. For the moment, let us consider the case where $a = 0$, and put $H_0 = \{ z \colon \operatorname{Im}(z/b) > 0 \}$, $b = \operatorname{cis} \beta$. If $z = r \operatorname{cis} \theta$ then $z/b = r \operatorname{cis}(\theta - \beta)$. Thus, z is in H_0 if and only if $\sin(\theta - \beta) > 0$; that is, when $\beta < \theta < \pi + \beta$. Hence H_0 is the half plane lying to the left of the line L if

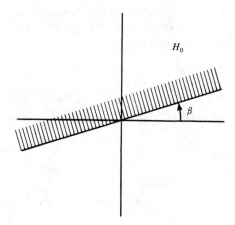

we are "walking along L in the direction of b." If we put

$$H_a = \left\{ z \colon \operatorname{Im}\left(\frac{z-a}{b}\right) > 0 \right\}$$

then it is easy to see that $H_a = a + H_0 \equiv \{ a + w \colon w \in H_0 \}$; that is, H_a is the translation of H_0 by a. Hence, H_a is the half plane lying to the left of L. Similarly,

$$K_a = \left\{ z \colon \operatorname{Im}\left(\frac{z-a}{b}\right) < 0 \right\}$$

is the half plane on the right of L.

Exercise

1. Let C be the circle $\{ z \colon |z-c| = r \}$, $r > 0$; let $a = c + r \operatorname{cis} \alpha$ and put

$$L_\beta = \left\{ z : \mathrm{Im}\left(\frac{z-a}{b}\right) = 0 \right\}$$

where $b = \mathrm{cis}\ \beta$. Find necessary and sufficient conditions in terms of β that L_β be tangent to C at a.

§6. The extended plane and its spherical representation

Often in complex analysis we will be concerned with functions that become infinite as the variable approaches a given point. To discuss this situation we introduce the *extended plane* which is $\mathbb{C} \cup \{\infty\} \equiv \mathbb{C}_\infty$. We also wish to introduce a distance function on $\mathbb{C}_\infty$ in order to discuss continuity properties of functions assuming the value infinity. To accomplish this and to give a concrete picture of $\mathbb{C}_\infty$ we represent $\mathbb{C}_\infty$ as the unit sphere in $\mathbb{R}^3$,

$$S = \{(x_1, x_2, x_3) \in \mathbb{R}^3 : x_1^2 + x_2^2 + x_3^2 = 1\}.$$

Let $N = (0, 0, 1)$; that is, N is the north pole on S. Also, identify $\mathbb{C}$ with $\{(x_1, x_2, 0) : x_1, x_2 \in \mathbb{R}\}$ so that $\mathbb{C}$ cuts S along the equator. Now for each point z in $\mathbb{C}$ consider the straight line in $\mathbb{R}^3$ through z and N. This intersects

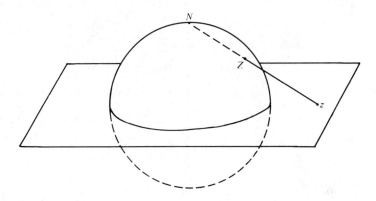

the sphere in exactly one point $Z \neq N$. If $|z| > 1$ then Z is in the northern hemisphere and if $|z| < 1$ then Z is in the southern hemisphere; also, for $|z| = 1$, $Z = z$. What happens to Z as $|z| \to \infty$? Clearly Z approaches N; hence, we identify N and the point ∞ in $\mathbb{C}_\infty$. Thus $\mathbb{C}_\infty$ is represented as the sphere S.

Let us explore this representation. Put $z = x + iy$ and let $Z = (x_1, x_2, x_3)$ be the corresponding point on S. We will find equations expressing x_1, x_2, and x_3 in terms of x and y. The line in $\mathbb{R}^3$ through z and N is given by $\{tN + (1-t)z : -\infty < t < \infty\}$, or by

6.1 $\{((1-t)x, (1-t)y, t) : -\infty < t < \infty\}.$

Hence, we can find the coordinates of Z if we can find the value of t at which this line intersects S. If t is this value then

$$1 = (1-t)^2x^2+(1-t)^2y^2+t^2$$
$$= (1-t)^2|z|^2+t^2$$

From which we get

$$1-t^2 = (1-t)^2|z|^2.$$

Since $t \neq 1$ ($z \neq \infty$) we arrive at

$$t = \frac{|z|^2-1}{|z|^2+1}.$$

Thus

6.2 $$x_1 = \frac{2x}{|z|^2+1}, \quad x_2 = \frac{2y}{|z|^2+1}, \quad x_3 = \frac{|z|^2-1}{|z|^2+1}$$

But this gives

6.3 $$x_1 = \frac{z+\bar{z}}{|z|^2+1} \quad x_2 = \frac{-i(z-\bar{z})}{|z|^2+1} \quad x_3 = \frac{|z|^2-1}{|z|^2+1}.$$

If the point Z is given ($Z \neq N$) and we wish to find z then by setting $t = x_3$ and using (6.1), we arrive at

6.4 $$z = \frac{x_1+ix_2}{1-x_3}$$

Now let us define a distance function between points in the extended plane in the following manner: for z, z' in $\mathbb{C}_\infty$ define the distance from z to z', $d(z, z')$, to be the distance between the corresponding points Z and Z' in $\mathbb{R}^3$. If $Z = (x_1, x_2, x_3)$ and $Z' = (x_1', x_2', x_3')$ then

6.5 $$d(z, z') = [(x_1-x_1')^2+(x_2-x_2')^2+(x_3-x_3')^2]^{\frac{1}{2}}$$

Using the fact that Z and Z' are on S, (6.5) gives

6.6 $$[d(z, z')]^2 = 2-2(x_1x_1'+x_2x_2'+x_3x_3')$$

By using equation (6.3) we get

6.7 $$d(z, z') = \frac{2|z-z'|}{[(1+|z|^2)(1+|z'|^2)]^{\frac{1}{2}}}, \quad (z, z' \in \mathbb{C})$$

In a similar manner we get for z in $\mathbb{C}$

6.8 $$d(z, \infty) = \frac{2}{(1+|z|^2)^{\frac{1}{2}}}$$

This correspondence between points of S and $\mathbb{C}_\infty$ is called the *stereographic projection*.

Exercises

1. Give the details in the derivation of (6.7) and (6.8).

2. For each of the following points in $\mathbb{C}$, give the corresponding point of S: 0, $1+i$, $3+2i$.

3. Which subsets of S correspond to the real and imaginary axes in $\mathbb{C}$.

4. Let Λ be a circle lying in S. Then there is a unique plane P in $\mathbb{R}^3$ such that $P \cap S = \Lambda$. Recall from analytic geometry that

$$P = \{(x_1, x_2, x_3): x_1\beta_1 + x_2\beta_2 + x_3\beta_3 = l\}$$

where $(\beta_1, \beta_2, \beta_3)$ is a vector orthogonal to P and l is some real number. It can be assumed that $\beta_1^2 + \beta_2^2 + \beta_3^2 = 1$. Use this information to show that if Λ contains the point N then its projection on $\mathbb{C}$ is a straight line. Otherwise, Λ projects onto a circle in $\mathbb{C}$.

5. Let Z and Z' be points on S corresponding to z and z' respectively. Let W be the point on S corresponding to $z+z'$. Find the coordinates of W in terms of the coordinates of Z and Z'.

Chapter II

Metric Spaces and the Topology of $\mathbb{C}$

§1. Definition and examples of metric spaces

A *metric space* is a pair (X, d) where X is a set and d is a function from $X \times X$ into $\mathbb{R}$, called a *distance function* or *metric*, which satisfies the following conditions for x, y, and z in X:

$$d(x, y) \geq 0$$

$$d(x, y) = 0 \text{ if and only if } x = y$$

$$d(x, y) = d(y, x) \text{ (symmetry)}$$

$$d(x, z) \leq d(x, y) + d(y, z) \text{ (triangle inequality)}$$

If x and $r > 0$ are fixed then define

$$B(x; r) = \{y \in X : d(x, y) < r\}$$

$$\bar{B}(x; r) = \{y \in X : d(x, y) \leq r\}.$$

$B(x; r)$ and $\bar{B}(x; r)$ are called the *open* and *closed balls*, respectively, with center x and radius r.

Examples

1.1 Let $X = \mathbb{R}$ or $\mathbb{C}$ and define $d(z, w) = |z - w|$. This makes both $(\mathbb{R}, d)$ and $(\mathbb{C}, d)$ metric spaces. In fact, $(\mathbb{C}, d)$ will be the example of principal interest to us. If the reader has never encountered the concept of a metric space before this, he should continually keep $(\mathbb{C}, d)$ in mind during the study of this chapter.

1.2 Let (X, d) be a metric space and let $Y \subset X$; then (Y, d) is also a metric space.

1.3 Let $X = \mathbb{C}$ and define $d(x + iy, a + ib) = |x - a| + |y - b|$. Then $(\mathbb{C}, d)$ is a metric space.

1.4 Let $X = \mathbb{C}$ and define $d(x + iy, a + ib) = \max \{|x - a|, |y - b|\}$.

1.5 Let X be any set and define $d(x, y) = 0$ if $x = y$ and $d(x, y) = 1$ if $x \neq y$. To show that the function d satisfies the triangle inequality one merely considers all possibilities of equality among x, y, and z. Notice here that $B(x; \epsilon)$ consists only of the point x if $\epsilon \leq 1$ and $B(x; \epsilon) = X$ if $\epsilon > 1$. This metric space does not appear in the study of analytic function theory.

1.6 Let $X = \mathbb{R}^n$ and for $x = (x_1, \ldots, x_n)$, $y = (y_1, \ldots, y_n)$ in $\mathbb{R}^n$ define

$$d(x, y) = \left[\sum_{j=1}^{n} (x_j - y_j)^2 \right]^{\frac{1}{2}}$$

11

1.7 Let S be any set and denote by $B(S)$ the set of all functions $f: S \to \mathbb{C}$ such that

$$\|f\|_\infty \equiv \sup \{|f(s)|: s \in S\} < \infty.$$

That is, $B(S)$ consists of all complex valued functions whose range is contained inside some disk of finite radius. For f and g in $B(S)$ define $d(f, g) = \|f-g\|_\infty$. We will show that d satisfies the triangle inequality. In fact if f, g, and h are in $B(S)$ and s is any point in S then $|f(s)-g(s)| = |f(s)-h(s)+h(s)-g(s)| \le |f(s)-h(s)|+|h(s)-g(s)| \le \|f-h\|_\infty + \|h-g\|_\infty$. Thus, when the supremum is taken over all s in S, $\|f-g\|_\infty \le \|f-h\|_\infty+\|h-g\|_\infty$, which is the triangle inequality for d.

1.8 Definition. For a metric space (X, d) a set $G \subset X$ is *open* if for each x in G there is an $\epsilon > 0$ such that $B(x; \epsilon) \subset G$.

Thus, a set in $\mathbb{C}$ is open if it has no "edge." For example, $G = \{z \in \mathbb{C}: a < \operatorname{Re} z < b\}$ is open; but $\{z: \operatorname{Re} z < 0\} \cup \{0\}$ is not because $B(0; \epsilon)$ is not contained in this set no matter how small we choose ϵ.

We denote the *empty set*, the set consisting of no elements, by $\square$.

1.9 Proposition. *Let (X, d) be a metric space; then:*
   (a) *The sets X and $\square$ are open;*
   (b) *If $G_1, \ldots, G_n$ are open sets in X then so is $\bigcap_{k=1}^{n} G_k$;*
   (c) *If $\{G_j: j \in J\}$ is a collection of open sets in X, J any indexing set, then $G = \cup \{G_j: j \in J\}$ is also open.*

Proof. The proof of (a) is a triviality. To prove (b) let $x \in G = \bigcap_{k=1}^{n} G_k$; then $x \in G_k$ for $k = 1, \ldots, n$. Thus, by the definition, for each k there is an $\epsilon_k > 0$ such that $B(x; \epsilon_k) \subset G_k$. But if $\epsilon = \min \{\epsilon_1, \epsilon_2, \ldots, \epsilon_n\}$ then for $1 \le k \le n$ $B(x; \epsilon) \subset B(x; \epsilon_k) \subset G_k$. Thus $B(x; \epsilon) \subset G$ and G is open.

The proof of (c) is left as an exercise for the reader. ∎

There is another class of subsets of a metric space which are distinguished. These are the sets which contain all their "edge"; alternately, the sets whose complements have no "edge."

1.10 Definition. A set $F \subset X$ is *closed* if its complement, $X-F$, is open.

The following proposition is the complement of Proposition 1.9. The proof, whose execution is left to the reader, is accomplished by applying de Morgan's laws to the preceding proposition.

1.11 Proposition. *Let (X, d) be a metric space. Then:*
   (a) *The sets X and $\square$ are closed;*
   (b) *If $F_1, \ldots, F_n$ are closed sets in X then so is $\bigcup_{k=1}^{n} F_k$;*
   (c) *If $\{F_j: j \in J\}$ is any collection of closed sets in X, J any indexing set, then $F = \cap \{F_j: j \in J\}$ is also closed.*

The most common error made upon learning of open and closed sets is to interpret the definition of closed set to mean that if a set is not open it is closed. This, of course, is false as can be seen by looking at $\{z \in \mathbb{C}: \operatorname{Re} z > 0\}$ $\cup \{0\}$; it is neither open nor closed.

1.12 Definition. Let A be a subset of X. Then the *interior* of A, int A, is the set $\bigcup \{G: G$ is open and $G \subset A\}$. The *closure* of A, A^-, is the set $\bigcap \{F: F$ is closed and $F \supset A\}$. Notice that int A may be empty and A^- may be X. If $A = \{a+bi: a$ and b are rational numbers$\}$ then simultaneously $A^- = \mathbb{C}$ and int $A = \square$. By Propositions 1.9 and 1.11 we have that A^- is closed and int A is open. The *boundary* of A is denoted by ∂A and defined by $\partial A = A^- \cap (X-A)^-$.

1.13 Proposition. *Let A and B be subsets of a metric space (X, d). Then:*
  (a) *A is open if and only if $A = $ int A;*
  (b) *A is closed if and only if $A = A^-$;*
  (c) *int $A = X-(X-A)^-$; $A^- = X-$int $(X-A)$; $\partial A = A^- -$int A;*
  (d) *$(A \cup B)^- = A^- \cup B^-$;*
  (e) *$x_0 \in$ int A if and only if there is an $\epsilon > 0$ such that $B(x_0; \epsilon) \subset A$;*
  (f) *$x_0 \in A^-$ if and only if for every $\epsilon > 0$, $B(x_0; \epsilon) \cap A \neq \square$.*

Proof. The proofs of (a)–(e) are left to the reader. To prove (f) assume $x_0 \in A^- = X-$int $(X-A)$; thus, $x_0 \notin$ int $(X-A)$. By part (e), for every $\epsilon > 0$ $B(x_0; \epsilon)$ is not contained in $X-A$. That is, there is a point $y \in B(x_0; \epsilon)$ which is not in $X-A$. Hence, $y \in B(x_0; \epsilon) \cap A$. Now suppose $x_0 \notin A^- = X-$int $(X-A)$. Then $x_0 \in$ int $(X-A)$ and, by (e), there is an $\epsilon > 0$ such that $B(x_0; \epsilon) \subset X-A$. That is, $B(x_0; \epsilon) \cap A = \square$ so that x_0 does not satisfy the condition. ∎

Finally, one last definition of a distinguished type of set.

1.14 Definition. A subset A of a metric space X is *dense* if $A^- = X$.

The set of rational numbers $\mathbb{Q}$ is dense in $\mathbb{R}$ and $\{x+iy: x, y \in \mathbb{Q}\}$ is dense in $\mathbb{C}$.

Exercises

1. Show that each of the examples of metric spaces given in (1.2)–(1.6) is, indeed, a metric space. Example (1.6) is the only one likely to give any difficulty. Also, describe $B(x;r)$ for each of these examples.
2. Which of the following subsets of $\mathbb{C}$ are open and which are closed: (a) $\{z: |z| < 1\}$; (b) the real axis; (c) $\{z: z^n = 1$ for some integer $n \geq 1\}$; (d) $\{z \in \mathbb{C}: z$ is real and $0 \leq z < 1\}$; (e) $\{z \in \mathbb{C}: z$ is real and $0 \leq z \leq 1\}$?
3. If (X, d) is any metric space show that every open ball is, in fact, an open set. Also, show that every closed ball is a closed set.
4. Give the details of the proof of (1.9c).
5. Prove Proposition 1.11.
6. Prove that a set $G \subset X$ is open if and only if $X-G$ is closed.
7. Show that $(\mathbb{C}_\infty, d)$ where d is given by (I. 6.7) and (I. 6.8) is a metric space.
8. Let (X, d) be a metric space and $Y \subset X$. Suppose $G \subset X$ is open; show

that $G \cap Y$ is open in (Y, d). Conversely, show that if $G_1 \subset Y$ is open in (Y, d), there is an open set $G \subset X$ such that $G_1 = G \cap Y$.

9. Do Exercise 8 with "closed" in place of "open."

10. Prove Proposition 1.13.

11. Show that $\{\text{cis} k : k \geq 0\}$ is dense in $T = \{z \in \mathbb{C} : |z| = 1\}$. For which values of θ is $\{\text{cis}(k\theta) : k \geq 0\}$ dense in T?

§2. Connectedness

Let us start this section by giving an example. Let $X = \{z \in \mathbb{C} : |z| \leq 1\}$ $\cup \{z : |z-3| < 1\}$ and give X the metric it inherits from $\mathbb{C}$. (Henceforward, whenever we consider subsets X of $\mathbb{R}$ or $\mathbb{C}$ as metric spaces we will assume, unless stated to the contrary, that X has the inherited metric $d(z, w) = |z-w|$.) Then the set $A = \{z : |z| \leq 1\}$ is simultaneously open and closed. It is closed because its complement in X, $B = X - A = \{z : |z-3| < 1\}$ is open; A is open because if $a \in A$ then $B(a; 1) \subset A$. (Notice that it may not happen that $\{z \in \mathbb{C} : |z-a| < 1\}$ is contained in A—for example, if $a = 1$. But the definition of $B(a; 1)$ is $\{z \in X : |z-a| < 1\}$ and this is contained in A.) Similarly B is also both open and closed in X.

This is an example of a non-connected space.

2.1 Definition. A metric space (X, d) is *connected* if the only subsets of X which are both open and closed are $\square$ and X. If $A \subset X$ then A is a *connected subset of X* if the metric space (A, d) is connected.

An equivalent formulation of connectedness is to say that X is not connected if there are disjoint open sets A and B in X, neither of which is empty, such that $X = A \cup B$. In fact, if this condition holds then $A = X - B$ is also closed.

2.2 Proposition. *A set* $X \subset \mathbb{R}$ *is connected iff* X *is an interval.*

Proof. Suppose $X = [a, b]$, a and b elements of $\mathbb{R}$. Let $A \subset X$ be an open subset of X such that $a \in A$, and $A \neq X$. We will show that A cannot also be closed—and hence, X must be connected. Since A is open and $a \in A$ there is an $\epsilon > 0$ such that $[a, a+\epsilon) \subset A$. Let

$$r = \sup \{\epsilon : [a, a+\epsilon) \subset A\}$$

Claim. $[a, a+r) \subset A$. In fact, if $a \leq x < a+r$ then, putting $h = a+r-x > 0$, the definition of supremum implies there is an ϵ with $r-h < \epsilon < r$ and $[a, a+\epsilon) \subset A$. But $a \leq x = a+(r-h) < a+\epsilon$ implies $x \in A$ and the claim is established.

However, $a+r \notin A$; for if, on the contrary, $a+r \in A$ then, by the openness of A, there is a $\delta > 0$ with $[a+r, a+r+\delta) \subset A$. But this gives $[a, a+r+\delta) \subset A$, contradicting the definition of r. Now if A were also closed then $a+r \in B = X - A$ which is open. Hence we could find a $\delta > 0$ such that $(a+r-\delta, a+r] \subset B$, contradicting the above claim.

The proof that other types of intervals are connected is similar and it will be left as an exercise.

The proof of the converse is Exercise 1. ∎

If w and z are in $\mathbb{C}$ then we denote the straight line segment from z to w by

$$[z, w] = \{tw + (1-t)z : 0 \le t \le 1\}$$

A *polygon* from a to b is a set $P = \bigcup\limits_{k=1}^{n} [z_k, w_k]$ where $z_1 = a$, $w_n = b$ and $w_k = z_{k+1}$ for $1 \le k \le n-1$; or, $P = [a, z_1 \ldots z_n, b]$.

2.3 Theorem. *An open set $G \subset \mathbb{C}$ is connected iff for any two points a, b in G there is a polygon from a to b lying entirely inside G.*

Proof. Suppose that G satisfies this condition and let us assume that G is not connected. We will obtain a contradiction. From the definition, $G = A \cup B$ where A and B are both open and closed, $A \cap B = \square$, and neither A nor B is empty. Let $a \in A$ and $b \in B$; by hypothesis there is a polygon P from a to b such that $P \subset G$. Now a moment's thought will show that one of the segments making up P will have one point in A and another in B. So we can assume that $P = [a, b]$. Define,

$$S = \{s \in [0, 1] : sb + (1-s)a \in A\}$$

$$T = \{t \in [0, 1] : tb + (1-t)a \in B\}$$

Then $S \cap T = \square$, $S \cup T = [0, 1]$, $0 \in S$ and $1 \in T$. However it can be shown that both S and T are open (Exercise 2), contradicting the connectedness of $[0, 1]$. Thus, G must be connected.

Now suppose that G is connected and fix a point a in G. To show how to construct a polygon (lying in G!) from a to a point b in G would be difficult. But we don't have to perform such a construction; we merely show that one exists. For a fixed a in G define

$$A = \{b \in G : \text{there is a polygon } P \subset G \text{ from } a \text{ to } b\}.$$

The plan is to show that A is simultaneously open and closed in G. Since $a \in A$ and G is connected this will give that $A = G$ and the theorem will be proved.

To show that A is open let $b \in A$ and let $P = [a, z_1, \ldots, z_n, b]$ be a polygon from a to b with $P \subset G$. Since G is open (this was not needed in the first half), there is an $\epsilon > 0$ such that $B(b; \epsilon) \subset G$. But if $z \in B(b; \epsilon)$ then $[b, z] \subset B(b; \epsilon) \subset G$. Hence the polygon $Q = P \cup [b, z]$ is inside G and goes from a to z. This shows that $B(b; \epsilon) \subset A$, and so A is open.

To show that A is closed suppose there is a point z in $G - A$ and let $\epsilon > 0$ be such that $B(z; \epsilon) \subset G$. If there is a point b in $A \cap B(z; \epsilon)$ then, as above, we can construct a polygon from a to z. Thus we must have that $B(z; \epsilon) \cap A = \square$, or $B(z; \epsilon) \subset G - A$. That is, $G - A$ is open so that A is closed. ∎

2.4 Corollary. *If $G \subset \mathbb{C}$ is open and connected and a and b are points in G then there is a polygon P in G from a to b which is made up of line segments parallel to either the real or imaginary axis.*

Proof. There are two ways of proving this corollary. One could obtain a polygon in G from a to b and then modify each of its line segments so that a new polygon is obtained with the desired properties. However, this proof is more easily executed using compactness (see Exercise 5.7). Another proof can be obtained by modifying the proof of Theorem 2.3. Define the set A as in the proof of (2.3) but add the restriction that the polygon's segments are all parallel to one of the axes. The remainder of the proof will be valid with one exception. If $z \in B(b; \epsilon)$ then $[b, z]$ may not be parallel to an axis. But it is easy to see that if $z = x+iy$, $b = p+iq$ then the polygon $[b, p+iy] \cup [p+iy, z] \subset B(b; \epsilon)$ and has segments parallel to an axis. ∎

It will now be shown that any set S in a metric space can be expressed, in a canonical way, as the union of connected pieces.

2.5 Definition. A subset D of a metric space X is a *component* of X if it is a maximal connected subset of X. That is, D is connected and there is no connected subset of X that properly contains D.

If the reader examines the example at the beginning of this section he will notice that both A and B are components and, furthermore, these are the only components of X. For another example let $X = \{0, 1, \frac{1}{2}, \frac{1}{3}, \ldots\}$. Then clearly every component of X is a point and each point is a component. Notice that while the components $\left\{\frac{1}{n}\right\}$ are all open in X, the component $\{0\}$ is not.

2.6 Lemma. *Let $x_0 \in X$ and let $\{D_j : j \in J\}$ be a collection of connected subsets of X such that $x_0 \in D_j$ for each j in J. Then $D = \bigcup \{D_j : j \in J\}$ is connected.*

Proof. Let A be a subset of the metric space (D, d) which is both open and closed and suppose that $A \neq \square$. Then $A \cap D_j$ is open in (D_j, d) for each j and it is also closed (Exercises 1.8 and 1.9). Since D_j is connected we get that either $A \cap D_j = \square$ or $A \cap D_j = D_j$. Since $A \neq \square$ there is at least one k such that $A \cap D_k \neq \square$; hence, $A \cap D_k = D_k$. In particular $x_0 \in A$ so that $x_0 \in A \cap D_j$ for every j. Thus $A \cap D_j = D_j$, or $D_j \subset A$, for each index j. This gives that $D = A$, so that D is connected. ∎

2.7 Theorem. *Let (X, d) be a metric space. Then:*
(a) *Each x_0 in X is contained in a component of X.*
(b) *Distinct components of X are disjoint.*
Note that part (a) says that X is the union of its components.

Proof. (a) Let $\mathscr{D}$ be the collection of connected subsets of X which contain the given point x_0. Notice that $\{x_0\} \in \mathscr{D}$ so that $\mathscr{D} \neq \square$. Also notice that the hypotheses of the preceding lemma apply to the collection $\mathscr{D}$. Hence $C = \bigcup \{D : D \in \mathscr{D}\}$ is connected and $x_0 \in C$. But C must be a component. In fact, if D is connected and $C \subset D$ then $x_0 \in D$ so that $D \in \mathscr{D}$; but then $D \subset C$, so that $C = D$. Thus C is maximal and part (a) is proved.

(b) Suppose C_1 and C_2 are components, $C_1 \neq C_2$, and suppose there is a point x_0 in $C_1 \cap C_2$. Again the lemma says that $C_1 \cup C_2$ is connected.

Since both C_1 and C_2 are components, this gives $C_1 = C_1 \cup C_2 = C_2$, a contradiction. ∎

2.8 Proposition. (a) *If $A \subset X$ is connected and $A \subset B \subset A^-$, then B is connected.*
(b) *If C is a component of X then C is closed.*
The proof is left as an exercise.

2.9 Theorem. *Let G be open in $\mathbb{C}$; then the components of G are open and there are only a countable number of them.*

Proof. Let C be a component of G and let $x_0 \in C$. Since G is open there is an $\epsilon > 0$ with $B(x_0; \epsilon) \subset G$. By Lemma 2.6, $B(x_0; \epsilon) \cup C$ is connected and so must be C. That is $B(x_0; \epsilon) \subset C$ and C is, therefore, open.

To see that the number of components is countable let $S = \{a+ib:$ a and b are rational and $a+bi \in G\}$. Then S is countable and each component of G contains a point of S, so that the number of components is countable. ∎

Exercises

1. The purpose of this exercise is to show that a connected subset of $\mathbb{R}$ is an interval.
(a) Show that a set $A \subset R$ is an interval iff for any two points a and b in A with $a < b$, the interval $[a, b] \subset A$.
(b) Use part (a) to show that if a set $A \subset R$ is connected then it is an interval.
2. Show that the sets S and T in the proof of Theorem 2.3 are open.
3. Which of the following subsets X of $\mathbb{C}$ are connected; if X is not connected, what are its components: (a) $X = \{z: |z| \leq 1\} \cup \{z: |z-2| < 1\}$. (b) $X =$ $[0, 1) \cup \left\{ 1 + \dfrac{1}{n} : n \geq 1 \right\}$. (c) $X = \mathbb{C} - (A \cup B)$ where $A = [0, \infty)$ and $B =$ $\{z = r \text{ cis } \theta: r = \theta, 0 \leq \theta \leq \infty\}$?
4. Prove the following generalization of Lemma 2.6. If $\{D_j: j \in J\}$ is a collection of connected subsets of X and if for each j and k in J we have $D_j \cap D_k \neq \square$ then $D = \bigcup \{D_j: j \in J\}$ is connected.
5. Show that if $F \subset X$ is closed and connected then for every pair of points a, b in F and each $\epsilon > 0$ there are points $z_0, z_1, \ldots, z_n$ in F with $z_0 = a$, $z_n = b$ and $d(z_{k-1}, z_k) < \epsilon$ for $1 \leq k \leq n$. Is the hypothesis that F be closed needed? If F is a set which satisfies this property then F is not necessarily connected, even if F is closed. Give an example to illustrate this.

§3. Sequences and completeness

One of the most useful concepts in a metric space is that of a convergent sequence. Their central role in calculus is duplicated in the study of metric spaces and complex analysis.

3.1 Definition. If $\{x_1, x_2, \ldots\}$ is a sequence in a metric space (X, d) then

$\{x_n\}$ *converges* to x—in symbols $x = \lim x_n$ or $x_n \to x$—if for every $\epsilon > 0$ there is an integer N such that $d(x, x_n) < \epsilon$ whenever $n \geq N$.

Alternately, $x = \lim x_n$ if $0 = \lim d(x, x_n)$.

If $X = \mathbb{C}$ then $z = \lim z_n$ means that for each $\epsilon > 0$ there is an N such that $|z - z_n| < \epsilon$ when $n \geq N$.

Many concepts in the theory of metric spaces can be phrased in terms of sequences. The following is an example.

3.2 Proposition. *A set $F \subset X$ is closed iff for each sequence $\{x_n\}$ in F with $x = \lim x_n$ we have $x \in F$.*

Proof. Suppose F is closed and $x = \lim x_n$ where each x_n is in F. So for every $\epsilon > 0$, there is a point x_n in $B(x; \epsilon)$; that is $B(x; \epsilon) \cap F \neq \square$, so that $x \in F^-$ $= F$ by Proposition 2.8.

Now suppose F is not closed; so there is a point x_0 in F^- which is not in F. By Proposition 1.13(f), for every $\epsilon > 0$ we have $B(x_0; \epsilon) \cap F \neq \square$. In particular for every integer n there is a point x_n in $B\left(x_0; \dfrac{1}{n}\right) \cap F$. Thus, $d(x_0, x_n) < \dfrac{1}{n}$ which implies that $x_n \to x_0$. Since $x_0 \notin F$, this says the condition fails. ∎

3.3 Definition. If $A \subset X$ then a point x in X is a *limit point* of A if there is a sequence $\{x_n\}$ of distinct points in A such that $x = \lim x_n$.

The reason for the word "distinct" in this definition can be illustrated by the following example. Let $X = \mathbb{C}$ and let $A = [0, 1] \cup \{i\}$; each point in $[0, 1]$ is a limit point of A but i is not. We do not wish to call a point such as i a limit point; but if "distinct" were dropped from the definition we could taken $x_n = i$ for each i and have $i = \lim x_n$.

3.4 Proposition. (a) *A set is closed iff it contains all its limit points.*

(b) *If $A \subset X$ then $A^- = A \cup \{x: x \text{ is a limit point of } A\}$.*

The proof is left as an exercise.

From real analysis we know that a basic property of $\mathbb{R}$ is that any sequence whose terms get closer together as n gets large, must be convergent. Such sequences are called Cauchy sequences. One of their attributes is that you know the limit will exist even though you can't produce it.

3.5 Definition. A sequence $\{x_n\}$ is called a *Cauchy sequence* if for every $\epsilon > 0$ there is an integer N such that $d(x_n, x_m) < \epsilon$ for all $n, m \geq N$.

If (X, d) has the property that each Cauchy sequence has a limit in X then (X, d) is *complete*.

3.6 Proposition. $\mathbb{C}$ *is complete.*

Proof. If $\{x_n + iy_n\}$ is a Cauchy sequence in $\mathbb{C}$ then $\{x_n\}$ and $\{y_n\}$ are Cauchy sequences in $\mathbb{R}$. Since $\mathbb{R}$ is complete, $x_n \to x$ and $y_n \to y$ for points x, y in $\mathbb{R}$. It follows that $x + iy = \lim (x_n + iy_n)$, and so $\mathbb{C}$ is complete. ∎

Consider $\mathbb{C}_\infty$ with its metric d (I. 6.7 and I. 6.8). Let z_n, z be points in $\mathbb{C}$;

it can be shown that $d(z_n, z) \to 0$ if and only if $|z_n - z| \to 0$. In spite of this, any sequence $\{z_n\}$ with $\lim |z_n| = \infty$ is Cauchy in $\mathbb{C}_\infty$, but, of course, is not Cauchy in $\mathbb{C}$.

If $A \subset X$ we define the *diameter* of A by diam $A = \sup \{d(x, y): x$ and y are in $A\}$.

3.7 Cantor's Theorem. *A metric space* (X, d) *is complete iff for any sequence* $\{F_n\}$ *of non-empty closed sets with* $F_1 \supset F_2 \supset \ldots$ *and* diam $F_n \to 0$, $\bigcap_{n=1}^{\infty} F_n$ *consists of a single point.*

Proof. Suppose (X, d) is complete and let $\{F_n\}$ be a sequence of closed sets having the properties: (i) $F_1 \supset F_2 \supset \ldots$ and (ii) $\lim$ diam $F_n = 0$. For each n, let x_n be an arbitrary point in F_n; if $n, m \geq N$ then x_n, x_m are in F_N so that, by definition, $d(x_n, x_m) \leq$ diam F_N. By the hypothesis N can be chosen sufficiently large that diam $F_N < \epsilon$; this shows that $\{x_n\}$ is a Cauchy sequence. Since X is complete, $x_0 = \lim x_n$ exists. Also, x_n is in F_N for all $n \geq N$ since $F_n \subset F_N$; hence, x_0 is in F_N for every N and this gives $x_0 \in \bigcap_{n=1}^{\infty} F_n = F$.
So F contains at least one point; if, also, y is in F then both x_0 and y are in F_n for each n and this gives $d(x_0, y) \leq$ diam $F_n \to 0$. Therefore $d(x_0, y) = 0$, or $x_0 = y$.

Now let us show that X is complete if it satisfies the stated condition. Let $\{x_n\}$ be a Cauchy sequence in X and put $F_n = \{x_n, x_{n+1}, \ldots\}^-$; then $F_1 \supset F_2 \supset \ldots$. If $\epsilon > 0$, choose N such that $d(x_n, x_m) < \epsilon$ for each n, $m \geq N$; this gives that diam $\{x_n, x_{n+1}, \ldots\} \leq \epsilon$ for $n \geq N$ and so diam $F_n \leq \epsilon$ for $n \geq N$ (Exercise 3). Thus diam $F_n \to 0$ and, by hypothesis, there is a point x_0 in X with $\{x_0\} = F_1 \cap F_2 \cap \ldots$. In particular x_0 is in F_n, and so $d(x_0, x_n) \leq$ diam $F_n \to 0$. Therefore, $x_0 = \lim x_n$. ■

There is a standard exercise associated with this theorem. It is to find a sequence of sets $\{F_n\}$ in $\mathbb{R}$ which satisfies two of the conditions:
(a) each F_n is closed,
(b) $F_1 \supset F_2 \supset \ldots$,
(c) diam $F_n \to 0$;
but which has $F = F_1 \cap F_2 \cap \ldots$ either empty or consisting of more than one point. Everyone should get examples satisfying the possible combinations.

3.8 Proposition. *Let* (X, d) *be a complete metric space and let* $Y \subset X$. *Then* (Y, d) *is a complete metric space iff* Y *is closed in* X.

Proof. It is left as an exercise to show that (Y, d) is complete whenever Y is a closed subset. Now assume (Y, d) to be complete; let x_0 be a limit point of Y. Then there is a sequence $\{y_n\}$ of points in Y such that $x_0 = \lim y_n$. Hence $\{y_n\}$ is a Cauchy sequence (Exercise 5) and must converge to a point y_0 in Y, since (Y, d) is complete. It follows that $y_0 = x_0$ and so Y contains all its limit points. Hence Y is closed by Proposition 3.4. ■

Exercises

1. Prove Proposition 3.4.
2. Furnish the details of the proof of Proposition 3.8.
3. Show that diam A = diam A^-.
4. Let z_n, z be points in $\mathbb{C}$ and let d be the metric on $\mathbb{C}_\infty$. Show that $|z_n - z| \to 0$ if and only if $d(z_n, z) \to 0$. Also show that if $|z_n| \to \infty$ then $\{z_n\}$ is Cauchy in $\mathbb{C}_\infty$. (Must $\{z_n\}$ converge in $\mathbb{C}_\infty$?)
5. Show that every convergent sequence in (X, d) is a Cauchy sequence.
6. Give three examples of non complete metric spaces.
7. Put a metric d on $\mathbb{R}$ such that $|x_n - x| \to 0$ if and only if $d(x_n, x) \to 0$, but that $\{x_n\}$ is a Cauchy sequence in $(\mathbb{R}, d)$ when $|x_n| \to \infty$. (Hint: Take inspiration from $\mathbb{C}_\infty$.)
8. Suppose $\{x_n\}$ is a Cauchy sequence and $\{x_{n_k}\}$ is a subsequence that is convergent. Show that $\{x_n\}$ must be convergent.

§4. Compactness

The concept of compactness is an extension of the benefits of finiteness to infinite sets. Most properties of compact sets are analogues of properties of finite sets which are quite trivial. For example, every sequence in a finite set has a convergent subsequence. This is quite trivial since there must be at least one point which is repeated an infinite number of times. However the same statement remains true if "finite" is replaced by "compact."

4.1 Definition. A subset K of a metric space X is *compact* if for every collection $\mathscr{G}$ of open sets in X with the property

4.2
$$K \subset \bigcup \{G : G \in \mathscr{G}\},$$

there is a finite number of sets $G_1, \ldots, G_n$ in $\mathscr{G}$ such that $K \subset G_1 \cup G_2 \cup \ldots \cup G_n$. A collection of sets $\mathscr{G}$ satisfying (4.2) is called a *cover* of K; if each member of $\mathscr{G}$ is an open set it is called an *open cover* of K.

Clearly the empty set and all finite sets are compact. An example of a non compact set is $D = \{z \in \mathbb{C} : |z| < 1\}$. If $G_n = \left\{ z : |z| < 1 - \dfrac{1}{n} \right\}$ for $n = 2, 3, \ldots$, then $\{G_2, G_3, \ldots\}$ is an open cover of D for which there is no finite subcover.

4.3 Proposition. *Let K be a compact subset of X; then:*
  (a) *K is closed;*
  (b) *If F is closed and $F \subset K$ then F is compact.*

Proof. To prove part (a) we will show that $K = K^-$. Let $x_0 \in K^-$; by Proposition 1.13(f), $B(x_0; \epsilon) \cap K \neq \square$ for each $\epsilon > 0$. Let $G_n = X - \bar{B}\left(x_0; \dfrac{1}{n} \right)$ and suppose that $x_0 \notin K$. Then each G_n is open and $K \subset \bigcup_{n=1}^{\infty} G_n \left(\text{because} \bigcap_{n=1}^{\infty} \right.$

$\bar{B}\left(x_0; \dfrac{1}{n}\right) = \{x_0\}$). Since K is compact there is an integer m such that $K \subset \bigcup\limits_{n=1}^{m} G_n$. But $G_1 \subset G_2 \subset \ldots$ so that $K \subset G_m = X - \bar{B}\left(x_0; \dfrac{1}{m}\right)$. But this gives that $B\left(x_0; \dfrac{1}{m}\right) \cap K = \square$, a contradiction. Thus $K = K^-$.

To prove part (b) let $\mathcal{G}$ be an open cover of F. Then, since F is closed, $\mathcal{G} \cup \{X - F\}$ is an open cover of K. Let $G_1, \ldots, G_n$ be sets in $\mathcal{G}$ such that $K \subset G_1 \cup \ldots \cup G_n \cup (X - F)$. Clearly, $F \subset G_1 \cup \ldots \cup G_n$ and so F is compact. ∎

If $\mathcal{F}$ is a collection of subsets of X we say that $\mathcal{F}$ has the *finite intersection property* (f.i.p.) if whenever $\{F_1, F_2, \ldots, F_n\} \subset \mathcal{F}$, $F_1 \cap F_2 \cap \ldots \cap F_n \neq \square$. An example of such a collection is $\{D - G_2, D - G_3, \ldots\}$ where the sets G_n are as in the example preceding Proposition 4.3.

4.4 Proposition. *A set $K \subset X$ is compact iff every collection $\mathcal{F}$ of closed subsets of K with the f.i.p. has $\bigcap \{F: F \in \mathcal{F}\} \neq \square$.*

Proof. Suppose K is compact and $\mathcal{F}$ is a collection of closed subsets of K having the f.i.p. Assume that $\bigcap \{F: F \in \mathcal{F}\} = \square$ and let $\mathcal{G} = \{X - F: F \in \mathcal{F}\}$. Then, $\bigcup \{X - F: F \in \mathcal{F}\} = X - \bigcap \{F: F \in \mathcal{F}\} = X$ by the assumption; in particular, $\mathcal{G}$ is an open cover of K. Thus, there are $F_1, \ldots,$ $F_n \in \mathcal{F}$ such that $K \subset \bigcup\limits_{k=1}^{n} (X - F_k) = X - \bigcap\limits_{k=1}^{n} F_k$. But this gives that $\bigcap\limits_{k=1}^{n} F_k \subset X - K$, and since each F_k is a subset of K it must be that $\bigcap\limits_{k=1}^{n} F_k = \square$. This contradicts the f.i.p.

The proof of the converse is left as an exercise. ∎

4.5 Corollary. *Every compact metric space is complete.*

Proof. This follows easily by applying the above proposition and Theorem 3.7. ∎

4.6. Corollary. *If X is compact then every infinite set has a limit point in X.*

Proof. Let S be an infinite subset of X and suppose S has no limit points. Let $\{a_1, a_2, \ldots\}$ be a sequence of distinct points in S; then $F_n = \{a_n, a_{n+1}, \ldots\}$ also has no limit points. But if a set has no limit points it contains all its limits points and must be closed! Thus, each F_n is closed and $\{F_n: n \geq 1\}$ has the f.i.p. However, since the points $a_1, a_2, \ldots$ are distinct, $\bigcap\limits_{n=1}^{\infty} F_n = \square$, contradicting the above proposition. ∎

4.7 Definition. A metric space (X, d) is *sequentially compact* if every sequence in X has a convergent subsequence.

It will be shown that compact and sequentially compact metric spaces are the same. To do this the following is needed.

4.8 Lebesgue's Covering Lemma. *If (X, d) is sequentially compact and*

$\mathcal{G}$ *is an open cover of* X *then there is an* $\epsilon > 0$ *such that if* x *is in* X, *there is a set* G *in* $\mathcal{G}$ *with* $B(x; \epsilon) \subset G$.

Proof. The proof is by contradiction; suppose that $\mathcal{G}$ is an open cover of X and no such $\epsilon > 0$ can be found. In particular, for every integer n there is a point x_n in X such that $B\left(x_n; \dfrac{1}{n}\right)$ is not contained in any set G in $\mathcal{G}$. Since X is sequentially compact there is a point x_0 in X and a subsequence $\{x_{n_k}\}$ such that $x_0 = \lim x_{n_k}$. Let $G_0 \in \mathcal{G}$ such that $x_0 \in G_0$ and choose $\epsilon > 0$ such that $B(x_0; \epsilon) \subset G_0$. Now let N be such that $d(x_0, x_{n_k}) < \epsilon/2$ for all $n_k \geq N$. Let n_k be any integer larger than both N and $2/\epsilon$, and let $y \in B(x_{n_k}; 1/n_k)$. Then $d(x_0, y) \leq d(x_0, x_{n_k}) + d(x_{n_k}, y) < \epsilon/2 + 1/n_k < \epsilon$. That is, $B(x_{n_k}; 1/n_k) \subset B(x_0; \epsilon) \subset G_0$, contradicting the choice of x_{n_k}. ∎

There are two common misinterpretations of Lebesgue's Covering Lemma; one implies that it says nothing and the other that it says too much. Since $\mathcal{G}$ is an open covering of X it follows that each x in X is contained in some G in $\mathcal{G}$. Thus there is an $\epsilon > 0$ such that $B(x; \epsilon) \subset G$ since G is open. The lemma, however, gives one $\epsilon > 0$ such that for any x, $B(x; \epsilon)$ is contained in some member of $\mathcal{G}$. The other misinterpretation is to believe that for the $\epsilon > 0$ obtained in the lemma, $B(x; \epsilon)$ is contained in each G in $\mathcal{G}$ such that $x \in G$.

4.9. Theorem. *Let* (X, d) *be a metric space; then the following are equivalent statements:*

(a) X *is compact;*
(b) *Every infinite set in* X *has a limit point;*
(c) X *is sequentially compact;*
(d) X *is complete and for every* $\epsilon > 0$ *there are a finite number of points* $x_1, \ldots, x_n$ *in* X *such that*

$$X = \bigcup_{k=1}^{n} B(x_k; \epsilon).$$

(The property mentioned in (d) is called *total boundedness*.)

Proof. That (a) implies (b) is the statement of Corollary 4.6.

(b) implies (c): Let $\{x_n\}$ be a sequence in X and suppose, without loss of generality, that the points $x_1, x_2, \ldots$ are all distinct. By (b), the set $\{x_1, x_2, \ldots\}$ has a limit point x_0. Thus there is a point $x_{n_1} \in B(x_0; 1)$; similarly, there is an integer $n_2 > n_1$ with $x_{n_2} \in B(x_0; 1/2)$. Continuing we get integers $n_1 < n_2 < \ldots$, with $x_{n_k} \in B(x_0; 1/k)$. Thus, $x_0 = \lim x_{n_k}$ and X is sequentially compact.

(c) implies (d): To see that X is complete let $\{x_n\}$ be a Cauchy sequence, apply the definition of sequential compactness, and appeal to Exercise 3.8.

Now let $\epsilon > 0$ and fix $x_1 \in X$. If $X = B(x_1; \epsilon)$ then we are done; otherwise choose $x_2 \in X - B(x_1; \epsilon)$. Again, if $X = B(x_1; \epsilon) \cup B(x_2; \epsilon)$ we are done;

if not, let $x_3 \in X - [B(x_1; \epsilon) \cup B(x_2; \epsilon)]$. If this process never stops we find a sequence $\{x_n\}$ such that

$$x_{n+1} \in X - \bigcup_{k=1}^{n} B(x_k; \epsilon).$$

But this implies that for $n \neq m$, $d(x_n, x_m) \geq \epsilon > 0$. Thus $\{x_n\}$ can have no convergent subsequence, contradicting (c).

(d) implies (c): This part of the proof will use a variation of the "pigeon hole principle." This principle states that if you have more objects than you have receptacles then at least one receptacle must hold more than one object. Moreover, if you have an infinite number of points contained in a finite number of balls then one ball contains infinitely many points. So part (d) says that for every $\epsilon > 0$ and any infinite set in X, there is a point $y \in X$ such that $B(y; \epsilon)$ contains infinitely many points of this set. Let $\{x_n\}$ be a sequence of distinct points. There is a point y_1 in X and a subsequence $\{x_n^{(1)}\}$ of $\{x_n\}$ such that $\{x_n^{(1)}\} \subset B(y_1; 1)$. Also, there is a point y_2 in X and a subsequence $\{x_n^{(2)}\}$ of $\{x_n^{(1)}\}$ such that $\{x_n^{(2)}\} \subset B(y_2; \frac{1}{2})$. Continuing, for each integer $k \geq 2$ there is a point y_k in X and a subsequence $\{x_n^{(k)}\}$ of $\{x_n^{(k-1)}\}$ such that $\{x_n^{(k)}\} \subset B(y_k; 1/k)$. Let $F_k = \{x_n^{(k)}\}^-$; then diam $F_k \leq 2/k$ and $F_1 \supset F_2 \supset \ldots$. By Theorem 3.7, $\bigcap_{k=1}^{\infty} F_k = \{x_0\}$. We claim that $x_k^{(k)} \to x_0$ (and $\{x_k^{(k)}\}$ is a subsequence of $\{x_n\}$). In fact, $x_0 \in F_k$ so that $d(x_0, x_k^{(k)}) \leq$ diam $F_k \leq 2/k$, and $x_0 = \lim x_k^{(k)}$.

(c) implies (a): Let $\mathcal{G}$ be an open cover of X. The preceding lemma gives an $\epsilon > 0$ such that for every $x \in X$ there is a G in $\mathcal{G}$ with $B(x; \epsilon) \subset G$. Now (c) also implies (d); hence there are points $x_1, \ldots, x_n$ in X such that $X = \bigcup_{k=1}^{n} B(x_k; \epsilon)$. Now for $1 \leq k \leq n$ there is a set $G_k \in \mathcal{G}$ with $B(x_k; \epsilon) \subset G_k$. Hence $X = \bigcup_{k=1}^{n} G_k$; that is, $\{G_1, \ldots, G_n\}$ is a finite subcover of $\mathcal{G}$. $\blacksquare$

4.10 Heine-Borel Theorem. *A subset K of $\mathbb{R}^n$ $(n \geq 1)$ is compact iff K is closed and bounded.*

Proof. If K is compact then K is totally bounded by part (d) of the preceding theorem. It follows that K must be closed (Proposition 4.3); also, it is easy to show that a totally bounded set is also bounded.

Now suppose that K is closed and bounded. Hence there are real numbers $a_1, \ldots, a_n$ and $b_1, \ldots, b_n$ such that $K \subset F = [a_1, b_1] \times \ldots \times [a_n, b_n]$. If it can be shown that F is compact then, because K is closed, it follows that K is compact (Proposition 4.3(b)). Since $\mathbb{R}^n$ is complete and F is closed it follows that F is complete. Hence, again using part (d) of the preceding theorem we need only show that F is totally bounded. This is easy although somewhat "messy" to write down. Let $\epsilon > 0$; we now will write F as the union of n-dimensional rectangles each of diameter less than ϵ. After doing this we will have $F \subset \bigcup_{k=1}^{m} B(x_k; \epsilon)$ where each x_k belongs to one of the aforementioned rectangles. The execution of the details of this strategy is left to the reader (Exercise 3). ∎

Exercises

1. Finish the proof of Proposition 4.4.
2. Let $p = (p_1, \ldots, p_n)$ and $q = (q_1, \ldots, q_n)$ be points in $\mathbb{R}^n$ with $p_k < q_k$ for each k. Let $R = [p_1, q_1] \times \ldots \times [p_n, q_n]$ and show that

$$\text{diam } R = d(p, q) = \left[\sum_{k=1}^{n} (q_k - p_k)^2 \right]^{\frac{1}{2}}.$$

3. Let $F = [a_1, b_1] \times \ldots \times [a_n, b_n] \subset \mathbb{R}^n$ and let $\epsilon > 0$; use Exercise 2 to show that there are rectangles $R_1, \ldots, R_m$ such that $F = \bigcup_{k=1}^{m} R_k$ and diam $R_k < \epsilon$ for each k. If $x_k \in R_k$ then it follows that $R_k \subset B(x_k; \epsilon)$.
4. Show that the union of a finite number of compact sets is compact.
5. Let X be the set of all bounded sequences of complex numbers. That is, $\{x_n\} \in X$ iff sup $\{|x_n|: n \geq 1\} < \infty$. If $x = \{x_n\}$ and $y = \{y_n\}$, define $d(x, y) = \sup \{|x_n - y_n|: n \geq 1\}$. Show that for each x in X and $\epsilon > 0$, $\bar{B}(x; \epsilon)$ is not totally bounded although it is complete. (Hint: you might have an easier time of it if you first show that you can assume $x = (0, 0, \ldots).$)
6. Show that the closure of a totally bounded set is totally bounded.

§5. Continuity

One of the most elementary properties of a function is continuity. The presence of continuity guarantees a certain degree of regularity and smoothness without which it is difficult to obtain any theory of functions on a metric space. Since the main subject of this book is the theory of functions of a complex variable which possess derivatives (and so are continuous), the study of continuity is basic.

5.1 Definition. Let (X, d) and (Ω, ρ) be metric spaces and let $f: X \to \Omega$ be a function. If $a \in X$ and $\omega \in \Omega$, then $\lim_{x \to a} f(x) = \omega$ if for every $\epsilon > 0$ there is a $\delta > 0$ such that $\rho(f(x), \omega) < \epsilon$ whenever $0 < d(x, a) < \delta$. The function f is *continuous at the point* a if $\lim_{x \to a} f(x) = f(a)$. If f is continuous at each point of X then f is a *continuous function from* X *to* Ω.

5.2 Proposition. *Let $f: (X, d) \to (\Omega, \rho)$ be a function and $a \in X$, $\alpha = f(a)$. The following are equivalent statements:*
   (a) *f is continuous at a;*
   (b) *For every $\epsilon > 0$, $f^{-1}(B(\alpha; \epsilon))$ contains a ball with center at a;*
   (c) $\alpha = \lim f(x_n)$ *whenever* $a = \lim x_n$.
The proof will be left as an exercise for the reader.

That was the last proposition concerning continuity of a function at a point. From now on we will concern ourselves only with functions continuous on all of X.

5.3 Proposition. *Let $f: (X, d) \to (\Omega, \rho)$ be a function. The following are equivalent statements:*

(a) *f is continuous;*

(b) *If Δ is open in Ω then $f^{-1}(\Delta)$ is open in X;*

(c) *If Γ is closed in Ω then $f^{-1}(\Gamma)$ is closed in X.*

Proof. (a) implies (b): Let Δ be open in Ω and let $x \in f^{-1}(\Delta)$. If $\omega = f(x)$ then ω is in Δ; by definition, there is an $\epsilon > 0$ with $B(\omega; \epsilon) \subset \Delta$. Since f is continuous, part (b) of the preceding proposition gives a $\delta > 0$ with $B(x; \delta) \subset f^{-1}(B(\omega; \epsilon)) \subset f^{-1}(\Delta)$. Hence, $f^{-1}(\Delta)$ is open.

(b) implies (c): If $\Gamma \subset \Omega$ is closed then let $\Delta = \Omega - \Gamma$. By (b), $f^{-1}(\Delta) = X - f^{-1}(\Gamma)$ is open, so that $f^{-1}(\Gamma)$ is closed.

(c) implies (a): Suppose there is a point x in X at which f is not continuous. Then there is an $\epsilon > 0$ and a sequence $\{x_n\}$ such that $\rho(f(x_n), f(x)) \geq \epsilon$ for every n while $x = \lim x_n$. Let $\Gamma = \Omega - B(f(x); \epsilon)$; then Γ is closed and each x_n is in $f^{-1}(\Gamma)$. Since (by (c)) $f^{-1}(\Gamma)$ is closed we have $x \in f^{-1}(\Gamma)$. But this implies $\rho(f(x), f(x)) \geq \epsilon > 0$, a contradiction. ∎

The following type of result is probably well understood by the reader and so the proof is left as an exercise.

5.4 Proposition. *Let f and g be continuous functions from X into $\mathbb{C}$ and let $\alpha, \beta \in \mathbb{C}$. Then $\alpha f + \beta g$ and fg are both continuous. Also, f/g is continuous provided $g(x) \neq 0$ for every x in X.*

5.5 Proposition. *Let $f: X \to Y$ and $g: Y \to Z$ be continuous functions. Then $g \circ f$ (where $g \circ f(x) = g(f(x))$) is a continuous function from X into Z.*

Proof. If U is open in Z then $g^{-1}(U)$ is open in Y; hence, $f^{-1}(g^{-1}(U)) = (g \circ f)^{-1}(U)$ is open in X. ∎

5.6 Definition. A function $f: (X, d) \to (\Omega, \rho)$ is *uniformly continuous* if for every $\epsilon > 0$ there is a $\delta > 0$ (depending only on ϵ) such that $\rho(f(x), f(y)) < \epsilon$ whenever $d(x, y) < \delta$. We say that f is a *Lipschitz function* if there is a constant $M > 0$ such that $\rho(f(x), f(y)) \leq M d(x, y)$ for all x and y in X.

It is easy to see that every Lipschitz function is uniformly continuous. In fact, if ϵ is given, take $\delta = \epsilon/M$. It is even easier to see that every uniformly continuous function is continuous. What are some examples of such functions? If $X = \Omega = \mathbb{R}$ then $f(x) = x^2$ is continuous but not uniformly continuous. If $X = \Omega = [0, 1]$ then $f(x) = x^{\frac{1}{2}}$ is uniformly continuous but is not a Lipschitz function. The following provides a wealthy supply of Lipschitz functions.

Let $A \subset X$ and $x \in X$; define the *distance from x to the set A, $d(x, A)$,* by

$$d(x, A) = \inf \{d(x, a): a \in A\}.$$

5.7 Proposition. *Let $A \subset X$; then:*

(a) *$d(x, A) = d(x, A^-)$;*

(b) $d(x, A) = 0$ iff $x \in A^-$;

(c) $|d(x, A) - d(y, A)| \leq d(x, y)$ for all x, y in X.

Proof. (a) If $A \subset B$ then it is clear from the definition that $d(x, B) \leq d(x, A)$. Hence, $d(x, A^-) \leq d(x, A)$. On the other hand, if $\epsilon > 0$ there is a point y in A^- such that $d(x, A^-) \geq d(x, y) - \epsilon/2$. Also, there is a point a in A with $d(y, a) < \epsilon/2$. But $|d(x, y) - d(x, a)| \leq d(y, a) < \epsilon/2$ by the triangle inequality. In particular, $d(x, y) > d(x, a) - \epsilon/2$. This gives, $d(x, A^-) \geq d(x, a) - \epsilon \geq d(x, A) - \epsilon$. Since ϵ was arbitrary $d(x, A^-) \geq d(x, A)$, so that (a) is proved.

(b) If $x \in A^-$ then $0 = d(x, A^-) = d(x, A)$. Now for any x in X there is a minimizing sequence $\{a_n\}$ in A such that $d(x, A) = \lim d(x, a_n)$. So if $d(x, A) = 0$, $\lim d(x, a_n) = 0$; that is, $x = \lim a_n$ and so $x \in A^-$.

(c) For a in A $d(x, a) \leq d(x, y) + d(y, a)$. Hence, $d(x, A) = \inf \{d(x, a): a \in A\} \leq \inf \{d(x, y) + d(y, a): a \in A\} = d(x, y) + d(y, A)$. This gives $d(x, A) - d(y, A) \leq d(x, y)$. Similarly $d(y, A) - d(x, A) \leq d(x, y)$ so the desired inequality follows. ■

Notice that part (c) of the proposition says that $f \colon X \to \mathbb{R}$ defined by $f(x) = d(x, A)$ is a Lipschitz function. If we vary the set A we get a large supply of these functions.

It is not true that the product of two uniformly continuous (Lipschitz) functions is again uniformly continuous (Lipschitz). For example, $f(x) = x$ is Lipschitz but $f \cdot f$ is not even uniformly continuous. However if both f and g are bounded then the conclusion is valid (see Exercise 3).

Two of the most important properties of continuous functions are contained in the following result.

5.8 Theorem. *Let* $f \colon (X, d) \to (\Omega, \rho)$ *be a continuous function.*

(a) *If* X *is compact then* $f(X)$ *is a compact subset of* Ω.

(b) *If* X *is connected then* $f(X)$ *is a connected subset of* Ω.

Proof. To prove (a) and (b) it may be supposed, without loss of generality, that $f(X) = \Omega$. (a) Let $\{\omega_n\}$ be a sequence in Ω; then there is, for each $n \geq 1$, a point x_n in X with $\omega_n = f(x_n)$. Since X is compact there is a point x in X and a subsequence $\{x_{n_k}\}$ such that $x = \lim x_{n_k}$. But if $\omega = f(x)$, then the continuity of f gives that $\omega = \lim \omega_{n_k}$; hence Ω is compact by Theorem 4.9. (b) Suppose $\Sigma \subset \Omega$ is both open and closed in Ω and that $\Sigma \neq \square$. Then, because $f(X) = \Omega$, $\square \neq f^{-1}(\Sigma)$; also, $f^{-1}(\Sigma)$ is both open and closed because f is continuous. By connectivity, $f^{-1}(\Sigma) = X$ and this gives $\Omega = \Sigma$. Thus, Ω is connected. ■

5.9 Corollary. *If* $f \colon X \to \Omega$ *is continuous and* $K \subset X$ *is either compact or connected in* X *then* $f(K)$ *is compact or connected, respectively, in* Ω.

5.10 Corollary. *If* $f \colon X \to R$ *is continuous and* X *is connected then* $f(X)$ *is an interval.*

This follows from the characterization of connected subsets of $\mathbb{R}$ as intervals.

5.11 Intermediate Value Theorem. *If* $f: [a, b] \to \mathbb{R}$ *is continuous and* $f(a) \leq \xi$ $\leq f(b)$ *then there is a point* x, $a \leq x \leq b$, *with* $f(x) = \xi$.

5.12 Corollary. *If* $f: X \to \mathbb{R}$ *is continuous and* $K \subset X$ *is compact then there are points* x_0 *and* y_0 *in* K *with* $f(x_0) = \sup \{f(x): x \in K\}$ *and* $f(y_0) = \inf \{f(x): x \in K\}$.

Proof. If $\alpha = \sup \{f(x): x \in K\}$ then α is in $f(K)$ because $f(K)$ is closed and bounded in $\mathbb{R}$. Similarly $\beta = \inf \{f(x): x \in K\}$ is in $f(K)$. ∎

5.13 Corollary. *If* $K \subset X$ *is compact and* $f: X \to \mathbb{C}$ *is continuous then there are points* x_0 *and* y_0 *in* K *with*

$$|f(x_0)| = \sup \{|f(x)|: x \in K\} \text{ and } |f(y_0)| = \inf \{|f(x)|: x \in K\}.$$

Proof. This corollary follows from the preceding one because $g(x) = |f(x)|$ defines a continuous function from X into $\mathbb{R}$.

5.14 Corollary. *If* K *is a compact subset of* X *and* x *is in* X *then there is a point* y *in* K *with* $d(x, y) = d(x, K)$.

Proof. Define $f: X \to \mathbb{R}$ by $f(y) = d(x, y)$. Then f is continuous and, by Corollary 5.12, assumes a minimum value on K. That is, there is a point y in K with $f(y) \leq f(z)$ for every $z \in K$. This gives $d(x, y) = d(x, K)$. ∎

The next two theorems are extremely important and will be used repeatedly throughout this book with no specific reference to the theorem numbers.

5.15. Theorem. *Suppose* $f: X \to \Omega$ *is continuous and* X *is compact; then* f *is uniformly continuous.*

Proof. Let $\epsilon > 0$; we wish to find a $\delta > 0$ such that $d(x, y) < \delta$ implies $\rho(f(x), f(y)) < \epsilon$. Suppose there is no such δ; in particular, each $\delta = 1/n$ will fail to work. Then for every $n \geq 1$ there are points x_n and y_n in X with $d(x_n, y_n) < 1/n$ but $\rho(f(x_n), f(y_n)) \geq \epsilon$. Since X is compact there is a subsequence $\{x_{n_k}\}$ and a point $x \in X$ with $x = \lim x_{n_k}$.

Claim. $x = \lim y_{n_k}$. In fact, $d(x, y_{n_k}) \leq d(x, x_{n_k}) + 1/n_k$ and this tends to zero as k goes to ∞.

But if $\omega = f(x)$, $\omega = \lim f(x_{n_k}) = \lim f(y_{n_k})$ so that

$$\epsilon \leq \rho(f(x_{n_k}), f(y_{n_k}))$$
$$\leq \rho(f(x_{n_k}), \omega) + \rho(\omega, f(y_{n_k}))$$

and the right hand side of this inequality goes to zero. This is a contradiction and completes the proof. ∎

5.16. Definition. If A and B are subsets of X then define the *distance from* A *to* B, $d(A, B)$, by

$$d(A, B) = \inf \{d(a, b): a \in A, b \in B\}.$$

Notice that if B is the single-point set $\{x\}$ then $d(A, \{x\}) = d(x, A)$. If

$A = \{y\}$ and $B = \{x\}$ then $d(\{x\}, \{y\}) = d(x, y)$. Also, if $A \cap B \neq \square$ then $d(A, B) = 0$, but we can have $d(A, B) = 0$ with A and B disjoint. The most popular type of example is to take $A = \{(x, 0): x \in \mathbb{R}\} \subset \mathbb{R}^2$ and $B = \{(x, e^x): x \in \mathbb{R}\}$. Notice that A and B are both closed and disjoint and still $d(A, B) = 0$.

5.17 Theorem. *If A and B are disjoint sets in X with B closed and A compact then $d(A, B) > 0$.*

Proof. Define $f: X \to \mathbb{R}$ by $f(x) = d(x, B)$. Since $A \cap B = \square$ and B is closed, $f(a) > 0$ for each a in A. But since A is compact there is a point a in A such that $0 < f(a) = \inf \{f(x): x \in A\} = d(A, B)$. ∎

Exercises

1. Prove Proposition 5.2.
2. Show that if f and g are uniformly continuous (Lipschitz) functions from X into $\mathbb{C}$ then so is $f+g$.
3. We say that $f: X \to \mathbb{C}$ is bounded if there is a constant $M > 0$ with $|f(x)| \leq M$ for all x in X. Show that if f and g are bounded uniformly continuous (Lipschitz) functions from X into $\mathbb{C}$ then so is fg.
4. Is the composition of two uniformly continuous (Lipschitz) functions again uniformly continuous (Lipschitz)?
5. Suppose $f: X \to \Omega$ is uniformly continuous; show that if $\{x_n\}$ is a Cauchy sequence in X then $\{f(x_n)\}$ is a Cauchy sequence in Ω. Is this still true if we only assume that f is continuous? (Prove or give a counterexample.)
6. Recall the definition of a dense set (1.14). Suppose that Ω is a complete metric space and that $f: (D, d) \to (\Omega; \rho)$ is uniformly continuous, where D is dense in (X, d). Use Exercise 5 to show that there is a uniformly continuous function $g: X \to \Omega$ with $g(x) = f(x)$ for every x in D.
7. Let G be an open subset of $\mathbb{C}$ and let P be a polygon in G from a to b. Use Theorems 5.15 and 5.17 to show that there is a polygon $Q \subset G$ from a to b which is composed of line segments which are parallel to either the real or imaginary axes.
8. Use Lebesgue's Covering Lemma (4.8) to give another proof of Theorem 5.15.
9. Prove the following converse to Exercise 2.5. Suppose (X, d) is a compact metric space having the property that for every $\epsilon > 0$ and for any points a, b in X, there are points $z_0, z_1, \ldots, z_n$ in X with $z_0 = a$, $z_n = b$, and $d(z_{k-1}, z_k) < \epsilon$ for $1 \leq k \leq n$. Then (X, d) is connected. (Hint: Use Theorem 5.17.)
10. Let f and g be continuous functions from (X, d) to (Ω, p) and let D be a dense subset of X. Prove that if $f(x) = g(x)$ for x in D then $f = g$. Use this to show that the function g obtained in Exercise 6 is unique.

§6. Uniform convergence

Let X be a set and (Ω, ρ) a metric space and suppose $f, f_1, f_2, \ldots$ are functions from X into Ω. The sequence $\{f_n\}$ *converges uniformly* to f—written

$f = u - \lim f_n$—if for every $\epsilon > 0$ there is an integer N (depending on ϵ alone) such that $\rho(f(x), f_n(x)) < \epsilon$ for all x in X, whenever $n \geq N$. Hence,

$$\sup \{\rho(f(x), f_n(x)) \colon x \in X\} \leq \epsilon$$

whenever $n \geq N$.

The first problem is this: If X is not just a set but a metric space and each f_n is continuous does it follow that f is continuous? The answer is yes.

6.1 Theorem. *Suppose $f_n \colon (X, d) \to (\Omega, \rho)$ is continuous for each n and that $f = u - \lim f_n$; then f is continuous.*

Proof. Fix x_0 in X and $\epsilon > 0$; we wish to find a $\delta > 0$ such that $\rho(f(x_0), f(x)) < \epsilon$ when $d(x_0, x) < \delta$. Since $f = u - \lim f_n$, there is a function f_n with $\rho(f(x), f_n(x)) < \epsilon/3$ for all x in X. Since f_n is continuous there is a $\delta > 0$ such that $\rho(f_n(x_0), f_n(x)) < \epsilon/3$ when $d(x_0, x) < \delta$. Therefore, if $d(x_0, x) < \delta$, $\rho(f(x_0), f(x)) \leq \rho(f(x_0), f_n(x_0)) + \rho(f_n(x_0), f_n(x)) + \rho(f_n(x), f(x)) < \epsilon$. ∎

Let us consider the special case where $\Omega = \mathbb{C}$. If $u_n \colon X \to \mathbb{C}$, let $f_n(x) = u_1(x) + \ldots + u_n(x)$. We say $f(x) = \sum_{n=1}^{\infty} u_n(x)$ iff $f(x) = \lim f_n(x)$ for each x in X. The *series* $\sum_{1}^{\infty} u_n$ is *uniformly* convergent to f iff $f = u - \lim f_n$.

6.2 Weierstrass M-Test. *Let $u_n \colon X \to \mathbb{C}$ be a function such that $|u_n(x)| \leq M_n$ for every x in X and suppose the constants satisfy $\sum_{n=1}^{\infty} M_n < \infty$. Then $\sum_{1}^{\infty} u_n$ is uniformly convergent.*

Proof. Let $f_n(x) = u_1(x) + \ldots + u_n(x)$. Then for $n > m$,

$$|f_n(x) - f_m(x)| = |u_{m+1}(x) + \ldots + u_n(x)| \leq \sum_{k=m+1}^{n} M_k \text{ for each } x. \text{ Since } \sum_{1}^{\infty} M_k$$

converges, $\{f_n(x)\}$ is a Cauchy sequence in $\mathbb{C}$. Thus there is a number $\xi \in \mathbb{C}$ with $\xi = \lim f_n(x)$. Define $f(x) = \xi$; this gives a function $f \colon X \to \mathbb{C}$. Now

$$|f(x) - f_n(x)| = |\sum_{k=n+1}^{\infty} u_n(x)| \leq \sum_{k=n+1}^{\infty} |u_n(x)| \leq \sum_{k=n+1}^{\infty} M_k;$$

since $\sum_{1}^{\infty} M_k$ is convergent, for any $\epsilon > 0$ there is an integer N such that $\sum_{k=n+1}^{\infty} M_k < \epsilon$ whenever $n \geq N$. This gives $|f(x) - f_n(x)| < \epsilon$ for all x in X when $n \geq N$. ∎

Exercise

1. Let $\{f_n\}$ in a sequence of uniformly continuous functions from (X, d) into (Ω, ρ) and suppose that $f = u - \lim f_n$ exists. Prove that f is uniformly continuous. If each f_n is a Lipschitz function with constant M_n and sup $M_n < \infty$, show that f is a Lipschitz function. If sup $M_n = \infty$, show that f may fail to be Lipschitz.

Chapter III

Elementary Properties and Examples of Analytic Functions

§1. Power series

In this section the definition and basic properties of a power series will be given. The power series will then be used to give examples of analytic functions. Before doing this it is necessary to give some elementary facts on infinite series in $\mathbb{C}$ whose statements for infinite series in $\mathbb{R}$ should be well known to the reader. If a_n is in $\mathbb{C}$ for every $n \geq 0$ then the series $\sum_{n=0}^{\infty} a_n$ *converges* to z iff for every $\epsilon > 0$ there is an integer N such that $|\sum_{n=0}^{m} a_n - z| < \epsilon$ whenever $m \geq N$. The series $\sum a_n$ *converges absolutely* if $\sum |a_n|$ converges.

1.1 Proposition. *If $\sum a_n$ converges absolutely then $\sum a_n$ converges.*

Proof. Let $\epsilon > 0$ and put $z_n = a_0 + a_1 + \ldots + a_n$. Since $\sum |a_n|$ converges there is an integer N such that $\sum_{n=N}^{\infty} |a_n| < \epsilon$. Thus, if $m > k \geq N$,

$$|z_m - z_k| = |\sum_{n=k+1}^{m} a_n| \leq \sum_{n=k+1}^{m} |a_n| \leq \sum_{n=N}^{\infty} |a_n| < \epsilon.$$

That is, $\{z_n\}$ is a Cauchy sequence and so there is a z in $\mathbb{C}$ with $z = \lim z_n$. Hence $\sum a_n = z$. ∎

Also recall the definitions of limit inferior and superior of a sequence in $\mathbb{R}$. If $\{a_n\}$ is a sequence in $\mathbb{R}$ then define

$$\liminf a_n = \lim_{n \to \infty} [\inf \{a_n, a_{n+1}, \ldots\}]$$

$$\limsup a_n = \lim_{n \to \infty} [\sup \{a_n, a_{n+1}, \ldots\}]$$

An alternate notation for $\liminf a_n$ and $\limsup a_n$ is $\underline{\lim} \, a_n$ and $\overline{\lim} \, a_n$. If $b_n = \inf \{a_n, a_{n+1}, \ldots\}$ then $\{b_n\}$ is an increasing sequence of real numbers or $\{-\infty\}$. Hence, $\liminf a_n$ always exists although it may be $\pm \infty$. Similarly $\limsup a_n$ always exists although it may be $\pm \infty$.

A number of properties of lim inf and lim sup are included in the exercises of this section.

A *power series* about a is an infinite series of the form $\sum_{n=0}^{\infty} a_n (z-a)^n$. One of the easiest examples of a power series (and one of the most useful) is the *geometric series* $\sum_{n=0}^{\infty} z^n$. For which values of z does this series converge and when does it diverge? It is easy to see that $1 - z^{n+1} = (1-z)(1+z+\ldots+z^n)$, so that

1.2
$$1 + z + \ldots + z^n = \frac{1 - z^{n+1}}{1 - z}.$$

If $|z| < 1$ then $0 = \lim z^n$ and so the geometric series is convergent with

$$\sum_0^\infty z^n = \frac{1}{1-z}.$$

If $|z| > 1$ then $\lim |z|^n = \infty$ and the series diverges. Not only is this result an archetype for what happens to a general power series, but it can be used to explore the convergence properties of power series.

1.3 Theorem. *For a given power series $\sum\limits_{n=0}^{\infty} a_n(z-a)^n$ define the number R, $0 \le R \le \infty$, by*

$$\frac{1}{R} = \limsup |a_n|^{1/n},$$

then:

(a) *if $|z - a| < R$, the series converges absolutely:*
(b) *if $|z - a| > R$, the terms of the series become unbounded and so the series diverges;*
(c) *if $0 < r < R$ then the series converges uniformly on $\{z : |z| \le r\}$.*
Moreover, the number R is the only number having properties (a) *and* (b).

Proof. We may suppose that $a = 0$. If $|z| < R$ there is an r with $|z| < r < R$. Thus, there is an integer N such that $|a_n|^{1/n} < \frac{1}{r}$ for all $n \ge N$ $\left(\text{because } \frac{1}{r} > \frac{1}{R}\right)$. But then $|a_n| < \frac{1}{r^n}$ and so $|a_n z^n| < \left(\frac{|z|}{r}\right)^n$ for all $n \ge N$. This says that the tail end $\sum\limits_{n=N}^{\infty} a_n z^n$ is dominated by the series $\sum \left(\frac{|z|}{r}\right)^n$, and since $\frac{|z|}{r} < 1$ the power series converges absolutely for each $|z| < R$.

Now suppose $r < R$ and choose ρ such that $r < \rho < R$. As above, let N be such that $|a_n| < \frac{1}{\rho^n}$ for all $n \ge N$. Then if $|z| \le r$, $|a_n z^n| \le \left(\frac{r}{\rho}\right)^n$ and $\left(\frac{r}{\rho}\right) < 1$. Hence the Weierstrass M-test gives that the power series converges uniformly on $\{z : |z| \le r\}$. This proves parts (a) and (c).

To prove (b), let $|z| > R$ and choose r with $|z| > r > R$. Hence $\frac{1}{r} < \frac{1}{R}$; from the definition of $\limsup$, this gives infinitely many integers n with $\frac{1}{r} < |a_n|^{1/n}$. It follows that $|a_n z^n| > \left(\frac{|z|}{r}\right)^n$ and, since $\left(\frac{|z|}{r}\right) > 1$, these terms become unbounded. ∎

The number R is called the *radius of convergence* of the power series.

1.4 Proposition. *If* $\sum a_n(z-a)^n$ *is a given power series with radius of convergence R, then*

$$R = \lim |a_n/a_{n+1}|$$

if this limit exists.

Proof. Again assume that $a = 0$ and let $\alpha = \lim |a_n/a_{n+1}|$, which we suppose to exist. Suppose that $|z| < r < \alpha$ and find an integer N such that $r < |a_n/a_{n+1}|$ for all $n \geq N$. Let $B = |a_N|r^N$; then $|a_{N+1}|r^{N+1} = |a_{N+1}|rr^N < |a_N|r^N = B$; $|a_{N+2}|r^{N+2} = |a_{N+2}|rr^{N+1} < |a_{N+1}|r^{N+1} < B$; continuing we get $|a_n r^n| \leq B$ for all $n \geq N$. But then $|a_n z^n| = |a_n r^n| \dfrac{|z|^n}{r^n} \leq B\dfrac{|z|^n}{r^n}$ for all $n \geq N$. Since $|z| < r$ we get that $\sum\limits_{n=1}^{\infty} |a_n z^n|$ is dominated by a convergent series and hence converges. Since $r < \alpha$ was arbitrary this gives that $\alpha \leq R$.

On the other hand if $|z| > r > \alpha$, then $|a_n| < r|a_{n+1}|$ for all n larger than some integer N. As before, we get $|a_n r^n| \geq B = |a_N r^N|$ for $n \geq N$. This gives $|a_n z^n| \geq B\dfrac{|z|^n}{|r|^n}$ which approaches ∞ as n does. Hence, $\sum a_n z^n$ diverges and so $R \leq \alpha$. Thus $R = \alpha$. ∎

Consider the series $\sum\limits_{n=0}^{\infty} \dfrac{z^n}{n!}$; by Proposition 1.4 we have that this series has radius of convergence ∞. Hence it converges at every complex number and the convergence is uniform on each compact subset of $\mathbb{C}$. Maintaining a parallel with calculus, we designate this series by

$$e^z = \exp z = \sum_{n=0}^{\infty} \frac{z^n}{n!},$$

the *exponential series* or *function.*

Recall the following proposition from the theory of infinite series (the proof will not be given).

1.5 Proposition. *Let* $\sum a_n$ *and* $\sum b_n$ *be two absolutely convergent series and put*

$$c_n = \sum_{k=0}^{n} a_k b_{n-k}.$$

Then $\sum c_n$ *is absolutely convergent with sum*

$$(\sum a_n)(\sum b_n).$$

1.6 Proposition. *Let* $\sum a_n(z-a)^n$ *and* $\sum b_n(z-a)^n$ *be power series with radius of convergence* $\geq r > 0$. *Put*

$$c_n = \sum_{k=0}^{n} a_k b_{n-k};$$

then both power series $\sum (a_n+b_n)(z-a)^n$ *and* $\sum c_n(z-a)^n$ *have radius of convergence* $\geq r$, *and*

$$\sum (a_n+b_n)(z-a)^n = \left[\sum a_n(z-a)^n + \sum b_n(z-a)^n\right]$$

$$\sum c_n(z-a)^n = \left[\sum a_n(z-a)^n\right]\left[\sum b_n(z-a)^n\right]$$

for $|z-a| < r$.

Proof. We only give an outline of the proof. If $0 < s < r$ then for $|z| \le s$, we get $\sum |a_n+b_n|\,|z|^n \le \sum |a_n|s^n + \sum |b_n|s^n < \infty$; $\sum |c_n|\,|z|^n \le (\sum |a_n|s^n) (\sum |b_n|s^n) < \infty$. From here the proof can easily be completed. ∎

Exercises

1. Prove Proposition 1.5.
2. Give the details of the proof of Proposition 1.6.
3. Prove that $\limsup (a_n+b_n) \le \limsup a_n + \limsup b_n$ and $\liminf (a_n+b_n) \ge \liminf a_n + \liminf b_n$ for $\{a_n\}$ and $\{b_n\}$ sequences of real numbers.
4. Show that $\liminf a_n \le \limsup a_n$ for any sequence in $\mathbb{R}$.
5. If $\{a_n\}$ is a convergent sequence in $\mathbb{R}$ and $a = \lim a_n$, show that $a = \liminf a_n = \limsup a_n$.
6. Find the radius of convergence for each of the following power series:

(a) $\sum_{n=0}^{\infty} a^n z^n$, $a \in \mathbb{C}$; (b) $\sum_{n=0}^{\infty} a^{n^2} z^n$, $a \in \mathbb{C}$; (c) $\sum_{n=0}^{\infty} k^n z^n$, k an integer $\ne 0$; (d) $\sum_{n=0}^{\infty} z^{n!}$.

7. Show that the radius of convergence of the power series

$$\sum_{n=1}^{\infty} \frac{(-1)^n}{n} z^{n(n+1)}$$

is 1, and discuss convergence for $z = 1$, -1, and i. (Hint: The nth co-efficient of this series is not $(-1)^n/n$.)

§2. Analytic functions

In this section analytic functions are defined and some examples are given. It is also shown that the Cauchy-Riemann equations hold for the real and imaginary parts of an analytic function.

2.1 Definition. If G is an open set in $\mathbb{C}$ and $f\colon G \to \mathbb{C}$ then f is *differentiable at a point a in G* if

$$\lim_{h \to 0} \frac{f(a+h)-f(a)}{h}$$

exists; the value of this limit is denoted by $f'(a)$ and is called the *derivative* of f at a. If f is differentiable at each point of G we say that f is *differentiable on G*. Notice that if f is differentiable on G then $f'(a)$ defines a function $f'\colon G \to \mathbb{C}$. If f' is continuous then we say that f is *continuously differentiable*. If f' is differentiable then f is *twice differentiable*; continuing, a differentiable function such that each successive derivative is again differentiable is called *infinitely differentiable*.

(Henceforward, all functions will be assumed to take their values in $\mathbb{C}$ unless it is stated to the contrary.)

The following was surely predicted by the reader.

2.2 Proposition. *If* $f: G \to \mathbb{C}$ *is differentiable at a point* a *in* G *then* f *is continuous at* a.

Proof. In fact,

$$\lim_{z \to a} |f(z) - f(a)| = \left[\lim_{z \to a} \frac{|f(z) - f(a)|}{|z - a|} \right] \cdot \left[\lim_{z \to a} |z - a| \right] = f'(a) \cdot 0 = 0. \blacksquare$$

2.3 Definition. A function $f: G \to \mathbb{C}$ is *analytic* if f is continuously differentiable on G.

It follows readily, as in calculus, that sums and products of functions analytic on G are analytic. Also, if f and g are analytic on G and G_1 is the set of points in G where g doesn't vanish, then f/g is analytic on G_1.

Since constant functions and the function z are clearly analytic it follows that all rational functions are analytic on the complement of the set of zeros of the denominator.

Moreover, the usual laws for differentiating sums, products, and quotients remain valid.

2.4 Chain Rule. *Let* f *and* g *be analytic on* G *and* Ω *respectively and suppose* $f(G) \subset \Omega$. *Then* $g \circ f$ *is analytic on* G *and*

$$(g \circ f)'(z) = g'(f(z))f'(z)$$

for all z *in* G.

Proof. Fix z_0 in G and choose a positive number r such that $B(z_0; r) \subset G$. We must show that if $0 < |h_n| < r$ and $\lim h_n = 0$ then $\lim \{ h_n^{-1}[g(f(z_0 + h_n)) - g(f(z_0))] \}$ exists and equals $g'(f(z_0))f'(z_0)$. (Why will this suffice for a proof?)

Case 1 Suppose $f(z_0) \neq f(z_0 + h_n)$ for all n.

In this case

$$\frac{g \circ f(z_0 + h_n) - g \circ f(z_0)}{h_n} = \frac{g(f(z_0 + h_n)) - g(f(z_0))}{f(z_0 + h_n) - f(z_0)} \cdot \frac{f(z_0 + h_n) - f(z_0)}{h_n}.$$

Since $\lim[f(z_0 + h_n) - f(z_0)] = 0$ by (2.2) we have that

$$\lim h_n^{-1} \left[g \circ f(z_0 + h_n) - g \circ f(z_0) \right] = g'(f(z_0))f'(z_0)$$

Case 2 $f(z_0) = f(z_0 + h_n)$ for infinitely many values of n.

Write $\{h_n\}$ as the union of two sequences $\{k_n\}$ and $\{l_n\}$ where $f(z_0) \neq f(z_0 + k_n)$ and $f(z_0) = f(z_0 + l_n)$ for all n. Since f is differentiable, $f'(z_0) = \lim l_n^{-1}[f(z_0 + l_n) - f(z_0)] = 0$. Also $\lim l_n^{-1}[g \circ f(z_0 + l_n) - g \circ f(z_0)] = 0$. By Case 1, $\lim k_n^{-1}[g \circ f(z_0 + k_n) - g \circ f(z_0)] = g'(f(z_0))f'(z_0) = 0$. Therefore $\lim h_n^{-1}[g \circ f(z_0 + h_n) - g \circ f(z_0)] = 0 = g'(f(z_0))f'(z_0)$.

The general case easily follows from the preceding two. $\blacksquare$

In order to define the derivative, the function was assumed to be defined on an open set. If we say f is analytic on a set A and A is not open, we mean that f is analytic on an open set containing A.

Perhaps the definition of analytic function has been anticlimatic to many readers. After seeing books written on analytic functions and year-long courses and seminars on the theory of analytic functions, one can excuse a certain degree of disappointment in discovering that the definition has already been encountered in calculus. Is this theory to be a simple generalization of calculus? The answer is a resounding no. To show how vastly different the two subjects are let us mention that we will show that *a differentiable function is analytic*. This is truly a remarkable result and one for which there is no analogue in the theory of functions of a real variable (e.g., consider $x^2 \sin \dfrac{1}{x}$). Another equally remarkable result is that *every analytic function is infinitely differentiable and, furthermore, has a power series expansion about each point of its domain*. How can such a humble hypothesis give such far-reaching results? One can get come indication of what produces this phenomenon if one considers the definition of derivative.

In the complex variable case there are an infinity of directions in which a variable can approach a point a. In the real case, however, there are only two avenues of approach. Continuity, for example, of a function defined on $\mathbb{R}$ can be discussed in terms of right and left continuity; this is far from the case for functions of a complex variable. So the statement that a function of a complex variable has a derivative is stronger than the same statement about a function of a real variable. Even more, if we consider a function f defined on $G \subset \mathbb{C}$ as a function of two real variables by putting $g(x, y) = f(x+iy)$ for $(x, y) \in G$, then requiring that f be Frechet differentiable will not ensure that f has a derivative in our sense.

In an exercise we ask the reader to show that $f(z) = |z|^2$ has a derivative only at $z = 0$; but, $g(x, y) = f(x+iy) = x^2+y^2$ is Frechet differentiable.

That differentiability implies analyticity is proved in Chapter IV; but right now we prove that power series are analytic functions.

2.5 Proposition. *Let* $f(z) = \displaystyle\sum_{n=0}^{\infty} a_n(z-a)^n$ *have radius of convergence* $R > 0$. *Then*:

(a) *For each* $k \geq 1$ *the series*

2.6
$$\sum_{n=k}^{\infty} n(n-1) \ldots (n-k+1)a_n(z-a)^{n-k}$$

has radius of convergence R;

(b) *The function* f *is infinitely differentiable on* $B(a; R)$ *and, furthermore,* $f^{(k)}(z)$ *is given by the series* (2.6) *for all* $k \geq 1$ *and* $|z-a| < R$;

(c) *For* $n \geq 0$,

2.7
$$a_n = \frac{1}{n!}f^{(n)}(a).$$

Proof. Again assume that $a = 0$.

(a) We first remark that if (a) is proved for $k = 1$ then the cases $k = 2, \ldots$ will follow. In fact, the case $k = 2$ can be obtained by applying part (a) for $k = 1$ to the series $\sum na_n(z-a)^{n-1}$. We have that $R^{-1} = \lim \sup |a_n|^{1/n}$; we wish to show that $R^{-1} = \limsup |na_n|^{1/(n-1)}$. Now it follows from l'Hôpital's rule that $\lim_{n \to \infty} \dfrac{\log n}{n-1} = 0$, so that $\lim_{n \to \infty} n^{1/(n-1)} = 1$. The result will follow from Exercise 2 if it can be shown that $\limsup |a_n|^{1/(n-1)} = R^{-1}$.

Let $(R')^{-1} = \limsup |a_n|^{1/(n-1)}$; then R' is the radius of convergence of $\sum_{1}^{\infty} a_n z^{n-1} = \sum_{0}^{\infty} a_{n+1} z^n$. Notice that $z \sum a_{n+1} z^n + a_0 = \sum a_n z^n$; hence if $|z| < R'$ then $\sum |a_n z^n| \le |a_0| + |z| \sum |a_{n+1} z^n| < \infty$. This gives $R' \le R$. If $|z| < R$ and $z \ne 0$ then $\sum |a_n z^n| < \infty$ and $\sum |a_{n+1} z^n| \le \dfrac{1}{|z|} \cdot \sum |a_n z^n| + \dfrac{1}{|z|} |a_0| < \infty$, so that $R \le R'$. This gives that $R = R'$ and completes the proof of part (a).

(b) For $|z| < R$ put $g(z) = \sum_{n=1}^{\infty} na_n z^{n-1}$, $s_n(z) = \sum_{k=0}^{n} a_k z^k$, and $R_n(z) = \sum_{k=n+1}^{\infty} a_k z^k$. Fix a point w in $B(0; R)$ and fix r with $|w| < r < R$; we wish to show that $f'(w)$ exists and is equal to $g(w)$. To do this let $\delta > 0$ be arbitrary except for the restriction that $\bar{B}(w; \delta) \subset B(0; r)$. (We will further restrict δ later in the proof.) Let $z \in B(w; \delta)$; then

2.8
$$\frac{f(z) - f(w)}{z - w} - g(w) = \left[\frac{s_n(z) - s_n(w)}{z - w} - s_n'(w) \right] + [s_n'(w) - g(w)]$$
$$+ \left[\frac{R_n(z) - R_n(w)}{z - w} \right]$$

Now
$$\frac{R_n(z) - R_n(w)}{z - w} = \frac{1}{z - w} \sum_{k=n+1}^{\infty} a_k(z^k - w^k)$$
$$= \sum_{k=n+1}^{\infty} a_k \left(\frac{z^k - w^k}{z - w} \right)$$

But
$$\frac{|z^k - w^k|}{|z - w|} = |z^{k-1} + z^{k-2}w + \dots + zw^{k-2} + w^{k-1}| \le kr^{k-1}.$$

Hence,
$$\left| \frac{R_n(z) - R_n(w)}{z - w} \right| \le \sum_{k=n+1}^{\infty} |a_k| kr^{k-1}$$

Since $r < R$, $\sum_{k=1}^{\infty} |a_k| kr^{k-1}$ converges and so for any $\epsilon > 0$ there is an integer N_1 such that for $n \ge N_1$
$$\left| \frac{R_n(z) - R_n(w)}{z - w} \right| < \frac{\epsilon}{3} \qquad (z \in B(w; \delta)).$$

Also, $\lim s_n'(w) = g(w)$ so there is an integer N_2 such that $|s_n'(w) - g(w)| < \dfrac{\epsilon}{3}$ whenever $n \ge N_2$. Let $n =$ the maximum of the two integers N_1 and N_2.

Then we can choose $\delta > 0$ such that

$$\left| \frac{s_n(z) - s_n(w)}{z - w} - s_n'(w) \right| < \frac{\epsilon}{3}$$

whenever $0 < |z - w| < \delta$. Putting these inequalities together with equation (2.8) we have that

$$\left| \frac{f(z) - f(w)}{z - w} - g(w) \right| < \epsilon$$

for $0 < |z - w| < \delta$. That is, $f'(w) = g(w)$.

(c) By a straightforward evaluation we get $f(0) = f^{(0)}(0) = a_0$. Using (2.6) (for $a = 0$), we get $f^{(k)}(0) = k! a_k$ and this gives formula (2.7). ∎

2.9 Corollary. *If the series $\sum\limits_{n=0}^{\infty} a_n(z - a)^n$ has radius of convergence $R > 0$ then*

$f(z) = \sum\limits_{n=0}^{\infty} a_n(z - a)^n$ *is analytic in $B(a; R)$.*

Hence, $\exp z = \sum\limits_{n=0}^{\infty} z^n / n!$ is analytic in $\mathbb{C}$. Before further examining the exponential function and defining $\cos z$ and $\sin z$, the following result must be proved.

2.10 Proposition. *If G is open and connected and $f: G \to \mathbb{C}$ is differentiable with $f'(z) = 0$ for all z in G, then f is constant.*

Proof. Fix z_0 in G and let $\omega_0 = f(z_0)$. Put $A = \{z \in G : f(z) = \omega_0\}$; we will show that $A = G$ by showing that A is both open and closed in G. Let $z \in G$ and let $\{z_n\} \subset A$ be such that $z = \lim z_n$. Since $f(z_n) = \omega_0$ for each $n \geq 1$ and f is continuous we get $f(z) = \omega_0$, or $z \in A$. Thus, A is closed in G. Now fix a in A, and let $\epsilon > 0$ be such that $B(a; \epsilon) \subset G$. If $z \in B(a; \epsilon)$, set $g(t) = f(tz + (1-t)a)$, $0 \leq t \leq 1$. Then

2.11 $$\frac{g(t) - g(s)}{t - s} = \frac{g(t) - g(s)}{(t-s)z + (s-t)a} \cdot \frac{(t-s)z + (s-t)a}{t - s}.$$

Thus, if we let $t \to s$ we get (A.4(b), Appendix A)

$$\lim_{t \to s} \frac{g(t) - g(s)}{t - s} = f'(sz + (1-s)a) \cdot (z - a) = 0.$$

That is, $g'(s) = 0$ for $0 \leq s \leq 1$, implying that g is a constant. Hence, $f(z) = g(1) = g(0) = f(a) = \omega_0$. That is, $B(a; \epsilon) \subset A$ and A is also open. ∎

Now differentiate $f(z) = e^z$; we do this by Proposition 2.5. This gives that

$$f'(z) = \sum_{n=1}^{\infty} \frac{n}{n!} z^{n-1} = \sum_{n=1}^{\infty} \frac{1}{(n-1)!} z^{n-1} = \sum_{n=0}^{\infty} \frac{1}{n!} z^n = f(z).$$

Thus the complex exponential function has the same property as its real counterpart. That is

2.12 $$\frac{d}{dz} e^z = e^z$$

Put $g(z) = e^z e^{a-z}$ for some fixed a in $\mathbb{C}$; then $g'(z) = e^z e^{a-z} + e^z(-e^{a-z}) = 0$. Hence $g(z) = \omega$ for all z in $\mathbb{C}$ and some constant ω. In particular, using $e^0 = 1$ we get $\omega = g(0) = e^a$. Then $e^z e^{a-z} = e^a$ for all z. Thus $e^{a+b} = e^a e^b$ for all a and b in $\mathbb{C}$. This also gives $1 = e^z e^{-z}$ which implies that $e^z \neq 0$ for any z and $e^{-z} = 1/e^z$. Returning to the power series expansion of e^z, since all the coefficients of this series are real we have $\exp \bar{z} = \overline{\exp z}$. In particular, for θ a real number we get $|e^{i\theta}|^2 = e^{i\theta} e^{-i\theta} = e^0 = 1$. More generally, $|e^z|^2 = e^z e^{\bar{z}} = e^{z+\bar{z}} = \exp(2\,\mathrm{Re}\,z)$. Thus,

2.13 $|\exp z| = \exp(\mathrm{Re}\,z)$.

We see, therefore, that e^z has the same properties that the real function e^x has. Again by analogy with the real power series we define the functions $\cos z$ and $\sin z$ by the power series

$$\cos z = 1 - \frac{z^2}{2!} + \frac{z^4}{4!} + \ldots + (-1)^n \frac{z^{2n}}{(2n)!} + \ldots$$

$$\sin z = z - \frac{z^3}{3!} + \frac{z^5}{5!} + \ldots + (-1)^n \frac{z^{2n+1}}{(2n+1)!} + \ldots$$

Each of the series has infinite radius of convergence and so $\cos z$ and $\sin z$ are analytic in $\mathbb{C}$. By using Proposition 2.5 we find that $(\cos z)' = -\sin z$ and $(\sin z)' = \cos z$. By manipulating power series (which is justified since these series converge absolutely)

2.14 $\cos z = \frac{1}{2}(e^{iz} + e^{-iz})$ $\sin z = \frac{1}{2i}(e^{iz} - e^{-iz})$

This gives for z in $\mathbb{C}$, $\cos^2 z + \sin^2 z = 1$ and

2.15 $e^{iz} = \cos z + i \sin z$.

In particular if we let $z = $ a real number θ in (2.15) we get $e^{i\theta} = \mathrm{cis}\,\theta$. Hence, for z in $\mathbb{C}$

2.16 $z = |z|e^{i\theta}$

where $\theta = \arg z$. Since $e^{x+iy} = e^x e^{iy}$ we have $|e^z| = \exp(\mathrm{Re}\,z)$ and $\arg e^z = \mathrm{Im}\,z$.

A function f is *periodic* with *period* c if $f(z+c) = f(z)$ for all z in $\mathbb{C}$. If c is a period of e^z then $e^z = e^{z+c} = e^z e^c$ implies that $e^c = 1$. Since $1 = |e^c| = \exp \mathrm{Re}(c)$, $\mathrm{Re}(c) = 0$. Thus $c = i\theta$ for some θ in $\mathbb{R}$. But $1 = e^c = e^{i\theta} = \cos\theta + i\sin\theta$ gives that the periods of e^z are the multiples of $2\pi i$. Thus, if we divide the plane into infinitely many horizontal strips by the lines $\mathrm{Im}\,z = \pi(2k-1)$, k any integer, the exponential function behaves the same in each of these strips. This property of periodicity is one which is not present in the real exponential function. Notice that by examining complex functions we have demonstrated a relationship (2.15) between the exponential function and the trigonometric functions which was not expected from our knowledge of the real case.

Now let us define log z. We could adopt the same procedure as before and let log z be the power series expansion of the real logarithm about some point. But this only gives log z in some disk. The method of defining the logarithm as the integral of t^{-1} from 1 to x, $x > 0$, is a possibility, but proves to be risky and unsatisfying in the complex case. Also, since e^z is not a one-one map as in the real case, log z cannot be defined as the inverse of e^z. We can, however, do something similar.

We want to define log w so that it satisfies $w = e^z$ when $z = $ log w. Now since $e^z \neq 0$ for any z we cannot define log 0. Therefore, suppose $e^z = w$ and $w \neq 0$; if $z = x+iy$ then $|w| = e^x$ and $y = $ arg $w+2\pi k$, for some k. Hence

2.17 $\{\log |w| + i(\text{arg } w + 2\pi k): k \text{ is any integer}\}$

is the solution set for $e^z = w$. (Note that log $|w|$ is the usual real logarithm.)

2.18 Definition. If G is an open connected set in $\mathbb{C}$ and $f: G \to \mathbb{C}$ is a continuous function such that $z = \exp f(z)$ for all z in G then f is a *branch of the logarithm.*

Notice that $0 \notin G$.

Suppose f is a given branch of the logarithm on the connected set G and suppose k is an integer. Let $g(z) = f(z)+2\pi ki$. Then $\exp g(z) = \exp f(z) = z$, so g is also a branch of the logarithm. Conversely, if f and g are both branches of log z then for each z in G, $f(z) = g(z)+2\pi ki$ for some integer k, where k depends on z. Does the same k work for each z in G? The answer is yes. In fact, if $h(z) = \dfrac{1}{2\pi i}[f(z)-g(z)]$ then h is continuous on G and $h(G) \subset \mathbb{Z}$, the integers. Since G is connected, $h(G)$ must also be connected (Theorem II. 5.8). Hence there is a k in $\mathbb{Z}$ with $f(z)+2\pi ki = g(z)$ for all z in G. This gives

2.19 Proposition. *If $G \subset \mathbb{C}$ is open and connected and f is a branch of* log z *on G then the totality of branches of* log z *are the functions $f(z)+2\pi ki$, $k \in \mathbb{Z}$.*

Now let us manufacture at least one branch of log z on some open connected set. Let

$$G = \mathbb{C} - \{z: z \leq 0\};$$

that is, "slit" the plane along the negative real axis. Clearly G is connected and each z in G can be uniquely represented by $z = |z|e^{i\theta}$ where $-\pi < \theta < \pi$. For θ in this range, define $f(re^{i\theta}) = \log r + i\theta$. We leave the proof of continuity to the reader (Exercise 9). It follows that f is a branch of the logarithm on G.

Is f analytic? To answer this we first prove a general fact.

2.20 Proposition. *Let G and Ω be open subsets of $\mathbb{C}$. Suppose that $f: G \to \mathbb{C}$ and $g: \Omega \to \mathbb{C}$ are continuous functions such that $f(G) \subset \Omega$ and $g(f(z)) = z$ for all z in G. If g is differentiable and $g'(z) \neq 0$, f is differentiable and*

$$f'(z) = \frac{1}{g'(f(z))}.$$

If g is analytic, f is analytic

Proof. Fix a in G and let $h \in \mathbb{C}$ such that $h \neq 0$ and $a+h \in G$. Hence $a = g(f(a))$ and $a+h = g(f(a+h))$ implies $f(a) \neq f(a+h)$. Also

$$1 = \frac{g(f(a+h)) - g(f(a))}{h}$$

$$= \frac{g(f(a+h)) - g(f(a))}{f(a+h) - f(a)} \cdot \frac{f(a+h) - f(a)}{h}.$$

Now the limit of the left hand side as $h \to 0$ is, of course, 1; so the limit of the right hand side exists. Since $\lim_{h \to 0} [f(a+h) - f(a)] = 0$,

$$\lim_{h \to 0} \frac{g(f(a+h)) - g(f(a))}{f(a+h) - f(a)} = g'(f(a)).$$

Hence we get that

$$\lim_{h \to 0} \frac{f(a+h) - f(a)}{h}$$

exists since $g'(f(a)) \neq 0$, and $1 = g'(f(a))f'(a)$.

Thus, $f'(z) = [g'(f(z))]^{-1}$. If g is analytic then g' is continuous and this gives that f is analytic. ∎

2.21 Corollary. *A branch of the logarithm function is analytic and its derivative is z^{-1}.*

We designate the particular branch of the logarithm defined above on $\mathbb{C} - \{z: z \leq 0\}$ to be the *principal branch of the logarithm*. If we write $\log z$ as a function we will always take it to be the principal branch of the logarithm unless otherwise stated.

If f is a branch of the logarithm on an open connected set G and if b in $\mathbb{C}$ is fixed then define $g: G \to \mathbb{C}$ by $g(z) = \exp(bf(z))$. If b is an integer, then $g(z) = z^b$. In this manner we define a branch of z^b, b in $\mathbb{C}$, for an open connected set on which there is a branch of $\log z$. If we write $g(z) = z^b$ as a function we will always understand that $z^b = \exp(b \log z)$ where $\log z$ is the principal branch of the logarithm; z^b is analytic since $\log z$ is.

As is evident from the considerations just concluded, connectedness plays an important role in analytic function theory. For example, Proposition 2.10 is false unless G is connected. This is analogous to the role played by intervals in calculus. Because of this it is convenient to introduce the term "region." A *region* is an open connected subset of the plane.

This section concludes with a discussion of the Cauchy-Riemann equations. Let $f: G \to \mathbb{C}$ be analytic and let $u(x, y) = \operatorname{Re} f(x+iy)$, $v(x, y) = \operatorname{Im} f(x+iy)$ for $x+iy$ in G. Let us evaluate the limit

$$f'(z) = \lim_{h \to 0} \frac{f(z+h) - f(z)}{h}$$

in two different ways. First let $h \to 0$ through real values of h. For $h \neq 0$ and h real we get

$$\frac{f(z+h)-f(z)}{h} = \frac{f(x+h+iy)-f(x+iy)}{h}$$

$$= \frac{u(x+h,\,y)-u(x,\,y)}{h} + i\,\frac{v(x+h,\,y)-v(x,\,y)}{h}$$

Letting $h \to 0$ gives

2.22
$$f'(z) = \frac{\partial u}{\partial x}(x,\,y) + i\,\frac{\partial v}{\partial x}(x,\,y)$$

Now let $h \to 0$ through purely imaginary values; that is, for $h \neq 0$ and h real,

$$\frac{f(z+ih)-f(z)}{ih} = -i\,\frac{u(x,\,y+h)-u(x,\,y)}{h} + \frac{v(x,\,y+h)-v(x,\,y)}{h}$$

Thus,

2.23
$$f'(z) = -i\,\frac{\partial u}{\partial y}(x,\,y) + \frac{\partial v}{\partial y}(x,\,y)$$

Equating the real and imaginary parts of (2.22) and (2.23) we get the *Cauchy-Riemann equations*

2.24
$$\frac{\partial u}{\partial x} = \frac{\partial v}{\partial y} \quad \text{and} \quad \frac{\partial u}{\partial y} = -\frac{\partial v}{\partial x}$$

Suppose that u and v have continuous second partial derivatives (we will eventually show that they are infinitely differentiable). Differentiating the Cauchy-Riemann equations again we get

$$\frac{\partial^2 u}{\partial x^2} = \frac{\partial^2 v}{\partial x \partial y} \quad \text{and} \quad \frac{\partial^2 u}{\partial y^2} = -\frac{\partial^2 v}{\partial y \partial x}$$

Hence,

2.25
$$\frac{\partial^2 u}{\partial x^2} + \frac{\partial^2 u}{\partial y^2} = 0.$$

Any function with continuous second derivatives satisfying (2.25) is said to be *harmonic*. In a similar fashion, v is also harmonic. We will study harmonic functions in Chapter X.

Let G be a region in the plane and let u and v be functions defined on G with continuous partial derivatives. Furthermore, suppose that u and v satisfy the Cauchy-Riemann equations. If $f(z) = u(z) + iv(z)$ then f can be shown to be analytic in G. To see this, let $z = x + iy \in G$ and let $B(z;r) \subset G$. If $h = s + it \in B(0;r)$ then

$$u(x+s,\,y+t)-u(x,\,y) = [u(x+s,\,y+t)-u(x,\,y+t)]+[u(x,\,y+t)-u(x,\,y)]$$

Applying the mean value theorem for the derivative of a function of one variable to each of these bracketed expressions, yields for each $s+it$ in $B(0;\,r)$ numbers s_1 and t_1 such that $|s_1| < |s|$ and $|t_1| < |t|$ and

2.26
$$\begin{cases} u(x+s,\,y+t)-u(x,\,y+t) = u_x(x+s_1,\,y+t)s \\ u(x,\,y+t)-u(x,\,y) = u_y(x,\,y+t_1)t \end{cases}$$

Letting

$$\varphi(s, t) = [u(x+s, y+t) - u(x, y)] - [u_x(x, y)s + u_y(x, y)t]$$

(2.26) gives that

$$\frac{\varphi(s,t)}{s+it} = \frac{s}{s+it}\left[u_x(x+s_1, y+t) - u_x(x,y)\right] + \frac{t}{s+it}\left[u_y(x,y+t_1) - u_y(x,y)\right]$$

But $|s| \le |s+it|$, $|t| \le |s+it|$, $|s_1| < |s|$, $|t_1| < |t|$, and the fact that u_x and u_y are continuous gives that

2.27 $$\lim_{s+it \to 0} \frac{\varphi(s, t)}{s+it} = 0$$

Hence

$$u(x+s, y+t) - u(x, y) = u_x(x, y)s + u_y(x, y)t + \varphi(s, t)$$

where φ satisfies (2.27). Similarly

$$v(x+s, y+t) - v(x, y) = v_x(x, y)s + v_y(x, y)t + \psi(s, t)$$

where ψ satisfies

2.28 $$\lim_{s+it \to 0} \frac{\psi(s, t)}{s+it} = 0$$

Using the fact that u and v satisfy the Cauchy-Riemann equations it is easy to see that

$$\frac{f(z+s+it) - f(z)}{s+it} = u_x(z) + iv_x(z) + \frac{\varphi(s, t) + \psi(x, t)}{s+it}$$

In light of (2.27) and (2.28), f is differentiable and $f'(z) = u_x(z) + iv_x(z)$. Since u_x and v_x are continuous, f' is continuous and f is analytic. These results are summarized as follows.

2.29. Theorem. *Let u and v be real-valued functions defined on a region G and suppose that u and v have continuous partial derivatives. Then $f: G \to \mathbb{C}$ defined by $f(z) = u(z) + iv(z)$ is analytic iff u and v satisfy the Cauchy-Riemann equations.*

Example. Is $u(x, y) = \log (x^2 + y^2)^{\frac{1}{2}}$ harmonic on $G = \mathbb{C} - \{0\}$? The answer is yes! This could be shown by differentiating u to see that it satisfies (2.25). However, it can also be shown by observing that in a neighborhood of each point of G, u is the real part of an analytic function defined in that neighborhood. (Which function?)

Another problem concerning harmonic functions which will be taken up in more detail in Section VIII. 3, is the following. Suppose G is a region in the plane and $u: G \to \mathbb{R}$ is harmonic. Does there exist a harmonic function $v: G \to \mathbb{R}$ such that $f = u + iv$ is analytic in G? If such a function v exists it is called a *harmonic conjugate* of u. If v_1 and v_2 are two harmonic conjugates of u then $i(v_1 - v_2) = (u + iv_1) - (u + iv_2)$ is analytic on G and only takes on purely imaginary values. It follows that two harmonic conjugates of a harmonic function differ by a constant (see Exercise 14).

Returning to the question of the existence of a harmonic conjugate, the above example $u(z) = \log|z|$ of a harmonic function on the region $G = \mathbb{C} - \{0\}$ has no harmonic conjugate. Indeed, if it did then it would be possible to define an analytic branch of the logarithm on G and this cannot be done. (Exercise 21.) However, there are some regions for which every harmonic function has a conjugate. In particular, it will now be shown that this is the case when G is any disk or the whole plane.

2.30 Theorem. *Let G be either the whole plane $\mathbb{C}$ or some open disk. If $u: G \to \mathbb{R}$ is a harmonic function then u has a harmonic conjugate.*

Proof. To carry out the proof of this theorem, Leibniz's rule for differentiating under the integral sign is needed (this is stated and proved in Proposition IV. 2.1). Let $G = B(0; R), 0 < R \le \infty$, and let $u: G \to \mathbb{R}$ be a harmonic function. The proof will be accomplished by finding a harmonic function v such that u and v satisfy the Cauchy-Riemann equations. So define

$$v(x, y) = \int_0^y u_x(x, t)dt + \varphi(x)$$

and determine φ so that $v_x = -u_y$. Differentiating both sides of this equation with respect to x gives

$$v_x(x, y) = \int_0^y u_{xx}(x, t)\, dt + \varphi'(x)$$

$$= -\int_0^y u_{yy}(x, t)\, dt + \varphi'(x)$$

$$= -u_y(x, y) + u_y(x, 0) + \varphi'(x)$$

So it must be that $\varphi'(x) = -u_y(x, 0)$. It is easily checked that u and

$$v(x, y) = \int_0^y u_x(x, t)dt - \int_0^x u_y(s, 0)ds$$

do satisfy the Cauchy-Riemann equations. ∎

Where was the fact that G is a disk or $\mathbb{C}$ used? Why can't this method of proof be doctored sufficiently that it holds for general regions G? Where does the proof break down when $G = \mathbb{C} - \{0\}$ and $u(z) = \log|z|$?

Exercises

1. Show that $f(z) = |z|^2 = x^2 + y^2$ has a derivative only at the origin.
2. Prove that if b_n, a_n are real and positive and $0 < b = \lim b_n$, $a = \lim \sup a_n$ then $ab = \lim \sup (a_n b_n)$. Does this remain true if the requirement of positivity is dropped?
3. Show that $\lim n^{1/n} = 1$.

4. Show that $(\cos z)' = -\sin z$ and $(\sin z)' = \cos z$.

5. Derive formulas (2.14).

6. Describe the following sets: $\{z: e^z = i\}$, $\{z: e^z = -1\}$, $\{z: e^z = -i\}$, $\{z: \cos z = 0\}$, $\{z: \sin z = 0\}$.

7. Prove formulas for $\cos(z+w)$ and $\sin(z+w)$.

8. Define $\tan z = \dfrac{\sin z}{\cos z}$; where is this function defined and analytic?

9. Suppose that z_n, $z \in G = \mathbb{C} - \{z: z \le 0\}$ and $z_n = r_n e^{i\theta_n}$, $z = re^{i\theta}$ where $-\pi < \theta, \theta_n < \pi$. Prove that if $z_n \to z$ then $\theta_n \to \theta$ and $r_n \to r$.

10. Prove the following generalization of Proposition 2.20. Let G and Ω be open in $\mathbb{C}$ and suppose f and h are functions defined on G, $g: \Omega \to \mathbb{C}$ and suppose that $f(G) \subset \Omega$. Suppose that g and h are analytic, $g'(\omega) \ne 0$ for any ω, that f is continuous, h is one-one, and that they satisfy $h(z) = g(f(z))$ for z in G. Show that f is analytic. Give a formula for $f'(z)$.

11. Suppose that $f: G \to \mathbb{C}$ is a branch of the logarithm and that n is an integer. Prove that $z^n = \exp(nf(z))$ for all z in G.

12. Show that the real part of the function $z^{\frac{1}{2}}$ is always positive.

13. Let $G = \mathbb{C} - \{z: z \le 0\}$ and let n be a positive integer. Find all analytic functions $f: G \to \mathbb{C}$ such that $z = (f(z))^n$ for all $z \in G$.

14. Suppose $f: G \to \mathbb{C}$ is analytic and that G is connected. Show that if $f(z)$ is real for all z in G then f is constant.

15. For $r > 0$ let $A = \left\{\omega: \omega = \exp\left(\dfrac{1}{z}\right) \text{ where } 0 < |z| < r\right\}$; determine the set A.

16. Find an open connected set $G \subset \mathbb{C}$ and two continuous functions f and g defined on G such that $f(z)^2 = g(z)^2 = 1 - z^2$ for all z in G. Can you make G maximal? Are f and g analytic?

17. Give the principal branch of $\sqrt{1-z}$.

18. Let $f: G \to \mathbb{C}$ and $g: G \to \mathbb{C}$ be branches of z^a and z^b respectively. Show that fg is a branch of z^{a+b} and f/g is a branch of z^{a-b}. Suppose that $f(G) \subset G$ and $g(G) \subset G$ and prove that both $f \circ g$ and $g \circ f$ are branches of z^{ab}.

19. Let G be a region and define $G^* = \{z: \bar{z} \in G\}$. If $f: G \to \mathbb{C}$ is analytic prove that $f^*: G^* \to C$, defined by $f^*(z) = \overline{f(\bar{z})}$, is also analytic.

20. Let $z_1, z_2, \ldots, z_n$ be complex numbers such that $\operatorname{Re} z_k > 0$ and $\operatorname{Re}(z_1 \ldots z_k) > 0$ for $1 \le k \le n$. Show that $\log(z, \ldots z_n) = \log z_1 + \ldots + \log z_n$, where $\log z$ is the principal branch of the logarithm. If the restrictions on the z_k are removed, does the formula remain valid?

21. Prove that there is no branch of the logarithm defined on $G = \mathbb{C} - \{0\}$. (Hint: Suppose such a branch exists and compare this with the principal branch.)

§3. Analytic functions as mappings. Möbius transformations

Consider the function defined by $f(z) = z^2$. If $z = x + iy$ and $\mu + iv = f(z)$ then $\mu = x^2 - y^2$, $v = 2xy$. Hence, the hyperbolas $x^2 - y^2 = c$ and $2xy = d$ are mapped by f into the straight lines $\mu = c$, $v = d$. One interesting fact is that for c and d not zero, these hyperbolas intersect at right angles, just as their images do. This is not an isolated phenomenon and this property will be explored in general later in this section.

Now examine what happens to the lines $x = c$ and $y = d$. First consider $x = c$ (y arbitrary); f maps this line into $\mu = c^2 - y^2$ and $v = 2cy$. Eliminating y we get that $x = c$ is mapped onto the parabola $v^2 = -4c^2(\mu - c^2)$. Similarly, f takes the line $y = d$ onto the parabola $v^2 = 4d^2(\mu + d^2)$. These parabolas intersect at $(c^2 - d^2, \pm 2|cd|)$. It is relevant to point out that as $c \to 0$ the parabola $v^2 = -4c^2(\mu - c^2)$ gets closer and closer to the negative real axis. This corresponds to the fact that the function $z^{\frac{1}{2}}$ maps $G = \mathbb{C} - \{z : z \le 0\}$ onto $\{z : \operatorname{Re} z > 0\}$. Notice also that $x = c$ and $x = -c$ (and $y = d$, $y = -d$) are mapped onto the same parabolas.

What happens to a circle centered at the origin? If $z = re^{i\theta}$ then $f(z) = r^2 e^{2i\theta}$; thus, the circle of radius r about the origin is mapped onto the circle of radius r^2 in a two to one fashion.

Finally, what happens to the sector $S(\alpha, \beta) = \{z : \alpha < \arg z < \beta\}$, for $\alpha < \beta$? It is easily seen that the image of $S(\alpha, \beta)$ is the sector $S(2\alpha, 2\beta)$. The restriction of f to $S(\alpha, \beta)$ will be one-one exactly when $\beta - \alpha < \pi$.

The above discussion sheds some light on the nature of $f(z) = z^2$ and, likewise, it is useful to study the mapping properties' other analytic functions. In the theory of analytic functions the following problem holds a paramount position: given two open connected sets G and Ω, is there an analytic function f defined on G such that $f(G) = \Omega$? Besides being intrinsically interesting, the solution (or rather, the information about the existence of a solution) of this problem is very useful.

3.1 Definition. A *path* in a region $G \subset \mathbb{C}$ is a continuous function $\gamma : [a,b] \to G$ for some interval $[a,b]$ in $\mathbb{R}$. If $\gamma'(t)$ exists for each t in $[a,b]$ and $\gamma' : [a,b] \to \mathbb{C}$ is continuous then γ is a *smooth path*. Also γ is *piecewise smooth* if there is a partition of $[a,b]$, $a = t_0 < t_1 < \ldots < t_n = b$, such that γ is smooth on each subinterval $[t_{j-1}, t_j]$, $1 \le j \le n$.

To say that a function $\gamma : [a,b] \to \mathbb{C}$ has a derivative $\gamma'(t)$ for each point t in $[a,b]$ means that

$$\lim_{h \to 0} \frac{\gamma(t+h) - \gamma(t)}{h} = \gamma'(t)$$

exists for $a < t < b$ and that the right and left sided limits exist for $t = a$ and $t = b$, respectively. This is, of course, equivalent to saying that $\operatorname{Re} \gamma$ and $\operatorname{Im} \gamma$ have a derivative (see Appendix A).

Suppose $\gamma : [a,b] \to G$ is a smooth path and that for some t_0 in (a,b), $\gamma'(t_0) \ne 0$. Then γ has a tangent line at the point $z_0 = \gamma(t_0)$. This line goes through the point z_0 in the direction of (the vector) $\gamma'(t_0)$; or, the slope of the line is $\tan(\arg \gamma'(t_0))$. If γ_1 and γ_2 are two smooth paths with $\gamma_1(t_1) = \gamma_2(t_2) = z_0$ and $\gamma_1'(t_1) \ne 0$, $\gamma_2'(t_2) \ne 0$, then define the *angle between the paths* γ_1 and γ_2 at z_0 to be

$$\arg \gamma_2'(t_2) - \arg \gamma_1'(t_1).$$

Suppose γ is a smooth path in G and $f: G \to \mathbb{C}$ is analytic. Then $\sigma = f \circ \gamma$ is also a smooth path and $\sigma'(t) = f'(\gamma(t))\gamma'(t)$. Let $z_0 = \gamma(t_0)$, and suppose that $\gamma'(t_0) \neq 0$ and $f'(z_0) \neq 0$; then $\sigma'(t_0) \neq 0$ and $\arg \sigma'(t_0) = \arg f'(z_0) + \arg \gamma'(t_0)$. That is,

3.2 $$\arg \sigma'(t_0) - \arg \gamma'(t_0) = \arg f'(z_0).$$

Now let γ_1 and γ_2 be smooth paths with $\gamma_1(t_1) = \gamma_2(t_2) = z_0$ and $\gamma_1'(t_1) \neq 0 \neq \gamma_2'(t_2)$; let $\sigma_1 = f \circ \gamma_1$ and $\sigma_2 = f \circ \gamma_2$. Also, suppose that the paths γ_1 and γ_2 are not tangent to each other at z_0; that is, suppose $\gamma_1'(t_1) \neq \gamma_2'(t_2)$. Equation (3.2) gives

3.3 $$\arg \gamma_2'(t_2) - \arg \gamma_1'(t_1) = \arg \sigma_2'(t_2) - \arg \sigma_1'(t_1).$$

This says that given any two paths through z_0, f maps these paths onto two paths through $\omega_0 = f(z_0)$ and, when $f'(z_0) \neq 0$, the angles between the curves are preserved both in *magnitude and direction*. This summarizes as follows.

3.4 Theorem. *If $f: G \to \mathbb{C}$ is analytic then f preserves angles at each point z_0 of G where $f'(z_0) \neq 0$.*

A function $f: G \to \mathbb{C}$ which has the angle preserving property and also has

$$\lim_{z \to a} \frac{|f(z) - f(a)|}{|z - a|}$$

existing is called a *conformal map*. If f is analytic and $f'(z) \neq 0$ for any z then f is conformal. The converse of this statement is also true.

If $f(z) = e^z$ then f is conformal throughout $\mathbb{C}$; let us look at the exponential function more closely. If $z = c + iy$ where c is fixed then $f(z) = re^{iy}$ for $r = e^c$. That is, f maps the line $x = c$ onto the circle with center at the origin and of radius e^c. Also, f maps the line $y = d$ onto the infinite ray $\{re^{id}: 0 < r < \infty\}$.

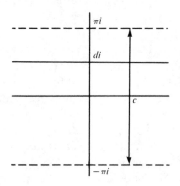
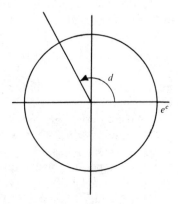

We have already seen that e^z is one-one on any horizontal strip of width $< 2\pi$. Let $G = \{z: -\pi < \operatorname{Im} z < \pi\}$. Then $f(G) = \Omega = \mathbb{C} - \{z: z \leq 0\}$; also

f maps the vertical segments $\{z = c+iy, -\pi < y < \pi\}$ onto the part of the circle $\{e^c e^{i\theta}: -\pi < \theta < \pi\}$, and the horizontal line $y = d$, $-\pi < d < \pi$, goes onto the ray making an angle d with the positive real axis.

Notice that log z, the principal branch of the logarithm, does the opposite. It maps Ω onto the strip G, circles onto vertical segments in G, rays onto horizontal lines in G.

The exploration of the mapping properties of cos z, sin z, and other analytic functions will be done in the exercises. We now proceed to an amazing class of mappings, the Möbius transformations.

3.5 Definition. A mapping of the form $S(z) = \dfrac{az+b}{cz+d}$ is called a *linear fractional transformation*. If a, b, c, and d also satisfy $ad-bc \neq 0$ then $S(z)$ is called a *Möbius transformation*.

If S is a Möbius transformation then $S^{-1}(z) = \dfrac{dz-b}{-cz+a}$ satisfies $S(S^{-1}(z)) = S^{-1}(S(z)) = z$; that is, S^{-1} is the inverse mapping of S. If S and T are both linear fractional transformations then it follows that $S \circ T$ is also. Hence, the set of Möbius maps forms a group under composition. Unless otherwise stated, the only linear fractional transformations we will consider are Möbius transformations.

Let $S(z) = \dfrac{az+b}{cz+d}$; if λ is any non-zero complex number, then

$$S(z) = \frac{(\lambda a)z+(\lambda b)}{(\lambda c)z+(\lambda d)}.$$

That is, the coefficients a, b, c, d are not unique (see Exercise 20).

We may also consider S as defined on $\mathbb{C}_\infty$ with $S(\infty) = a/c$ and $S(-d/c) = \infty$. (Notice that we cannot have $a = 0 = c$ or $d = 0 = c$ since either situation would contradict $ad-bc \neq 0$.) Since S has an inverse it maps $\mathbb{C}_\infty$ onto $\mathbb{C}_\infty$.

If $S(z) = z+a$ then S is called a *translation*; if $S(z) = az$ with $a \neq 0$ then S is a *dilation*; if $S(z) = e^{i\theta}z$ then it is a *rotation*; finally, if $S(z) = 1/z$ it is the *inversion*.

3.6 Proposition. *If S is a Möbius transformation then S is the composition of translations, dilations, and the inversion.* (Of course, some of these may be missing.)

Proof. First, suppose $c = 0$. Hence $S(z) = (a/d)z+(b/d)$ so if $S_1(z) = (a/d)z$, $S_2(z) = z+(b/d)$, then $S_2 \circ S_1 = S$ and we are done.

Now let $c \neq 0$ and put $S_1(z) = z+d/c$, $S_2(z) = 1/z$, $S_3(z) = \dfrac{(bc-ad)}{c^2}z$, $S_4(z) = z+a/c$. Then $S = S_4 \circ S_3 \circ S_2 \circ S_1$. ∎

What are the finite fixed points of S? That is, what are the points z satisfying $S(z)=z$. If z satisfies this condition then

$$cz^2+(d-a)z-b = 0.$$

Hence, a Möbius transformation can have at most two fixed points unless $S(z) = z$ for all z. (Why?)

Now let S be a Möbius transformation and let a, b, c be distinct points in $\mathbb{C}_\infty$ with $\alpha = S(a)$, $\beta = S(b)$, $\gamma = S(c)$. Suppose that T is another map with this property. Then $T^{-1} \circ S$ has a, b, and c as fixed points and, therefore, $T^{-1} \circ S = I = $ the identity. That is, $S = T$. Hence, a Möbius map is uniquely determined by its action on any three given points in $\mathbb{C}_\infty$.

Let z_2, z_3, z_4 be points in $\mathbb{C}_\infty$. Define $S: \mathbb{C}_\infty \to \mathbb{C}_\infty$ by

$$S(z) = \left(\frac{z - z_3}{z - z_4} \right) \bigg/ \left(\frac{z_2 - z_3}{z_2 - z_4} \right) \quad \text{if} \quad z_2, z_3, z_4 \in \mathbb{C};$$

$$S(z) = \frac{z - z_3}{z - z_4} \quad \text{if} \quad z_2 = \infty;$$

$$S(z) = \frac{z_2 - z_4}{z - z_4} \quad \text{if} \quad z_3 = \infty;$$

$$S(z) = \frac{z - z_3}{z_2 - z_3} \quad \text{if} \quad z_4 = \infty.$$

In any case $S(z_2) = 1$, $S(z_3) = 0$, $S(z_4) = \infty$ and S is the only transformation having this property.

3.7 Definition. If $z_1 \in \mathbb{C}_\infty$ then (z_1, z_2, z_3, z_4). (The *cross ratio* of z_1, z_2, z_3, and z_4) is the image of z_1 under the unique Möbius transformation which takes z_2 to 1, z_3 to 0, and z_4 to ∞.

For example: $(z_2, z_2, z_3, z_4) = 1$ and $(z, 1, 0, \infty) = z$. Also, if M is any Möbius map and w_2, w_3, w_4 are the points such that $Mw_2 = 1$, $Mw_3 = 0$, $Mw_4 = \infty$ then $Mz = (z, w_2, w_3, w_4)$.

3.8 Proposition. *If z_2, z_3, z_4 are distinct points and T is any Möbius transformation then*

$$(z_1, z_2, z_3, z_4) = (Tz_1, Tz_2, Tz_3, Tz_4)$$

for any point z_1.

Proof. Let $Sz = (z, z_2, z_3, z_4)$; then S is a Möbius map. If $M = ST^{-1}$ then $M(Tz_2) = 1$, $M(Tz_3) = 0$, $M(Tz_4) = \infty$; hence, $ST^{-1}z = (z, Tz_2, Tz_3, Tz_4)$ for all z in $\mathbb{C}_\infty$. In particular, if $z = Tz_1$ the desired result follows. ∎

3.9 Proposition. *If z_2, z_3, z_4 are distinct points in $\mathbb{C}_\infty$ and ω_2, ω_3, ω_4 are also distinct points of $\mathbb{C}_\infty$, then there is one and only one Möbius transformation S such that $Sz_2 = \omega_2$, $Sz_3 = \omega_3$, $Sz_4 = \omega_4$*

Proof. Let $Tz = (z, z_2, z_3, z_4)$, $Mz = (z, \omega_2, \omega_3, \omega_4)$ and put $S = M^{-1}T$. Clearly S has the desired property. If R is another Möbius map with $Rz_j = \omega_j$ for $j = 2, 3, 4$ then $R^{-1} \circ S$ has three fixed points (z_2, z_3, and z_4). Hence $R^{-1} \circ S = I$, or $S = R$. ∎

It is well known from high school geometry that three points in the plane determine a circle. (Recall that a circle in $\mathbb{C}_\infty$ passing through ∞ corresponds to a straight line in $\mathbb{C}$. Hence there is no need to inject in the previous statement the word "non-colinear.") A straight line in the plane will be called a circle.) The next result explains when four points lie on a circle.

3.10 Proposition. *Let z_1, z_2, z_3, z_4 be four distinct points in $\mathbb{C}_\infty$. Then (z_1, z_2, z_3, z_4) is a real number iff all four points lie on a circle.*

Proof. Let $S: \mathbb{C}_\infty \to \mathbb{C}_\infty$ be defined by $Sz = (z, z_2, z_3, z_4)$; then $S^{-1}(\mathbb{R}) =$ the set of z such that (z, z_2, z_3, z_4) is real. Hence, we will be finished if we can show that the image of $\mathbb{R}_\infty$ under a Möbius transformation is a circle.

Let $Sz = \dfrac{az+b}{cz+d}$; if $z = x \in \mathbb{R}$ and $\omega = S^{-1}(x)$ then $x = S\omega$ implies that $S(\omega) = \overline{S(\omega)}$. That is,

$$\frac{a\omega+b}{c\omega+d} = \frac{\bar{a}\bar{\omega}+\bar{b}}{\bar{c}\bar{\omega}+\bar{d}}$$

Cross multiplying this gives

3.11 $\qquad (a\bar{c}-\bar{a}c)|\omega|^2 + (a\bar{d}-\bar{b}c)\omega + (b\bar{c}-d\bar{a})\bar{\omega} + (b\bar{d}-\bar{b}d) = 0.$

If $a\bar{c}$ is real then $a\bar{c}-\bar{a}c = 0$; putting $\alpha = 2(a\bar{d}-\bar{b}c)$, $\beta = i(b\bar{d}-\bar{b}d)$ and multiplying (3.11) by i gives

3.12 $\qquad\qquad 0 = \text{Im}\,(\alpha\omega)-\beta = \text{Im}\,(\alpha\omega-\beta)$

since β is real. That is, ω lies on the line determined by (3.12) for fixed α and β. If $a\bar{c}$ is not real then (3.11) becomes

$$|\omega|^2 + \bar{\gamma}\omega + \gamma\bar{\omega} - \delta = 0$$

for some constants γ in $\mathbb{C}$, δ in $\mathbb{R}$. Hence,

3.13 $\qquad\qquad\qquad |\omega+\gamma| = \lambda$

where

$$\lambda = (|\gamma|^2+\delta)^{\frac{1}{2}} = \left|\frac{ad-bc}{\bar{a}c-a\bar{c}}\right| > 0.$$

Since γ and λ are independent of x and since (3.13) is the equation of a circle, the proof is finished. ∎

3.14 Theorem. *A Möbius transformation takes circles onto circles.*

Proof. Let Γ be any circle in $\mathbb{C}_\infty$ and let S be any Möbius transformation. Let z_2, z_3, z_4 be three distinct points on Γ and put $\omega_j = Sz_j$ for $j = 2, 3, 4$. Then $\omega_2, \omega_3, \omega_4$ determine a circle Γ'. We claim that $S(\Gamma) = \Gamma'$. In fact, for any z in $\mathbb{C}_\infty$

3.15 $\qquad\qquad (z, z_2, z_3, z_4) = (Sz, \omega_2, \omega_3, \omega_4)$

by Proposition 3.8. By the preceding proposition, if z is on Γ then both sides of (3.15) are real. But this says that $Sz \in \Gamma'$. ∎

Now let Γ and Γ' be two circles in $\mathbb{C}_\infty$ and let $z_2, z_3, z_4 \in \Gamma$; $\omega_2, \omega_3, \omega_4 \in \Gamma'$. Put $Rz = (z, z_2, z_3, z_4)$, $Sz = (z, \omega_2, \omega_3, \omega_4)$. Then $T = S^{-1} \circ R$ maps

Γ onto Γ'. In fact, $Tz_j = \omega_j$ for $j = 2, 3, 4$ and, as in the above proof, it follows that $T(\Gamma) = \Gamma'$.

3.16 Proposition. *For any given circles Γ and Γ' in $\mathbb{C}_\infty$ there is a Möbius transformation T such that $T(\Gamma) = \Gamma'$. Furthermore we can specify that T take any three points on Γ onto any three points of Γ'. If we do specify Tz_j for $j = 2, 3, 4$ (distinct z_j in Γ) then T is unique.*

Proof. The proof, except for the uniqueness statement, is given in the previous paragraph. The uniqueness part is a trivial exercise for the reader. ∎

Now that we know that a Möbius map takes circles to circles, the next question is: What happens to the inside and the outside of these circles? To answer this we introduce some new concepts.

3.17 Definition. Let Γ be a circle through points z_2, z_3, z_4. The points z, z^* in $\mathbb{C}_\infty$ are said to be *symmetric* with respect to Γ if

3.18 $(z^*, z_2, z_3, z_4) = \overline{(z, z_2, z_3, z_4)}.$

As it stands, this definition not only depends on the circle but also on the points z_2, z_3, z_4. It is left as an exercise for the reader to show that symmetry is independent of the points chosen (Exercise 11).

Also, by Proposition 3.10 z is symmetric to itself with respect to Γ if and only if $z \in \Gamma$.

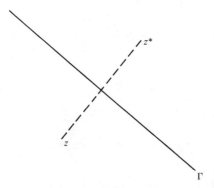

Let us investigate what it means for z and z^* to be symmetric. If Γ is a straight line then our linguistic prejudices lead us to believe that z and z^* are symmetric with respect to Γ if the line through z and z^* is perpendicular to Γ and z and z^* are the same distance from Γ but on opposite sides of Γ. This is indeed the case.

If Γ is a straight line then, choosing $z_4 = \infty$, equation (3.18) becomes

$$\frac{z^* - z_3}{z_2 - z_3} = \frac{\bar{z} - \bar{z}_3}{\bar{z}_2 - \bar{z}_3}.$$

This gives $|z^* - z_3| = |z - z_3|$; since z_3 was not specified, we have that z

and z^* are equidistant from each point on Γ. Also

$$\operatorname{Im}\frac{z^*-z_3}{z_2-z_3} = \operatorname{Im}\frac{\bar{z}-\bar{z}_3}{\bar{z}_2-\bar{z}_3}$$

$$= -\operatorname{Im}\frac{z-z_3}{z_2-z_3}$$

Hence, we have (unless $z \in \Gamma$) that z and z^* lie in different half planes determined by Γ. It now follows that $[z, z^*]$ is perpendicular to Γ.

Now suppose that $\Gamma = \{z: |z-a| = R\}$ $(0 < R < \infty)$. Let z_2, z_3, z_4 be points in Γ; using (3.18) and Proposition 3.8 for a number of Möbius transformations gives

$$(z^*, z_2, z_3, z_4) = \overline{(z, z_2, z_3, z_4)}$$

$$= \overline{(z-a, z_2-a, z_3-a, z_4-a)}$$

$$= \left(\bar{z}-\bar{a}, \frac{R^2}{z_2-a}, \frac{R^2}{z_3-a}, \frac{R^2}{z_4-a}\right)$$

$$= \left(\frac{R^2}{\bar{z}-\bar{a}}, z_2-a, z_3-a, z_4-a\right)$$

$$= \left(\frac{R^2}{\bar{z}-\bar{a}} + a, z_2, z_3, z_4\right)$$

Hence, $z^* = a+R^2(\bar{z}-\bar{a})^{-1}$ or $(z^*-a)(\bar{z}-\bar{a}) = R^2$. From this it follows that

$$\frac{z^*-a}{z-a} = \frac{R^2}{|z-a|^2} > 0,$$

so that z^* lies on the ray $\{a+t(z-a): 0 < t < \infty\}$ from a through z. Using the fact that $|z-a|\,|z^*-a| = R^2$ we can obtain z^* from z (if z lies inside Γ) as in the figure below. That is: Let L be the ray from a through z. Construct

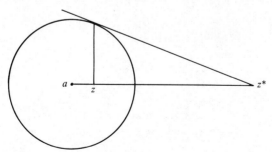

a line P perpendicular to L at z and at the point where P intersects Γ construct the tangent to Γ. The point of intersection of this tangent with L is the point z^*. Thus, the points a and ∞ are symmetric with respect to Γ.

3.19 Symmetry Principle. *If a Möbius transformation T takes a circle Γ_1 onto the circle Γ_2 then any pair of points symmetric with respect to Γ_1 are mapped by T onto a pair of points symmetric with respect to Γ_2.*

Proof. Let z_2, z_3, $z_4 \in \Gamma_1$; it follows that if z and z^* are symmetric with respect to Γ_1 then

$$(Tz^*, Tz_2, Tz_3, Tz_4) = (z^*, z_2, z_3, z_4)$$
$$= \overline{(z, z_2, z_3, z_4)}$$
$$= \overline{(Tz, Tz_2, Tz_3, Tz_4)}$$

by Proposition 3.8. Hence Tz^* and Tz are symmetric with respect to Γ_2. ∎

Now we will discuss orientation for circles in $\mathbb{C}_\infty$; this will enable us to distinguish between the "inside" and "outside" of a circle in $\mathbb{C}_\infty$. Notice that on $\mathbb{C}_\infty$ (the sphere) there is no obvious choice for the inside and outside of a circle.

3.20 Definition. If Γ is a circle then an *orientation* for Γ is an ordered triple of points (z_1, z_2, z_3) such that each z_j is in Γ.

Intuitively, these three points give a direction to Γ. That is we "go" from z_1 to z_2 to z_3. If only two points were given, this would, of course, be ambiguous.

Let $\Gamma = \mathbb{R}$ and let z_1, z_2, $z_3 \in \mathbb{R}$; also, put $Tz = (z, z_1, z_2, z_3) = \dfrac{az+b}{cz+d}$.

Since $T(\mathbb{R}_\infty) = \mathbb{R}_\infty$ it follows that a, b, c, d can be chosen to be real numbers (see Exercise 8). Hence,

$$Tz = \frac{az+b}{cz+d}$$
$$= \frac{az+b}{|cz+d|^2} \cdot (c\bar{z}+d)$$
$$= \frac{1}{|cz+d|^2}[ac|z|^2 + bd + bc\bar{z} + adz]$$

Hence,

$$\mathrm{Im}\,(z, z_1, z_2, z_3) = \frac{(ad-bc)}{|cz+d|^2}\,\mathrm{Im}\,z.$$

Thus, $\{z: \mathrm{Im}\,(z, z_1, z_2, z_3) < 0\}$ is either the upper or lower half plane depending on whether $(ad-bc) > 0$ or $(ad-bc) < 0$. (Note that $ad-bc$ is the "determinant" of T.)

Now let Γ be arbitrary, and suppose that z_1, z_2, z_3 are on Γ; for any Möbius transformation S we have (by Proposition 3.8)

$$\{z: \mathrm{Im}\,(z, z_1, z_2, z_3) > 0\} = \{z: \mathrm{Im}\,(Sz, Sz_1, Sz_2, Sz_3) > 0\}$$
$$= S^{-1}\{z: \mathrm{Im}\,(z, Sz_1, Sz_2, Sz_3) > 0\}$$

In particular, if S is chosen so that S maps Γ onto $\mathbb{R}_\infty$, then $\{z: \mathrm{Im}\,(z, z_1, z_2, z_3) > 0\}$ is equal to S^{-1} of either the upper or lower half plane.

If (z_1, z_2, z_3) is an orientation of Γ then we define the *right side of* Γ (with respect to (z_1, z_2, z_3)) to be

$$\{z: \mathrm{Im}\,(z, z_1, z_2, z_3) > 0\}.$$

Similarly, we define the *left side of* Γ to be

$$\{z \colon \operatorname{Im} (z, z_1, z_2, z_3) < 0\}.$$

The proof of the following theorem is left as an exercise.

3.21 Orientation Principle. *Let* Γ_1 *and* Γ_2 *be two circles in* $\mathbb{C}_\infty$ *and let* T *be a Möbius transformation such that* $T(\Gamma_1) = \Gamma_2$. *Let* (z_1, z_2, z_3) *be an orientation for* Γ_1. *Then* T *takes the right side and the left side of* Γ_1 *onto the right side and left side of* Γ_2 *with respect to the orientation* (Tz_1, Tz_2, Tz_3).

Consider the orientation $(1, 0, \infty)$ of $\mathbb{R}$. By the definition of the cross ratio, $(z, 1, 0, \infty) = z$. Hence, the right side of $\mathbb{R}$ with respect to $(1, 0, \infty)$ is the upper half plane. This fits our intuition that the right side lies on our right as we walk along $\mathbb{R}$ from 1 to 0 to ∞.

As an example consider the following problem: Find an analytic function $f \colon G \to \mathbb{C}$, where $G = \{z \colon \operatorname{Re} z > 0\}$, such that $f(G) = D = \{z \colon |z| < 1\}$. We solve this problem by finding a Möbius transformation which takes the imaginary axis onto the unit circle and, by the Orientation Principle, takes G onto D (that is, we must choose this map carefully in order that it does not send G onto $\{z \colon |z| > 1\}$).

If we give the imaginary axis the orientation $(-i, 0, i)$ then $\{z \colon \operatorname{Re} z > 0\}$ is on the right of this axis. In fact,

$$(z, -i, 0, i) = \frac{2z}{z-i}$$

$$= \frac{2z}{z-i} \cdot \frac{\bar{z}+i}{\bar{z}+i}$$

$$= \frac{2}{|z-i|^2} \cdot (|z|^2 + iz)$$

Hence, $\{z \colon \operatorname{Im} (z, -i, 0, i) > 0\} = \{z \colon \operatorname{Im} (iz) > 0\} = \{z \colon \operatorname{Re} z > 0\}$. Giving Γ the orientation $(-i, -1, i)$ we have that D lies on the right of Γ. Also,

$$(z, -i, -1, i) = \frac{2i}{i-1} \cdot \frac{z+1}{z-i}.$$

If

$$Sz = \frac{2z}{z-i} \quad \text{and} \quad Rz = \left(\frac{2i}{i-1}\right)\left(\frac{z+1}{z-i}\right)$$

then $T = R^{-1}S$ maps G onto D (and the imaginary axis onto Γ). By algebraic manipulations we have

$$Tz = \frac{z-1}{z+1}$$

Combining this with previous results we have that $g(z) = \dfrac{e^z-1}{e^z+1}$ maps the infinite strip $\{z \colon |\operatorname{Im} z| < \pi/2\}$ onto the open unit disk D. (It is worth mentioning that $\dfrac{e^z-1}{e^z+1} = \tanh (z/2)$.)

Let G_1, G_2 be open connected sets; to try to find an analytic function f such that $f(G_1) = G_2$ we try to map both G_1 and G_2 onto the open unit disk. If this can be done, f can be obtained by taking the composition of one function with the inverse of the other.

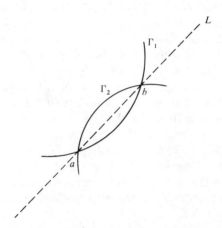

As an example, let G be the open set inside two circles Γ_1 and Γ_2, intersecting at points a and b ($a \neq b$). Let L be the line passing through a and b and give L the orientation (∞, a, b). Then $Tz = (z, \infty, a, b) = \left(\dfrac{z-a}{z-b}\right)$ maps L onto the real axis ($T\infty = 1$, $Ta = 0$, $Tb = \infty$). Since T must map circles onto circles, T maps Γ_1 and Γ_2 onto circles through 0 and ∞. That is, $T(\Gamma_1)$ and $T(\Gamma_2)$ are straight lines. By the use of orientation we have that $T(G) = \{\omega - \alpha < \arg \omega < \alpha\}$ for some $\alpha > 0$, or the complement of some such closed sector. By the use of an appropriate power of z and possibly a rotation we can map this wedge onto the right half plane. Now, composing with the map $(z-1)(z+1)^{-1}$ gives a map of G onto $D = \{z: |z| < 1\}$.

Exercises

1. Find the image of $\{z: \operatorname{Re} z < 0, |\operatorname{Im} z| < \pi\}$ under the exponential function.
2. Do exercise 1 for the set $\{z: |\operatorname{Im} z| < \pi/2\}$.
3. Discuss the mapping properties of $\cos z$ and $\sin z$.
4. Discuss the mapping properties of z^n and $z^{1/n}$ for $n \geq 2$. (Hint: use polar coordinates.)
5. Find the fixed points of a dilation, a translation and the inversion on $\mathbb{C}_\infty$.
6. Evaluate the following cross ratios: (a) $(7+i, 1, 0, \infty)$ (b) $(2, 1-i, 1, 1+i)$ (c) $(0, 1, i, -1)$ (d) $(i-1, \infty, 1+i, 0)$.
7. If $Tz = \dfrac{az+b}{cz+d}$ find z_2, z_3, z_4 (in terms of a, b, c, d) such that $Tz = (z, z_2, z_3, z_4)$.

8. If $Tz = \dfrac{az+b}{cz+d}$ show that $T(\mathbb{R}_\infty) = \mathbb{R}_\infty$ iff we can choose a, b, c, d to be real numbers.

9. If $Tz = \dfrac{az+b}{cz+d}$, find necessary and sufficient conditions that $T(\Gamma) = \Gamma$ where Γ is the unit circle $\{z: |z| = 1\}$.

10. Let $D = \{z: |z| < 1\}$ and find all Möbius transformations T such that $T(D) = D$.

11. Show that the definition of symmetry (3.17) does not depend on the choice of points z_2, z_3, z_4. That is, show that if $\omega_2, \omega_3, \omega_4$ are also in Γ then equation (3.18) is satisfied iff $(z^*, \omega_2, \omega_3, \omega_4) = \overline{(z, \omega_2, \omega_3, \omega_4)}$. (Hint: Use Exercise 8.)

12. Prove Theorem 3.4.

13. Give a discussion of the mapping $f(z) = \frac{1}{2}(z+1/z)$.

14. Suppose that one circle is contained inside another and that they are tangent at the point a. Let G be the region between the two circles and map G conformally onto the open unit disk. (Hint: first try $(z-a)^{-1}$.)

15. Can you map the open unit disk conformally onto $\{z: 0<|z|<1\}$?

16. Map $G = \mathbb{C} - \{z: -1 \le z \le 1\}$ onto the open unit disk by an analytic function f. Can f be one-one?

17. Let G be a region and suppose that $f: G \to \mathbb{C}$ is analytic such that $f(G)$ is a subset of a circle. Show that f is constant.

18. Let $-\infty < a < b < \infty$ and put $Mz = \dfrac{z - ia}{z - ib}$. Define the lines $L_1 = \{z: \operatorname{Im} z = b\}$, $L_2 = \{z: \operatorname{Im} z = a\}$ and $L_3 = \{z: \operatorname{Re} z = 0\}$. Determine which of the regions A, B, C, D, E, F in Figure 1, are mapped by M onto the regions U, V, W, X, Y, Z in Figure 2.

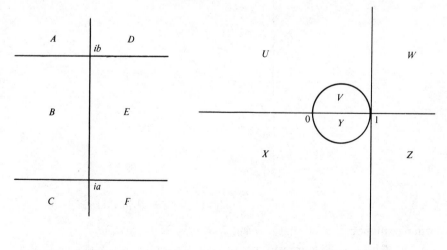

Figure 1   Figure 2

19. Let a, b, and M be as in Exercise 18 and let log be the principal branch of the logarithm.

(a) Show that log (Mz) is defined for all z except $z = ic$, $a \le c \le b$; and if $h(z) = $ Im [log Mz] then $0 < h(z) < \pi$ for Re $z > 0$.

(b) Show that log $(z - ic)$ is defined for Re $z > 0$ and any real number c; also prove that $|$Im log $(z - ic)| < \dfrac{\pi}{2}$ if Re $z > 0$.

(c) Let h be as in (a) and prove that $h(z) = $ Im [log $(z - ia) - $ log $(z - ib)$].

(d) Show that

$$\int_a^b \frac{dt}{z - it} = i[\log (z - ib) - \log (z - ia)]$$

(Hint: Use the Fundamental Theorem of Calculus.)

(e) Combine (c) and (d) to get that

$$h(x + iy) = \int_a^b \frac{x}{x^2 + (y - t)^2}\, dt = \arctan\left(\frac{y - a}{x}\right) - \arctan\left(\frac{y - b}{x}\right)$$

(f) Interpret part (e) geometrically and show that for Re $z > 0$ $h(z)$ is the angle depicted in the figure.

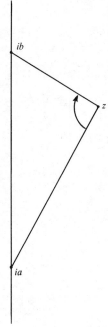

20. Let $Sz = \dfrac{az + b}{cz + d}$ and $Tz = \dfrac{\alpha z + \beta}{\gamma z + \delta}$; show that $S = T$ iff there is a non zero complex number λ such that $\alpha = \lambda a$, $\beta = \lambda b$, $\gamma = \lambda c$, $\delta = \lambda d$.

21. Let T be a Möbius transformation with fixed points z_1 and z_2. If S is a Möbius transformation show that $S^{-1}TS$ has fixed points $S^{-1}z_1$ and $S^{-1}z_2$.

22. (a) Show that a Möbius transformation has 0 and ∞ as its only fixed points iff it is a dilation.

(b) Show that a Möbius transformation has ∞ as its only fixed point iff it is a translation.

23. Show that a Möbius transformation T satisfies $T(0) = \infty$ and $T(\infty) = 0$ iff $Tz = az^{-1}$ for some a in $\mathbb{C}$.

24. Let T be a Möbius transformation, $T \neq$ the identity. Show that a Möbius transformation S commutes with T if S and T have the same fixed points. (Hint: Use Exercises 21 and 22.)

25. Find all the abelian subgroups of the group of Möbius transformations.

26. (a) Let $GL_2(\mathbb{C}) =$ all invertible 2×2 matrices with entries in $\mathbb{C}$ and let $\mathcal{M}$ be the group of Möbius transformations. Define $\varphi: GL_2(\mathbb{C}) \to \mathcal{M}$ by $\varphi\begin{pmatrix} a & b \\ c & d \end{pmatrix} = \dfrac{az+b}{cz+d}$. Show that φ is a group homomorphism of $GL_2(\mathbb{C})$ onto $\mathcal{M}$. Find the kernel of φ.

(b) Let $SL_2(\mathbb{C})$ be the subgroup of $GL_2(\mathbb{C})$ consisting of all matrices of determinant 1. Show that the image of $SL_2(\mathbb{C})$ under φ is all of $\mathcal{M}$. What part of the kernel of φ is in $SL_2(\mathbb{C})$?

27. If $\mathcal{G}$ is a group and $\mathcal{N}$ is a subgroup then $\mathcal{N}$ is said to be a *normal subgroup* of $\mathcal{G}$ if $S^{-1}TS \in \mathcal{N}$ whenever $T \in \mathcal{N}$ and $S \in \mathcal{G}$. $\mathcal{G}$ is a *simple group* if the only normal subgroups of $\mathcal{G}$ are $\{I\}$ ($I =$ the identity of $\mathcal{G}$) and $\mathcal{G}$ itself. Prove that the group $\mathcal{M}$ of Möbius transformations is a simple group.

28. Discuss the mapping properties of $(1-z)^i$.

29. For complex numbers α and β with $|\alpha|^2 + |\beta|^2 = 1$

$$u_{\alpha,\beta}(z) = \frac{\alpha z - \bar{\beta}}{\beta z - \bar{\alpha}} \quad \text{and let} \quad U = \left\{ u_{\alpha,\beta} : |\alpha|^2 + |\beta|^2 = 1 \right\}.$$

(a) Show that U is a group under composition.

(b) If SU_2 is the set of all unitary matrices with determinant 1, show that SU_2 is a group under matrix multiplication and that for each A in SU_2 there are unique complex numbers α and β with $|\alpha|^2 + |\beta|^2 = 1$ and

$$A = \begin{pmatrix} \alpha & \beta \\ -\bar{\beta} & \bar{\alpha} \end{pmatrix}.$$

(c) Show that $\begin{pmatrix} \alpha & \beta \\ -\bar{\beta} & \bar{\alpha} \end{pmatrix} \mapsto u_{\alpha,\beta}$ is a homomorphism of the group SU_2 onto U. What is its kernel?

(d) If $l \in \{0, \frac{1}{2}, 1, \frac{3}{2}, \ldots\}$ let $H_l =$ all the polynomials of degree $\leq 2l$. For $u_{\alpha,\beta} = u$ in U define $T_u^{(l)} : H_l \to H_l$ by $(T_u^{(l)}f)(z) = (\beta z + \bar{\alpha})^{2l} f(u(z))$. Show that $T_u^{(l)}$ is an invertible linear transformation on H_l and $u \mapsto T_u^{(l)}$ is an injective homomorphism of U into the group of invertible linear transformations of H_l onto H_l.

30. For $|z| < 1$ define $f(z)$ by

$$f(z) = \exp\left\{ -i \log\left[i\left(\frac{1+z}{1-z} \right) \right]^{1/2} \right\}.$$

(a) Show that f maps $D = \{z : |z| < 1\}$ conformally onto an annulus G.

(b) Find all Mobius transformations $S(z)$ that map D onto D and such that $f(S(z)) = f(z)$ when $|z| < 1$.

Chapter IV

Complex Integration

In this chapter results are derived which are fundamental in the study of analytic functions. The theorems presented here constitute one of the pillars of Mathematics and have far ranging applications.

§1. Riemann-Stieltjes integrals

We will begin by defining the Riemann-Stieltjes integral in order to define the integral of a function along a path in $\mathbb{C}$. The discussion of this integral is by no means complete, but is limited to those results essential to a cogent exposition of line integrals.

1.1 Definition. A function $\gamma: [a, b] \to \mathbb{C}$, for $[a, b] \subset \mathbb{R}$, is of *bounded variation* if there is a constant $M > 0$ such that for any partition $P = \{a = t_0 < t_1 < \ldots < t_m = b\}$ of $[a, b]$

$$v(\gamma; P) = \sum_{k=1}^{m} |\gamma(t_k) - \gamma(t_{k-1})| \le M.$$

The *total variation* of γ, $V(\gamma)$, is defined by

$$V(\gamma) = \sup \{v(\gamma; P): P \text{ a partition of } [a, b]\}.$$

Clearly $V(\gamma) \le M < \infty$.

It is easily shown that γ is of bounded variation if and only if Re γ and Im γ are of bounded variation. If γ is real valued and is non-decreasing then γ is of bounded variation and $V(\gamma) = \gamma(b) - \gamma(a)$. (Exercise 1) Other examples will be given, but first let us give some easily deduced properties of these functions.

1.2 Proposition. *Let $\gamma: [a, b] \to \mathbb{C}$ be of bounded variation. Then:*
   (a) *If P and Q are partitions of $[a, b]$ and $P \subset Q$ then $v(\gamma; P) \le v(\gamma; Q)$;*
   (b) *If $\sigma: [a, b] \to \mathbb{C}$ is also of bounded variation and $\alpha, \beta \in \mathbb{C}$ then $\alpha\gamma + \beta\sigma$ is of bounded variation and $V(\alpha\gamma + \beta\sigma) \le |\alpha| V(\gamma) + |\beta| V(\sigma)$.*
   The proof is left to the reader.

The next proposition gives a wealthy collection of functions of bounded variation. In actuality this is the set of functions which is of principal concern to us.

1.3 Proposition. *If $\gamma: [a,b] \to \mathbb{C}$ is piecewise smooth then γ is of bounded variation and*

$$V(\gamma) = \int_a^b |\gamma'(t)| \, dt$$

Proof. Assume that γ is smooth (the complete proof is easily deduced from this). Recall that when we say that γ is smooth this means γ' is continuous.

Let $P = \{a = t_0 < t_1 < \ldots < t_m = b\}$. Then, from the definition,

$$v(\gamma; P) = \sum_{k=1}^{m} |\gamma(t_k) - \gamma(t_{k-1})|$$

$$= \sum_{k=1}^{m} \left| \int_{t_{k-1}}^{t_k} \gamma'(t)\, dt \right|$$

$$\leq \sum_{k=1}^{m} \int_{t_{k-1}}^{t_k} |\gamma'(t)|\, dt$$

$$= \int_{a}^{b} |\gamma'(t)|\, dt$$

Hence $V(\gamma) \leq \int_a^b |\gamma'(t)|\, dt$, so that γ is of bounded variation.

Since γ' is continuous it is uniformly continuous; so if $\epsilon > 0$ is given we can choose $\delta_1 > 0$ such that $|s - t| < \delta_1$ implies that $|\gamma'(s) - \gamma'(t)| < \epsilon$. Also, we may choose $\delta_2 > 0$ such that if $P = \{a = t_0 < t_1 < \ldots < t_m = b\}$ and $\|P\| = \max \{(t_k - t_{k-1}): 1 \leq k \leq m\} < \delta_2$ then

$$\left| \int_a^b |\gamma'(t)|\, dt - \sum_{k=1}^{m} |\gamma'(\tau_k)|\, (t_k - t_{k-1}) \right| < \epsilon$$

where τ_k is any point in $[t_{k-1}, t_k]$. Hence

$$\int_a^b |\gamma'(t)|\, dt \leq \epsilon + \sum_{k=1}^{m} |\gamma'(\tau_k)|\, (t_k - t_{k-1})$$

$$= \epsilon + \sum_{k=1}^{m} \left| \int_{t_{k-1}}^{t_k} \gamma'(\tau_k)\, dt \right|$$

$$\leq \epsilon + \sum_{k=1}^{m} \left| \int_{t_{k-1}}^{t_k} [\gamma'(\tau_k) - \gamma'(t)]\, dt \right| + \sum_{k=1}^{m} \left| \int_{t_{k-1}}^{t_k} \gamma'(t)\, dt \right|$$

If $\|P\| < \delta = \min(\delta_1, \delta_2)$ then $|\gamma'(\tau_k) - \gamma'(t)| < \epsilon$ for t in $[t_{k-1}, t_k]$ and

$$\int_a^b |\gamma'(t)|\, dt \leq \epsilon + \epsilon(b - a) + \sum_{k=1}^{m} |\gamma(t_k) - \gamma(t_{k-1})|$$

$$= \epsilon[1 + (b - a)] + v(\gamma; P)$$

$$\leq \epsilon[1 + (b - a)] + V(\gamma).$$

Letting $\epsilon \to 0+$, gives

$$\int_a^b |\gamma'(t)|\, dt \le V(\gamma),$$

which yields equality. ∎

1.4 Theorem. *Let* $\gamma\colon [a,\, b] \to \mathbb{C}$ *be of bounded variation and suppose that* $f\colon [a,\, b] \to \mathbb{C}$ *is continuous. Then there is a complex number* I *such that for every* $\epsilon > 0$ *there is a* $\delta > 0$ *such that when* $P = \{t_0 < t_1 < \ldots < t_m\}$ *is a partition of* $[a,\, b]$ *with* $\|P\| = \max\,\{(t_k - t_{k-1})\colon 1 \le k \le m\} < \delta$ *then*

$$\left| I - \sum_{k=1}^m f(\tau_k)\,[\gamma(t_k) - \gamma(t_{k-1})]\right| < \epsilon$$

for whatever choice of points τ_k, $t_{k-1} \le \tau_k \le t_k$.

This number I is called the *integral of* f *with respect to* γ over $[a,\, b]$ and is designated by

$$I = \int_a^b f\, d\gamma = \int_a^b f(t)\, d\gamma(t).$$

Proof. Since f is continuous it is uniformly continuous; thus, we can find (inductively) positive numbers $\delta_1 > \delta_2 > \delta_3 > \ldots$ such that if $|s - t| < \delta_m$, $|f(s) - f(t)| < \dfrac{1}{m}$. For each $m \ge 1$ let $\mathscr{P}_m =$ the collection of all partitions P of $[a,\, b]$ with $\|P\| < \delta_m$; so $\mathscr{P}_1 \supset \mathscr{P}_2 \supset \cdots$. Finally define F_m to be the closure of the set:

1.5 $\quad \left\{ \displaystyle\sum_{k=1}^n f(\tau_k)[\gamma(t_k) - \gamma(t_{k-1})]\colon P \in \mathscr{P}_m \quad \text{and} \quad t_{k-1} \le \tau_k \le t_k \right\}.$

The following are claimed to hold:

1.6 $\quad \begin{cases} F_1 \supset F_2 \supset \cdots \text{ and} \\[2mm] \operatorname{diam} F_m \le \dfrac{2}{m}\, V(\gamma) \end{cases}$

If this is done then, by Cantor's Theorem (II. 3.7), there is exactly one complex number I such that $I \in F_m$ for every $m \ge 1$. Let us show that this will complete the proof. If $\epsilon > 0$ let $m > (2/\epsilon)\, V(\gamma)$; then $\epsilon > (2/m)\, V(\gamma) \ge \operatorname{diam} F_m$. Since $I \in F_m$, $F_m \subset B(I;\, \epsilon)$. Thus, if $\delta = \delta_m$ the theorem is proved.

Now to prove (1.6). The fact that $F_1 \supset F_2 \supset \ldots$ follows trivially from the fact that $\mathscr{P}_1 \supset \mathscr{P}_2 \supset \ldots$. To show that $\operatorname{diam} F_m \le \dfrac{2}{m}\, V(\gamma)$ it suffices to show that the diameter of the set in (1.5) is $\le \dfrac{2}{m}\, V(\gamma)$. This is done in two stages, each of which is easy although the first is tedious.

If $P = \{t_0 < \ldots < t_n\}$ is a partition we will denote by $S(P)$ a sum of the form $\sum f(\tau_k)[\gamma(t_k) - \gamma(t_{k-1})]$ where τ_k is any point with $t_{k-1} \le \tau_k \le t_k$.

Fix $m \geq 1$ and let $P \in \mathscr{P}_m$; the first step will be to show that if $P \subset Q$ (and hence $Q \in \mathscr{P}_m$) then $|S(P) - S(Q)| < \dfrac{1}{m} V(\gamma)$. We only give the proof for the case where Q is obtained by adding one extra partition point to P. Let $1 \leq p \leq m$ and let $t_{p-1} < t^* < t_p$; suppose that $P \cup \{t'\} = Q$. If $t_{p-1} \leq \sigma \leq t^*$, $t^* \leq \sigma' \leq t_p$, and

$$S(Q) = \sum_{k \neq p} f(\sigma_k) \left[\gamma(t_k) - \gamma(t_{k-1}) \right] + f(\sigma) \left[\gamma(t^*) - \gamma(t_{p-1}) \right]$$

$$+ f(\sigma') \left[\gamma(t_p) - \gamma(t^*) \right],$$

then, using the fact that $|f(\tau) - f(\sigma)| < \dfrac{1}{m}$ for $|\tau - \sigma| < \delta_m$,

$$|S(P) - S(Q)| = \left| \sum_{k \neq p} [f(\tau_k) - f(\sigma_k)] \, [\gamma(t_k) - \gamma(t_{k-1})] + f(\tau_p) \, [\gamma(t_p) - \gamma(t_{p-1})] \right.$$

$$\left. - f(\sigma) \, [\gamma(t^*) - \gamma(t_{p-1})] - f(\sigma') \, [\gamma(t_p) - \gamma(t^*)] \right|$$

$$\leq \frac{1}{m} \sum_{k \neq p} |\gamma(t_k) - \gamma(t_{k-1})| + |[f(\tau_p) - f(\sigma)] \, [\gamma(t^*) - \gamma(t_{p-1})]$$

$$+ [f(\tau_p) - f(\sigma')] \, [\gamma(t_p) - \gamma(t^*)]|$$

$$\leq \frac{1}{m} \sum_{k \neq p} |\gamma(t_k) - \gamma(t_{k-1})| + \frac{1}{m} |\gamma(t^*) - \gamma(t_{p-1})|$$

$$+ \frac{1}{m} |\gamma(t_p) - \gamma(t^*)|$$

$$\leq \frac{1}{m} V(\gamma)$$

For the second and final stage let P and R be any two partitions in $\mathscr{P}_m$. Then $Q = P \cup R$ is a partition and contains both P and R. Using the first part we get

$$|S(P) - S(R)| \leq |S(P) - S(Q)| + |S(Q) - S(R)| \leq \frac{2}{m} V(\gamma).$$

It now follows that the diameter of (1.5) is $\leq \dfrac{2}{m} V(\gamma)$. ∎

The next result follows from the definitions by a routine $\epsilon - \delta$ argument.

1.7 Proposition. *Let f and g be continuous functions on $[a, b]$ and let γ and σ be functions of bounded variation on $[a, b]$. Then for any scalars α and β:*
   (a) $\int_a^b (\alpha f + \beta g) \, d\gamma = \alpha \int_a^b f d\gamma + \beta \int_a^b g d\gamma$
   (b) $\int_a^b f d(\alpha \gamma + \beta \sigma) = \alpha \int_a^b f d\gamma + \beta \int_a^b f d\sigma$.
   The following is a very useful result in calculating these integrals.

1.8 Proposition. *Let $\gamma: [a, b] \to \mathbb{C}$ be of bounded variation and let $f: [a, b] \to \mathbb{C}$ be continuous. If $a = t_0 < t_1 < \cdots < t_n = b$ then*

$$\int_a^b f d\gamma = \sum_{k=1}^n \int_{t_{k-1}}^{t_k} f d\gamma.$$

The proof is left as an exercise.

As was mentioned before, we will mainly be concerned with those γ which are piecewise smooth. The following theorem says that in this case we can find $\int f d\gamma$ by the methods of integration learned in calculus.

1.9 Theorem. *If γ is piecewise smooth and $f:[a,b]\to\mathbb{C}$ is continuous then*

$$\int_a^b f d\gamma = \int_a^b f(t)\gamma'(t)\,dt.$$

Proof. Again we only consider the case where γ is smooth. Also, by looking at the real and imaginary parts of γ, we reduce the proof to the case where $\gamma([a,b])\subset\mathbb{R}$. Let $\epsilon>0$ and choose $\delta>0$ such that if $P=\{a=t_0<\ldots<t_n=b\}$ has $\|P\|<\delta$ then

1.10
$$\left|\int_a^b f d\gamma - \sum_{k=1}^n f(\tau_k)\,[\gamma(t_k)-\gamma(t_{k-1})]\right| < \tfrac{1}{2}\epsilon$$

and

1.11
$$\left|\int_a^b f(t)\gamma'(t)\,dt - \sum_{k=1}^n f(\tau_k)\gamma'(\tau_k)\,(t_k-t_{k-1})\right| < \tfrac{1}{2}\epsilon$$

for any choice of τ_k in $[t_{k-1}, t_k]$. If we apply the Mean Value Theorem for derivatives we get that there is a τ_k in $[t_{k-1}, t_k]$ with $\gamma'(\tau_k) = [\gamma(t_k)-\gamma(t_{k-1})]$ $(t_k-t_{k-1})^{-1}$. (Note that the fact that γ is real valued is needed to apply the Mean Value Theorem.) Thus,

$$\sum_{k=1}^n f(\tau_k)\,[\gamma(t_k)-\gamma(t_{k-1})] = \sum_{k=1}^n f(\tau_k)\gamma'(\tau_k)\,(t_k-t_{k-1}).$$

Combining this with inequalities (1.10) and (1.11) gives

$$\left|\int_a^b f d\gamma - \int_a^b f(t)\gamma'(t)\,dt\right| < \epsilon.$$

Since $\epsilon > 0$ was arbitrary, this completes the proof of the theorem. ∎

We have already defined a path as a continuous function $\gamma: [a, b] \to \mathbb{C}$. If $\gamma: [a, b] \to \mathbb{C}$ is a path then the set $\{\gamma(t): a \le t \le b\}$ is called the *trace* of γ and is denoted it by $\{\gamma\}$. Notice that the trace of a path is always a compact set. γ is a *rectifiable path* if γ is a function of bounded variation. If P is a partition of $[a, b]$ then $v(\gamma; P)$ is exactly the sum of lengths of line segments connecting points on the trace of γ. To say that γ is rectifiable is to say that γ has finite length and its length is $V(\gamma)$. In particular, if γ is piecewise smooth then γ is rectifiable and its length is $\int_a^b |\gamma'|\, dt$.

If $\gamma: [a, b] \to \mathbb{C}$ is a rectifiable path with $\{\gamma\} \subset E \subset \mathbb{C}$ and $f: E \to \mathbb{C}$ is a continuous function then $f \circ \gamma$ is a continuous function on $[a, b]$. With this in mind the following definition makes sense.

1.12 Definition. If γ: $[a, b] \to \mathbb{C}$ is a rectifiable path and f is a function defined and continuous on the trace of γ then the *(line) integral of f along γ* is

$$\int_a^b f(\gamma(t)) \, d\gamma(t).$$

This line integral is also denoted by $\int_\gamma f = \int_\gamma f(z) \, dz$.

As an example let us take γ: $[0, 2\pi] \to \mathbb{C}$ to be $\gamma(t) = e^{it}$ and define $f(z) = \dfrac{1}{z}$ for $z \neq 0$. Now γ is differentiable so, by Theorem 1.9 we have

$\int_\gamma \dfrac{1}{z} \, dz = \int_0^{2\pi} e^{-it}(ie^{it}) \, dt = 2\pi i$.

Using the same definition of γ and letting m be any integer ≥ 0 gives $\int_\gamma z^m \, dz = \int_0^{2\pi} e^{imt}(ie^{it}) \, dt = i \int_0^{2\pi} \exp\,(i(m+1)t) \, dt = i \int_0^{2\pi} \cos\,(m+1)t \, dt - \int_0^{2\pi} \sin\,(m+1)t \, dt = 0$.

Now let a, b $\mathbb{C}$ and put $\gamma(t) = tb + (1-t)a$ for $0 \leq t \leq 1$. Then $\gamma'(t) = b - a$, and using the Fundamental Theorem of Calculus we get that for

$n \geq 0$, $\int_\gamma z^n \, dz = (b-a) \int_0^1 [tb + (1-t)a]^n \, dt = \dfrac{1}{n+1} (b^{n+1} - a^{n+1})$.

There are more examples in the exercises, but now we will prove a certain "invariance" result which, besides being useful in computations, forms the basis for our definition of a curve.

If γ: $[a, b] \to \mathbb{C}$ is a rectifiable path and φ: $[c, d] \to [a, b]$ is a continuous non-decreasing function whose image is all of $[a, b]$ (i.e., $\varphi(c) = a$ and $\varphi(d) = b$) then $\gamma \circ \varphi$: $[c, d] \to \mathbb{C}$ is a path with the same trace as γ. Moreover, $\gamma \circ \varphi$ is rectifiable because if $c = s_0 < s_1 < \cdots < s_n = d$ then $a = \varphi(s_0) \leq \varphi(s_1) \leq \cdots \leq \varphi(s_n) = b$ is a partition of $[a, b]$. Hence

$$\sum_{k=1}^n |\gamma(\varphi(s_k)) - \gamma(\varphi(s_{k-1}))| \leq V(\gamma)$$

so that $V(\gamma \circ \varphi) \leq V(\gamma) < \infty$. So if f is continuous on $\{\gamma\} = \{\gamma \circ \varphi\}$ then $\int_{\gamma \circ \varphi} f$ is well defined.

1.13 Proposition. *If γ: $[a, b] \to \mathbb{C}$ is a rectifiable path and φ: $[c, d] \to [a, b]$ is a continuous non-decreasing function with $\varphi(c) = a$, $\varphi(d) = b$; then for any function f continuous on $\{\gamma\}$*

$$\int_\gamma f = \int_{\gamma \circ \varphi} f$$

Proof. Let $\epsilon > 0$ and choose $\delta_1 > 0$ such that for $\{s_0 < s_1 < \cdots < s_n\}$, a partition of $[c, d]$ with $(s_k - s_{k-1}) < \delta_1$, and $s_{k-1} \leq \sigma_k \leq s_k$ we have

1.14
$$\left| \int_{\gamma \circ \varphi} f - \sum_{k=1}^n f(\gamma \circ \varphi(\sigma_k)) \left[\gamma \circ \varphi(s_k) - \gamma \circ \varphi(s_{k-1}) \right] \right| < \tfrac{1}{2}\epsilon$$

Similarly choose $\delta_2 > 0$ such that if $\{t_0 < t_1 < \cdots < t_n\}$ is a partition of $[a, b]$ with $(t_k - t_{k-1}) < \delta_2$ and $t_{k-1} \leq \tau_k \leq t_k$, then

1.15 $$\left| \int_\gamma f - \sum_{k=1}^n f(\gamma(\tau_k))\, [\gamma(t_k) - \gamma(t_{k-1})] \right| < \tfrac{1}{2}\epsilon.$$

But φ is uniformly continuous on $[c, d]$; hence there is a $\delta > 0$, which can be chosen with $\delta < \delta_1$, such that $|\varphi(s) - \varphi(s')| < \delta_2$ whenever $|s - s'| < \delta$. So if $\{s_0 < s_1 < \cdots < s_n\}$ is a partition of $[c, d]$ with $(s_k - s_{k-1}) < \delta < \delta_1$ and $t_k = \varphi(s_k)$, then $\{t_0 \le t_1 \le \cdots \le t_n\}$ is a partition of $[a, b]$ with $(t_k - t_{k-1}) < \delta_2$. If $s_{k-1} \le \sigma_k \le s_k$ and $\tau_k = \varphi(\sigma_k)$ then both (1.14) and (1.15) hold. Moreover, the right hand parts of these two differences are equal! It follows that

$$\left| \int_\gamma f - \int_{\gamma \circ \varphi} f \right| < \epsilon$$

Since $\epsilon > 0$ was arbitrary, equality is proved. ∎

We wish to define an equivalence relation on the collection of rectifiable paths so that each member of an equivalence class has the same trace and so that the line integral of a function continuous on this trace is the same for each path in the class. It would seem that we should define σ and γ to be equivalent if $\sigma = \gamma \circ \varphi$ for some function φ as above. However, this is not an equivalence relation!

1.16 Definition. Let $\sigma: [c, d] \to \mathbb{C}$ and $\gamma: [a, b] \to \mathbb{C}$ be rectifiable paths. The path σ is *equivalent* to γ if there is a function $\varphi: [c, d] \to [a, b]$ which is continuous, strictly increasing, and with $\varphi(c) = a$, $\varphi(d) = b$; such that $\sigma = \gamma \circ \varphi$. We call the function φ a *change of parameter*.

A *curve* is an equivalence class of paths. The trace of a curve is the trace of any one of its members. If f is continuous on the trace of the curve then the integral of f over the curve is the integral of f over any member of the curve.

A curve is smooth (piecewise smooth) if and only if some one of its representatives is smooth (piecewise smooth).

Henceforward, we will not make this distinction between a curve and its representative. In fact, expressions such as "let γ be the unit circle traversed once in the counter-clockwise direction" will be used to indicate a curve. The reader is asked to trust that a result for curves which is, in fact, a result only about paths will not be stated.

Let $\gamma: [a, b] \to \mathbb{C}$ be a rectifiable path and for $a \le t \le b$, let $|\gamma|(t)$ be $V(\gamma; [a, t])$. That is,

$$|\gamma|(t) = \sup \left\{ \sum_{k=1}^n |\gamma(t_k) - \gamma(t_{k-1})| : \{t_0, \ldots, t_n\} \text{ is a partition of } [a, t] \right\}.$$

Clearly $|\gamma|(t)$ is increasing and so $|\gamma|: [a, b] \to \mathbb{C}$ is of bounded variation. If f is continuous on $\{\gamma\}$ define

$$\int_\gamma f\, |dz| = \int_a^b f(\gamma(t))\, d|\gamma|(t).$$

If γ is a rectifiable curve then denote by $-\gamma$ the curve defined by $(-\gamma)(t) =$

$\gamma(-t)$ for $-b \leq t \leq -a$. Another notation for this is γ^{-1}. Also if $c \in \mathbb{C}$ let $\gamma+c$ denote the curve defined by $(\gamma+c)(t) = \gamma(t)+c$. The following proposition gives many basic properties of the line integral.

1.17 Proposition. *Let γ be a rectifiable curve and suppose that f is a function continuous on $\{\gamma\}$. Then:*

(a) $\int_\gamma f = -\int_{-\gamma} f$;
(b) $|\int_\gamma f| \leq \int_\gamma |f| \, |dz| \leq V(\gamma) \sup [|f(z)|: z \in \{\gamma\}]$;
(c) If $c \in \mathbb{C}$ then $\int_\gamma f(z) \, dz = \int_{\gamma+c} f(z-c) \, dz$.

The proof is left as an exercise.

The next result is the analogue of the Fundamental Theorem of Calculus for line integrals.

1.18 Theorem. *Let G be open in $\mathbb{C}$ and let γ be a rectifiable path in G with initial and end points α and β respectively. If $f: G \to \mathbb{C}$ is a continuous function with a primitive $F: G \to \mathbb{C}$, then*

$$\int_\gamma f = F(\beta) - F(\alpha)$$

(Recall that F is a *primitive* of f when $F' = f$.)

The following useful fact will be needed in the proof of this theorem.

1.19 Lemma. *If G is an open set in $\mathbb{C}$, $\gamma:[a,b] \to G$ is a rectifiable path, and $f: G \to \mathbb{C}$ is continuous then for every $\epsilon > 0$ there is a polygonal path Γ in G such that $\Gamma(a) = \gamma(a)$, $\Gamma(b) = \gamma(b)$, and $|\int_\gamma f - \int_\Gamma f| < \epsilon$.*

Proof. Case I. Suppose G is an open disk. Since $\{\gamma\}$ is a compact set, $d = \text{dist}(\{\gamma\}, \partial G) > 0$. It follows that if $G = B(c;r)$ then $\{\gamma\} \subset B(c;\rho)$ where $\rho = r - \frac{1}{2}d$. The reason for passing to this smaller disk is that f is uniformly continuous in $\bar{B}(c;\rho) \subset G$. Hence without loss of generality it can be assumed that f is uniformly continuous on G. Choose $\delta > 0$ such that $|f(z) - f(w)| < \epsilon$ whenever $|z - w| < \delta$. If $\gamma:[a,b] \to \mathbb{C}$ then γ is uniformly continuous so there is a partition $\{t_0 < t_1 < \ldots < t_n\}$ of $[a,b]$ such that

1.19a $$|\gamma(s) - \gamma(t)| < \delta$$

if $t_{k-1} \leq s, t \leq t_k$; and such that for $t_{k-1} \leq \tau_k \leq t_k$ we have

1.20 $$\left| \int_\gamma f - \sum_{k=1}^{n} f(\gamma(\tau_k)) [\gamma(t_k) - \gamma(t_{k-1})] \right| < \epsilon.$$

Define $\Gamma:[a,b] \to \mathbb{C}$ by

$$\Gamma(t) = \frac{1}{t_k - t_{k-1}} [(t_k - t)\gamma(t_{k-1}) + (t - t_{k-1})\gamma(t_k)]$$

if $t_{k-1} \leq t \leq t_k$. So on $[t_{k-1}, t_k]$, $\Gamma(t)$ traces out the straight line segment from $\gamma(t_{k-1})$ to $\gamma(t_k)$; that is, Γ is a polygonal path in G. From (1.19a)

1.21 $\qquad\qquad |\Gamma(t) - \gamma(\tau_k)| < \delta \quad \text{for } t_{k-1} \le t \le t_k.$

Since $\int_\Gamma f = \int_a^b f(\Gamma(t))\Gamma'(t)\,dt$ it follows that

$$\int_\Gamma f = \sum_{k=1}^n \frac{\gamma(t_k) - \gamma(t_{k-1})}{t_k - t_{k-1}} \int_{t_{k-1}}^{t_k} f(\Gamma(t))\,dt.$$

Using (1.20) we obtain

$$\left| \int_\gamma f - \int_\Gamma f \right| \le \epsilon + \left| \sum_{k=1}^n f(\gamma(\tau_k))[\gamma(t_k) - \gamma(t_{k-1})] - \int_\Gamma f \right|$$

$$\le \epsilon + \sum_{k=1}^n |\gamma(t_k) - \gamma(t_{k-1})|(t_k - t_{k-1})^{-1} \int_{t_{k-1}}^{t_k} |f(\Gamma(t)) - f(\gamma(\tau_k))|\,dt.$$

Applying (1.21) to the integrand gives

$$\left| \int_\gamma f - \int_\Gamma f \right| \le \epsilon + \epsilon \sum_{k=1}^n |\gamma(t_k) - \gamma(t_{k-1})| \le \epsilon(1 + V(\gamma)).$$

The proof of Case I now follows.

Case II. G is arbitrary. Since $\{\gamma\}$ is compact there is a number r with $0 < r < \text{dist}(\{\gamma\}, \partial G)$. Choose $\delta > 0$ such that $|\gamma(s) - \gamma(t)| < r$ when $|s - t| < \delta$. If $P = \{t_0 < t_1 < \ldots < t_n\}$ is a partition of $[a,b]$ with $\|P\| < \delta$ then $|\gamma(t) - \gamma(t_{k-1})| < r$ for $t_{k-1} \le t \le t_k$. That is if $\gamma_k : [t_{k-1}, t_k] \to G$ is defined by $\gamma_k(t) = \gamma(t)$ then $\{\gamma_k\} \subset B(\gamma(t_{k-1}); r)$ for $1 \le k \le n$. By Case I there is a polygonal path $\Gamma_k : [t_{k-1}, t_k] \to B(\gamma(t_{k-1}); r)$ such that $\Gamma_k(t_{k-1}) = \gamma(t_{k-1})$, $\Gamma_k(t_k) = \gamma(t_k)$, and $|\int_{\gamma_k} f - \int_{\Gamma_k} f| < \epsilon/n$. If $\Gamma(t) = \Gamma_k(t)$ on $[t_{k-1}, t_k]$ then Γ has the required properties. ∎

Proof of Theorem 1.18. Case I. $\gamma : [a,b] \to \mathbb{C}$ is piecewise smooth. Then $\int_\gamma f = \int_a^b f(\gamma(t))\gamma'(t)\,dt = \int_a^b F'(\gamma(t))\gamma'(t)\,dt = \int_a^b (F \circ \gamma)'(t)\,dt = F(\gamma(b)) - F(\gamma(a)) = F(\beta) - F(\alpha)$ by the Fundamental Theorem of Calculus.

Case II The General Case. If $\epsilon > 0$ then Lemma 1.19 implies there is a polygonal path Γ from α to β such that $|\int_\gamma f - \int_\Gamma f| < \epsilon$. But Γ is piecewise smooth, so by Case I $\int_\Gamma f = F(\beta) - F(\alpha)$. Hence $|\int_\gamma f - [F(\beta) - F(\alpha)]| < \epsilon$; since $\epsilon > 0$ is arbitrary, the desired equality follows. ∎

The use of Lemma 1.19 in the proof of Theorem 1.18 to pass from the piecewise smooth case to the rectifiable case is typical of many proofs of results on line integrals. We shall see applications of Lemma 1.19 in the future.

A curve $\gamma: [a, b] \to \mathbb{C}$ is said to be *closed* if $\gamma(a) = \gamma(b)$.

1.22 Corollary. *Let G, γ, and f satisfy the same hypothesis as in Theorem 1.18. If γ is a closed curve then*

$$\int_\gamma f = 0$$

The Fundamental Theorem of Calculus says that each continuous function has a primitive. This is far from being true for functions of a complex variable. For example let $f(z) = |z|^2 = x^2 + y^2$. If F is a primitive of f then F is analytic. So if $F = U + iV$ then $x^2 + y^2 = F'(x+iy)$. Now, using the Cauchy-Riemann equations,

$$\frac{\partial U}{\partial x} = \frac{\partial V}{\partial y} = x^2 + y^2 \quad \text{and} \quad \frac{\partial U}{\partial y} = \frac{\partial V}{\partial x} = 0.$$

But $\dfrac{\partial U}{\partial y} = 0$ implies that $U(x, y) = u(x)$ for some differentiable function u.

But this gives $x^2 + y^2 = \dfrac{\partial U}{\partial x} = u'(x)$, a clear contradiction. Another way to see that $|z|^2$ does not have a primitive is to apply Theorem 1.18 (see Exercise 8).

Exercises

1. Let $\gamma: [a, b] \to \mathbb{R}$ be non decreasing. Show that γ is of bounded variation and $V(\gamma) = \gamma(b) - \gamma(a)$.
2. Prove Proposition 1.2.
3. Prove Proposition 1.7.
4. Prove Proposition 1.8 (Use induction).
5. Let $\gamma(t) = \exp((-1+i)t^{-1})$ for $0 < t \le 1$ and $\gamma(0) = 0$. Show that γ is a rectifiable path and find $V(\gamma)$. Give a rough sketch of the trace of γ.
6. Show that if $\gamma; [a, b] \to \mathbb{C}$ is a Lipschitz function then γ is of bounded variation.
7. Show that $\gamma: [0, 1] \to \mathbb{C}$, defined by $\gamma(t) = t + it \sin \dfrac{1}{t}$ for $t \ne 0$ and $\gamma(0) = 0$, is a path but is not rectifiable. Sketch this path.
8. Let γ and σ be the two polygons $[1, i]$ and $[1, 1+i, i]$. Express γ and σ as paths and calculate $\int_\gamma f$ and $\int_\sigma f$ where $f(z) = |z|^2$.
9. Define $\gamma: [0, 2\pi] \to \mathbb{C}$ by $\gamma(t) = \exp(int)$ where n is some integer (positive, negative, or zero). Show that $\displaystyle\int_\gamma \frac{1}{z}\, dz = 2\pi in$.
10. Define $\gamma(t) = e^{it}$ for $0 \le t \le 2\pi$ and find $\int_\gamma z^n\, dz$ for every integer n.
11. Let γ be the closed polygon $[1-i, 1+i, -1+i, -1-i, 1-i]$. Find $\displaystyle\int_\gamma \frac{1}{z}\, dz$.
12. Let $I(r) = \displaystyle\int_\gamma \frac{e^{iz}}{z}\, dz$ where $\gamma: [0, \pi] \to \mathbb{C}$ is defined by $\gamma(t) = re^{it}$. Show that $\lim_{r \to \infty} I(r) = 0$.
13. Find $\int_\gamma z^{-\frac{1}{2}}\, dz$ where: (a) γ is the upper half of the unit circle from $+1$ to -1: (b) γ is the lower half of the unit circle from $+1$ to -1.

14. Prove that if $\varphi: [a, b] \to [c, d]$ is continuous and $\varphi(a) = c$, $\varphi(b) = d$ then φ is one-one iff φ is strictly increasing.

15. Show that the relation in Definition 1.16 is an equivalence relation.

16. Show that if γ and σ are equivalent rectifiable paths then $V(\gamma) = V(\sigma)$.

17. Show that if $\gamma: [a, b] \to \mathbb{C}$ is a path then there is an equivalent path $\sigma: [0, 1] \to \mathbb{C}$.

18. Prove Proposition 1.17.

19. In the proof of Case I of Lemma 1.19, where was the assumption that γ lies in a disk used?

20. Let $\gamma(t) = 1 + e^{it}$ for $0 \le t \le 2\pi$ and find $\int_\gamma (z^2 - 1)^{-1} \, dz$.

21. Let $\gamma(t) = 2e^{it}$ for $-\pi \le t \le \pi$ and find $\int_\gamma (z^2 - 1)^{-1} \, dz$.

22. Show that if F_1 and F_2 are primitives for $f: G \to \mathbb{C}$ and G is connected then there is a constant c such that $F_1(z) = c + F_2(z)$ for each z in G.

23. Let γ be a closed rectifiable curve in G and $a \notin G$. Show that for $n \ge 2$, $\int_\gamma (z - a)^{-n} \, dz = 0$.

24. Prove the following integration by parts formula. Let f and g be analytic in G and let γ be a rectifiable curve from a to b in G. Then $\int_\gamma fg' = f(b)g(b) - f(a)g(a) - \int_\gamma f'g$.

§2. *Power series representation of analytic functions*

In this section we will see that a function f, analytic in an open set G, has a power series expansion about each point of G. In particular, an analytic function is infinitely differentiable.

We begin by proving Leibniz's rule from Advanced Calculus.

2.1 Proposition. *Let $\varphi: [a, b] \times [c, d] \to \mathbb{C}$ be a continuous function and define $g: [c, d] \to \mathbb{C}$ by*

2.2
$$g(t) = \int_a^b \varphi(s, t) \, ds.$$

Then g is continuous. Moreover, if $\dfrac{\partial \varphi}{\partial t}$ exists and is a continuous function on $[a, b] \times [c, d]$ then g is continuously differentiable and

2.3
$$g'(t) = \int_a^b \frac{\partial \varphi}{\partial t}(s, t) \, ds.$$

Proof. The proof that g is continuous is left as an exercise. Notice that if we prove that g is differentiable with g' given by formula (2.3) then it will follow from the first part that g' is continuous since $\dfrac{\partial \varphi}{\partial t}$ is continuous. Hence, we need only verify formula (2.3).

Fix a point t_0 in $[c, d]$ and let $\epsilon > 0$. Denote $\dfrac{\partial \varphi}{\partial t}$ by φ_2; it follows that φ_2 must be uniformly continuous on $[a, b] \times [c, d]$. Thus, there is a $\delta > 0$ such that $|\varphi_2(s', t') - \varphi_2(s, t)| < \epsilon$ whenever $(s - s')^2 + (t - t')^2 < \delta^2$. In particular

2.4
$$|\varphi_2(s, t) - \varphi_2(s, t_0)| < \epsilon$$

whenever $|t - t_0| < \delta$ and $a \le s \le b$. This gives that for $|t - t_0| < \delta$ and $a \le s \le b$,

2.5
$$\left| \int_{t_0}^{t} [\varphi_2(s, \tau) - \varphi_2(s, t_0)] \, d\tau \right| \le \epsilon |t - t_0|.$$

But for a fixed s in $[a, b]$ $\Phi(t) = \varphi(s, t) - t\varphi_2(s, t_0)$ is a primitive of $\varphi_2(s, t) - \varphi_2(s, t_0)$. By combining the Fundamental Theorem of Calculus with inequality (2.5), it follows that

$$|\varphi(s, t) - \varphi(s, t_0) - (t - t_0)\varphi_2(s, t_0)| \le \epsilon |t - t_0|$$

for any s when $|t - t_0| < \delta$. But from the definition of g this gives

$$\left| \frac{g(t) - g(t_0)}{t - t_0} - \int_{a}^{b} \varphi_2(s, t_0) \, ds \right| \le \epsilon(b - a)$$

when $0 < |t - t_0| < \delta$. ∎

This result can be used to prove that

$$\int_{0}^{2\pi} \frac{e^{is}}{e^{is} - z} \, ds = 2\pi \quad \text{if} \quad |z| < 1.$$

Actually, we will need this formula in the proof of the next proposition. Let $\varphi(s, t) = \dfrac{e^{is}}{e^{is} - tz}$ for $0 \le t \le 1, 0 \le s \le 2\pi$; (Note that φ is continuously differentiable because $|z| < 1$.) Hence $g(t) = \int_0^{2\pi} \varphi(s, t) \, ds$ is continuously differentiable. Also, $g(0) = 2\pi$; so if it can be shown that g is a constant, then $2\pi = g(1)$ and the desired result is obtained.

Now

$$g'(t) = \int_{0}^{2\pi} \frac{ze^{is}}{(e^{is} - tz)^2} \, ds;$$

but for t fixed, $\Phi(s) = zi(e^{is} - tz)^{-1}$ has $\Phi'(s) = -zi(e^{is} - tz)^{-2}(ie^{is}) = ze^{is}(e^{is} - tz)^{-2}$. Hence $g'(t) = \Phi(2\pi) - \Phi(0) = 0$, so g must be a constant.

The next result, although very important, is transitory. We will see a much more general result than this—Cauchy's Integral Formula; a formula which is one of the essential facts of the theory.

2.6 Proposition. *Let* $f: G \to \mathbb{C}$ *be analytic and suppose* $\bar{B}(a; r) \subset G (r > 0)$. *If* $\gamma(t) = a+re^{it}$, $0 \leq t \leq 2\pi$, *then*

$$f(z) = \frac{1}{2\pi i} \int_{\gamma} \frac{f(w)}{w-z} \, dw$$

for $|z-a| < r$.

Proof. By considering $G_1 = \left\{ \frac{1}{r}(z-a): z \in G \right\}$ and the function $g(z) = f(a+rz)$ we see that, without loss of generality, it may be assumed that $a = 0$ and $r = 1$. That is we may assume that $\bar{B}(0; 1) \subset G$.

Fix z, $|z| < 1$; it must be shown that

$$f(z) = \frac{1}{2\pi i} \int_{\gamma} \frac{f(w)}{w-z} \, dw$$

$$= \frac{1}{2\pi} \int_{0}^{2\pi} \frac{f(e^{is})e^{is}}{e^{is}-z} \, ds;$$

that is, we want to show that

$$0 = \int_{0}^{2\pi} \frac{f(e^{is})e^{is}}{e^{is}-z} \, ds - 2\pi f(z)$$

$$= \int_{0}^{2\pi} \left[\frac{f(e^{is})e^{is}}{e^{is}-z} - f(z) \right] ds$$

We will apply Leibniz's rule by letting

$$\varphi(s, t) = \frac{f(z+t(e^{is}-z))e^{is}}{e^{is}-z} - f(z),$$

for $0 \leq t \leq 1$ and $0 \leq s \leq 2\pi$. Since $|z+t(e^{is}-z)| = |z(1-t)+te^{is}| < 1$, φ is well defined and is continuously differentiable. Let $g(t) = \int_{0}^{2\pi} \varphi(s, t) \, ds$; so g has a continuous derivative.

The proposition will be proved if it can be shown that $g(1) = 0$; this is done by showing that $g(0) = 0$ and that g is constant. To see that $g(0) = 0$

compute:

$$g(0) = \int_0^{2\pi} \varphi(s, 0) \, ds$$

$$= \int_0^{2\pi} \left[\frac{f(z)e^{is}}{e^{is} - z} - f(z) \right] ds$$

$$= f(z) \int_0^{2\pi} \frac{e^{is}}{e^{is} - z} \, ds - 2\pi f(z)$$

$$= 0,$$

since we showed that $\int_0^{2\pi} \frac{e^{is}}{e^{is} - z} \, ds = 2\pi$ prior to the statement of this proposition.

To show that g is constant compute g'. By Leibniz's rule, $g'(t) = \int_0^{2\pi} \varphi_2(s, t) \, ds$ where

$$\varphi_2(s, t) = e^{is} f'(z + t(e^{is} - z)).$$

However, for $0 < t \le 1$ we have that $\Phi(s) = -it^{-1} f(z + t(e^{is} - z))$ is a primitive of $\varphi_2(s, t)$. So $g'(t) = \Phi(2\pi) - \Phi(0) = 0$ for $0 < t \le 1$. Since g' is continuous we have $g' = 0$ and g must be a constant. ∎

How is this result used to get the power series expansion? The answer is that we use a geometric series. Let $|z - a| < r$ and suppose that w is on the circle $|w - a| = r$. Then

$$\frac{1}{w - z} = \frac{1}{w - a} \cdot \frac{1}{1 - \left[\dfrac{z - a}{w - a} \right]} = \frac{1}{(w - a)} \sum_{n=0}^{\infty} \left(\frac{z - a}{w - a} \right)^n$$

since $|z - a| < r = |w - a|$. Now, multiplying both sides by $[f(w)/2\pi i]$ and integrating around the circle $\gamma: |w - a| = r$, the left hand side yields $f(z)$ by the preceding proposition. The right hand side becomes—what? To find the answer we must know that we can distribute the integral through the infinite sum.

2.7 Lemma. *Let γ be a rectifiable curve in $\mathbb{C}$ and suppose that F_n and F are continuous functions on $\{\gamma\}$. If $F = u - \lim F_n$ on $\{\gamma\}$ then*

$$\int_\gamma F = \lim \int_\gamma F_n.$$

Proof. Let $\epsilon > 0$; then there is an integer N such that $|F_n(w) - F(w)| < \epsilon/V(\gamma)$ for all w on $\{\gamma\}$ and $n \geq N$. But this gives, by Proposition 1.17(b),

$$\left| \int_\gamma F - \int_\gamma F_n \right| = \left| \int_\gamma (F - F_n) \right|$$

$$\leq \int_\gamma |F(w) - F_n(w)| \, |dw|$$

$$\leq \epsilon$$

whenever $n \geq N$. ∎

2.8 Theorem. *Let f be analytic in $B(a; R)$; then $f(z) = \sum\limits_{n=0}^{\infty} a_n(z-a)^n$ for $|z-a| < R$ where $a_n = \dfrac{1}{n!} f^{(n)}(a)$ and this series has radius of convergence $\geq R$.*

Proof. Let $0 < r < R$ so that $\bar{B}(a; r) \subset B(a; R)$. If $\gamma(t) = a + re^{it}, 0 \leq t \leq 2\pi$, then by Proposition 2.6,

$$f(z) = \frac{1}{2\pi i} \int_\gamma \frac{f(w)}{w - z} \, dw \quad \text{for} \quad |z-a| < r.$$

But, since $|z-a| < r$ and w is on the circle $\{\gamma\}$,

$$\frac{|f(w)| \, |z-a|^n}{|w-a|^{n+1}} \leq \frac{M}{r} \left(\frac{|z-a|}{r} \right)^n$$

where $M = \max \, \{|f(w)|; \, |w-a| = r\}$. Since $\dfrac{|z-a|}{r} < 1$, the Weierstrass M-test gives that $\sum f(w) \, (z-a)^n/(w-a)^{n+1}$ converges uniformly for w on $\{\gamma\}$. By Lemma 2.7 and the discussion preceding it

2.9 $$f(z) = \sum_{n=0}^{\infty} \left[\frac{1}{2\pi i} \int_\gamma \frac{f(w)}{(w-a)^{n+1}} \, dw \right] (z-a)^n$$

If we set

$$a_n = \frac{1}{2\pi i} \int_\gamma \frac{f(w)}{(w-a)^{n+1}} \, dw$$

then a_n is independent of z, and so (2.9) is a power series which converges for $|z-a| < r$.

It follows (Proposition III. 2.5) that $a_n = \dfrac{1}{n!} f^{(n)}(a)$, so that the value of a_n is independent of γ; that is, it is independent of r. So

2.10 $$f(z) = \sum_{n=0}^{\infty} a_n(z-a)^n$$

for $|z-a| < r$. Since r was chosen arbitrarily, $r < R$, we have that (2.10) holds for $|z-a| < R$; giving that the radius of convergence of (2.10) must be at least R. ∎

2.11 Corollary. *If $f: G \to \mathbb{C}$ is analytic and $a \in G$ then $f(z) = \sum\limits_{0}^{\infty} a_n(z-a)^n$ for $|z-a| < R$ where $R = d(a, \partial G)$.*

Proof. Since $R = d(a, \partial G)$, $B(a; R) \subset G$ so that f is analytic on $B(a; R)$. The result now follows from the theorem. ∎

2.12 Corollary. *If $f: G \to \mathbb{C}$ is analytic then f is infinitely differentiable.*

2.13 Corollary. *If $f: G \to \mathbb{C}$ is analytic and $\bar{B}(a; r) \subset G$ then*

$$f^{(n)}(a) = \frac{n!}{2\pi i} \int_{\gamma} \frac{f(w)}{(w-a)^{n+1}} \, dw$$

where $\gamma(t) = a + re^{it}$, $0 \le t \le 2\pi$.

2.14 Cauchy's Estimate. *Let f be analytic in $B(a; R)$ and suppose $|f(z)| \le M$ for all z in $B(a; R)$. Then*

$$|f^{(n)}(a)| \le \frac{n!M}{R^n} \, .$$

Proof. Since Corollary 2.13 applies with $r < R$, Proposition 1.17 implies that

$$|f^{(n)}(a)| \le \left(\frac{n!}{2\pi}\right)\frac{M}{r^{n+1}} \cdot 2\pi r = \frac{n!M}{r^n}$$

Since $r < R$ is arbitrary, the result follows by letting $r \to R-$. ∎

We will conclude this section by proving a proposition which is a special case of a more general result which will be presented later in this chapter.

2.15 Proposition. *Let f be analytic in the disk $B(a; R)$ and suppose that γ is a closed rectifiable curve in $B(a; R)$. Then*

$$\int_{\gamma} f = 0.$$

Proof. This is proved by showing that f has a primitive (Corollary 1.22). Now, by Theorem 2.8, $f(z) = \sum a_n(z-a)^n$ for $|z-a| < R$. Let

$$F(z) = \sum_{n=0}^{\infty} \left(\frac{a_n}{n+1}\right)(z-a)^{n+1} = (z-a)\sum_{0}^{\infty}\left(\frac{a_n}{n+1}\right)(z-a)^n.$$

Since $\lim (n+1)^{1/n} = 1$, it follows that this power series has the same radius of convergence as $\sum a_n(z-a)^n$. Hence, F is defined on $B(a; R)$. Moreover, $F'(z) = f(z)$ for $|z-a| < R$. ∎

Exercises

1. Show that the function defined by (2.2) is continuous.
2. Prove the following analogue of Leibniz's rule (this exercise will be

frequently used in the later sections.) Let G be an open set and let γ be a rectifiable curve in $\mathbb{C}$. Suppose that $\varphi : \{\gamma\} \times G \to \mathbb{C}$ is a continuous function and define $g : G \to \mathbb{C}$ by

$$g(z) = \int_{\gamma} \varphi(w, z)\, dw$$

then g is continuous. If $\dfrac{\partial \varphi}{\partial z}$ exists for each (w, z) in $\{\gamma\} \times G$ and is continuous then g is analytic and

$$g'(z) = \int_{\gamma} \frac{\partial \varphi}{\partial z}(w, z)\, dw.$$

3. Suppose that γ is a rectifiable curve in $\mathbb{C}$ and φ is defined and continuous on $\{\gamma\}$. Use Exercise 2 to show that

$$g(z) = \int_{\gamma} \frac{\varphi(w)}{w - z}\, dw$$

is analytic on $\mathbb{C} - \{\gamma\}$ and

$$g^{(n)}(z) = n! \int_{\gamma} \frac{\varphi(w)}{(w - z)^{n+1}}\, dw.$$

4. (a) Prove Abel's Theorem: Let $\sum a_n (z - a)^n$ have radius of convergence 1 and suppose that $\sum a_n$ converges to A. Prove that

$$\lim_{r \to 1-} \sum a_n r^n = A.$$

(Hint: Find a summation formula which is the analogue of integration by parts.)

(b) Use Abel's Theorem to prove that $\log 2 = 1 - \frac{1}{2} + \frac{1}{3} - \dots$

5. Give the power series expansion of $\log z$ about $z = i$ and find its radius of convergence.

6. Give the power series expansion of $\sqrt{z}$ about $z = 1$ and find its radius of convergence.

✓ 7. Use the results of this section to evaluate the following integrals:

(a) $\displaystyle\int_{\gamma} \frac{e^{iz}}{z^2}\, dz$, $\qquad \gamma(t) = e^{it}$, $\qquad 0 \le t \le 2\pi$;

(b) $\displaystyle\int_{\gamma} \frac{dz}{z - a}$, $\qquad \gamma(t) = a + re^{it}$, $\qquad 0 \le t \le 2\pi$;

(c) $\displaystyle\int_{\gamma} \frac{\sin z}{z^3}\, dz$, $\qquad \gamma(t) = e^{it}$, $\qquad 0 \le t \le 2\pi$;

(d) $\displaystyle\int_{\gamma} \frac{\log z}{z^n}\, dz$, $\qquad \gamma(t) = 1 + \frac{1}{2}e^{it}$, $\qquad 0 \le t \le 2\pi$ and $n \ge 0$.

8. Use a Mobuis transformation to show that Proposition 2.15 holds if the disk $B(a; R)$ is replaced by a half plane.

9. Use Corollary 2.13 to evaluate the following integrals:

(a) $\int_\gamma \dfrac{e^z - e^{-z}}{z^n}\, dz$ where n is a positive integer and $\gamma(t) = e^{it}, 0 \le t \le 2\pi$;

(b) $\int_\gamma \dfrac{dz}{\left(z - \frac{1}{2}\right)^n}$ where n is a positive integer and $\gamma(t) = \frac{1}{2} + e^{it}, 0 \le t \le 2\pi$;

(c) $\int_\gamma \dfrac{dz}{z^2 + 1}$ where $\gamma(t) = 2e^{it}, 0 \le t \le 2\pi$. (Hint: expand $(z^2 + 1)^{-1}$ by means of partial fractions);

(d) $\int_\gamma \dfrac{\sin z}{z}\, dz$ where $\gamma(t) = e^{it}, 0 \le t \le 2\pi$;

(e) $\int_\gamma \dfrac{z^{1/m}}{(z-1)^m}\, dz$ where $\gamma(t) = 1 + \frac{1}{2}e^{it}, 0 \le t \le 2\pi$.

10. Evaluate $\int_\gamma \dfrac{z^2 + 1}{z(z^2 + 4)}\, dz$ where $\gamma(t) = re^{it}, 0 \le t \le 2\pi$, for all possible values of r, $0 < r < 2$ and $2 < r < \infty$.

11. Find the domain of analyticity of

$$f(z) = \frac{1}{2i} \log\left(\frac{1+iz}{1-iz}\right);$$

also, show that $\tan f(z) = z$ (i.e., f is a branch of arctan z). Show that

$$f(z) = \sum_{k=0}^{\infty} (-1)^k \frac{z^{2k+1}}{2k+1} \quad \text{for } |z| < 1$$

(Hint: see Exercise III. 3.19.)

12. Show that

$$\sec z = 1 + \sum_{k=1}^{\infty} \frac{E_{2k}}{(2k)!} z^{2k}$$

for some constants $E_2, E_4, \cdots$. These numbers are called Euler's numbers. What is the radius of convergence of this series? Use the fact that $1 = \cos z$ sec z to show that

$$E_{2n} - \binom{2n}{2n-2} E_{2n-2} + \binom{2n}{2n-4} E_{2n-4} + \cdots + (-1)^{n-1}\binom{2n}{2} E_2 + (-1)^n = 0.$$

Evaluate E_2, E_4, E_6, E_8. ($E_{10} = 50521$ and $E_{12} = 2702765$).

⨉

13. Find the series expansion of $\dfrac{e^z - 1}{z}$ about zero and determine its radius

of convergence. Consider $f(z) = \dfrac{z}{e^z - 1}$ and let

$$f(z) = \sum_{k=0}^{\infty} \frac{a_k}{k!} z^k$$

be its power series expansion about zero. What is the radius of convergence?
Show that

$$0 = a_0 + \binom{n+1}{1} a_1 + \cdots + \binom{n+1}{n} a_n.$$

Using the fact that $f(z) + \frac{1}{2}z$ is an even function show that $a_k = 0$ for k odd
and $k > 1$. The numbers $B_{2n} = (-1)^{n-1} a_{2n}$ are called the Bernoulli numbers
for $n \geq 1$. Calculate $B_2, B_4, \cdots B_{10}$.

14. Find the power series expansion of $\tan z$ about $z = 0$, expressing the
coefficients in terms of Bernoulli numbers. (Hint: use Exercise 13 and the
formula $\cot 2z = \frac{1}{2} \cot z - \frac{1}{2} \tan z$.)

§3. Zeros of an analytic function

If $p(z)$ and $q(z)$ are two polynomials then it is well known that
$p(z) = s(z)q(z) + r(z)$ where $s(z)$ and $r(z)$ are also polynomials and the degree
of $r(z)$ is less than the degree of $q(z)$. In particular, if a is such that $p(a) = 0$
then choose $(z-a)$ for $q(z)$. Hence, $p(z) = (z-a)s(z) + r(z)$ and $r(z)$ must
be a constant polynomial. But letting $z = a$ gives $0 = p(a) = r(a)$. Thus,
$p(z) = (z-a)s(z)$. If we also have that $s(a) = 0$ we can factor $(z-a)$ from
$s(z)$. Continuing we get $p(z) = (z-a)^m t(z)$ where $1 \leq m \leq$ degree of $p(z)$,
and $t(z)$ is a polynomial such that $t(a) \neq 0$. Also, degree $t(z) =$ degree
$p(z) - m$.

3.1 Definition. If $f: G \to \mathbb{C}$ is analytic and a in G satisfies $f(a) = 0$ then a
is a *zero of f of multiplicity* $m \geq 1$ if there is an analytic function $g: G \to \mathbb{C}$
such that $f(z) = (z-a)^m g(z)$ where $g(a) \neq 0$.

Returning to the discussion of polynomials, we have that the multiplicity
of a zero of a polynomial must be less than the degree of the polynomial.
If $n =$ the degree of the polynomial $p(z)$ and $a_1, \ldots, a_k$ are all the distinct
zeros of $p(z)$ then $p(z) = (z-a_1)^{m_1} \cdots (z-a_k)^{m_k} s(z)$ where $s(z)$ is a polynomial
with no zeros. Now the Fundamental Theorem of Algebra says that a
polynomial with no zeros is constant. Hence, if we can prove this result we
will have succeeded in completely factoring $p(z)$ into the product of first
degree polynomials. The reader might be pleasantly surprised to know that
after many years of studying Mathematics he is right now on the threshold
of proving the Fundamental Theorem of Algebra. But first it is necessary to
prove a famous result about analytic functions. It is also convenient to
introduce some new terminology.

3.2 Definition. An *entire function* is a function which is defined and analytic in the whole complex plane $\mathbb{C}$. (The term "integral function" is also used.)

The following result follows from Theorem 2.8 and the fact that $\mathbb{C}$ contains $B(0; R)$ for arbitrarily large R.

3.3 Proposition. *If f is an entire function then f has a power series expansion*

$$f(z) = \sum_{n=0}^{\infty} a_n z^n$$

with infinite radius of convergence.

In light of the preceding proposition, entire functions can be considered as polynomials of "infinite degree". So the question arises: can the theory of polynomials be generalized to entire functions? For example, can an entire function be factored? The answer to this is difficult and is postponed to Section VII. 5. Another property of polynomials is that no non constant polynomial is bounded. Indeed, if $p(z) = z^n + a_{n-1} z^{n-1} + \cdots + a_0$ then $\lim_{z \to \infty} p(z) = \lim_{z \to \infty} z^n [1 + a_{n-1} z^{-1} + \cdots + a_0 z^{-n}] = \infty$. The fact that this also holds for entire functions is an extremely useful result.

3.4 Liouville's Theorem. *If f is a bounded entire function then f is constant.*
Proof. Suppose $|f(z)| \le M$ for all z in $\mathbb{C}$. We will show that $f'(z) = 0$ for all z in $\mathbb{C}$. To do this use Cauchy's Estimate (Corollary 2.14). Since f is analytic in any disk $B(z; R)$ we have that $|f'(z)| \le M/R$. Since R was arbitrary, it follows that $f'(z) = 0$ for each z in $\mathbb{C}$. ∎

The reader should not be deceived into thinking that this theorem is insignificant because it has such a short proof. We have expended a great deal of effort building up machinery and increasing our knowledge of analytic functions. We have plowed, planted, and fertilized; we shouldn't be surprised if, occasionally, something is available for easy picking.

Liouville's Theorem will be better appreciated in the following application.

3.5 Fundamental Theorem of Algebra. *If p(z) is a non constant polynomial then there is a complex number a with p(a) = 0.*

Proof. Suppose $p(z) \ne 0$ for all z and let $f(z) = [p(z)]^{-1}$; then f is an entire function. If p is not constant then, as was shown above, $\lim_{z \to \infty} p(z) = \infty$; so $\lim_{z \to \infty} f(z) = 0$. In particular, there is a number $R > 0$ such that $|f(z)| < 1$ if $|z| > R$. But f is continuous on $\bar{B}(0; R)$ so there is a constant M such that $|f(z)| \le M$ for $|z| \le R$. Hence f is bounded and, therefore, must be constant by Liouville's theorem. It follows that p must be constant, contradicting our assumption. ∎

3.6 Corollary. *If p(z) is a polynomial and $a_1, \ldots, a_m$ are its zeros with a_j having multiplicity k_j then $p(z) = c(z - a_1)^{k_1} \cdots (z - a_m)^{k_m}$ for some constant c and $k_1 + \cdots + k_m$ is the degree of p.*

Returning to the analogy between entire functions and polynomials, the reader should be warned that this cannot be taken too far. For example, if p is a polynomial and $a \in \mathbb{C}$ then there is a number z with $p(z) = a$. In fact, this follows from the Fundamental Theorem of Algebra by considering the polynomial $p(z) - a$. However the exponential function fails to have this property since it does not assume the value zero. (Nevertheless, we are able to show that this is the worst that can happen. That is, a function analytic in $\mathbb{C}$ omits at most one value. This is known as Picard's Little Theorem and will be proved later.) Moreover, no one should begin to make an analogy between analytic functions in an open set G and a polynomial p defined on $\mathbb{C}$; rather, you should only think of the polynomials as defined on G.

For example, let

$$f(z) = \cos\left(\frac{1+z}{1-z}\right), \; |z| < 1.$$

Notice that $\dfrac{1+z}{1-z}$ maps $D = \{z: |z| < 1\}$ onto $G = \{z: \operatorname{Re} z > 0\}$. The zeros

of f are the points $\left\{\dfrac{n\pi - 2}{n\pi + 2}: n \text{ is odd}\right\}$; so f has infinitely many zeros. However, as $n \to \infty$ the zeros approach 1 which is not in the domain of analyticity D. This is the story for the most general case.

3.7 Theorem. *Let G be a connected open set and let $f: G \to \mathbb{C}$ be an analytic function. Then the following are equivalent statements:*
   (a) *$f \equiv 0$;*
   (b) *there is a point a in G such that $f^{(n)}(a) = 0$ for each $n \geq 0$;*
   (c) *$\{z \in G: f(z) = 0\}$ has a limit point in G.*

Proof. Clearly (a) implies both (b) and (c). (c) *implies* (b): Let $a \in G$ and a limit point of $Z = \{z \in G: f(z) = 0\}$, and let $R > 0$ be such that $B(a; R) \subset G$. Since a is a limit point of Z and f is continuous it follows that $f(a) = 0$. Suppose there is an integer $n \geq 1$ such that $f(a) = f'(a) = \cdots = f^{(n-1)}(a) = 0$ and $f^{(n)}(a) \neq 0$. Expanding f in power series about a gives that

$$f(z) = \sum_{k=n}^{\infty} a_k(z-a)^k$$

for $|z-a| < R$. If

$$g(z) = \sum_{k=n}^{\infty} a_k(z-a)^{k-n}$$

then g is analytic in $B(a; R)$, $f(z) = (z-a)^n g(z)$, and $g(a) = a_n \neq 0$. Since g is analytic (and therefore continuous) in $B(a; R)$ we can find an r, $0 < r < R$, such that $g(z) \neq 0$ for $|z-a| < r$. But since a is a limit point of Z there is a point b with $f(b) = 0$ and $0 < |b-a| < r$. This gives $0 = (b-a)^n g(b)$ and so $g(b) = 0$, a contradiction. Hence no such integer n can be found; this proves part (b).

   (b) *implies* (a): Let $A = \{z \in G: f^{(n)}(z) = 0 \text{ for all } n \geq 0\}$. From the hypothesis of (b) we have that $A \neq \square$. We will show that A is both open and closed in G; by the connectedness of G it will follow that A must be G and so $f \equiv 0$. To see that A is closed let $z \in A^-$ and let z_k be a sequence in A such that $z = \lim z_k$. Since each $f^{(n)}$ is continuous it follows that $f^{(n)}(z) = \lim f^{(n)}(z_k) = 0$. So $z \in A$ and A is closed.

To see that A is open, let $a \in A$ and let $R > 0$ be such that $B(a; R) \subset G$. Then $f(z) = \Sigma a_n(z-a)^n$ for $|z-a| < R$ where $a_n = \dfrac{1}{n!} f^{(n)}(a) = 0$ for each $n \geq 0$. Hence $f(z) = 0$ for all z in $B(a; R)$ and, consequently, $B(a; R) \subset A$. Thus A is open and this completes the proof of the theorem. ∎

3.8 Corollary. *If f and g are analytic on a region G then $f \equiv g$ iff $\{z \in G: f(z) = g(z)\}$ has a limit point in G.*

This follows by applying the preceding theorem to the analytic function $f - g$.

3.9 Corollary. *If f is analytic on an open connected set G and f is not identically zero then for each a in G with $f(a) = 0$ there is an integer $n \geq 1$ and an analytic function $g: G \to \mathbb{C}$ such that $g(a) \neq 0$ and $f(z) = (z-a)^n g(z)$ for all z in G. That is, each zero of f has finite multiplicity.*

Proof. Let n be the largest integer (≥ 1) such that $f^{(n-1)}(a) = 0$ and define $g(z) = (z-a)^{-n}f(z)$ for $z \neq a$ and $g(a) = \dfrac{1}{n!} f^{(n)}(a)$. Then g is clearly analytic in $G - \{a\}$; to see that g is analytic in G it need only be shown to be analytic in a neighborhood of a. This is accomplished by using the method of the proof that (c) implies (b) in the theorem. ∎

3.10 Corollary. *If $f: G \to \mathbb{C}$ is analytic and not constant, $a \in G$, and $f(a) = 0$ then there is an $R > 0$ such that $B(a; R) \subset G$ and $f(z) \neq 0$ for $0 < |z-a| < R$.*

Proof. By the above theorem the zeros of f are isolated. ∎

There is one instance where the analogy between polynomials and analytic functions works in reverse. That is, there is a property of analytic functions which is not so transparent for polynomials.

3.11 Maximum Modulus Theorem. *If G is a region and $f: G \to \mathbb{C}$ is an analytic function such that there is a point a in G with $|f(a)| \geq |f(z)|$ for all z in G, then f is constant.*

Proof. Let $\bar{B}(a; r) \subset G$, $\gamma(t) = a + re^{it}$ for $0 \leq t \leq 2\pi$; according to Proposition 2.6

$$f(a) = \frac{1}{2\pi i} \int_\gamma \frac{f(w)}{w-a}\, dw$$

$$= \frac{1}{2\pi} \int_0^{2\pi} f(a+re^{it})\, dt$$

Hence

$$|f(a)| \leq \frac{1}{2\pi} \int_0^{2\pi} |f(a+re^{it})| \, dt \leq |f(a)|$$

since $|f(a+re^{it})| \leq |f(a)|$ for all t. This gives that

$$0 = \int_0^{2\pi} [|f(a)| - |f(a+re^{it})|] \, dt;$$

but since the integrand is non-negative it follows that $|f(a)| = |f(a+re^{it})|$ for all t. Moreover, since r was arbitrary, we have that f maps any disk $B(a; R) \subset G$ into the circle $|z| = |\alpha|$ where $\alpha = f(a)$. But this implies that f is constant on $B(a; R)$ (Exercise III. 3.17). In particular $f(z) = \alpha$ for $|z-a| < R$. According to Corollary 3.8, $f \equiv \alpha$. ∎

According to the Maximum Modulus Theorem, a non-constant analytic function on a region cannot assume its maximum modulus; this fact is far from obvious even in the case of polynomials. The consequences of this theorem are far reaching; some of these, along with a closer examination of the Maximum Modulus Theorem, are presented in Chapter VI. (Actually, the reader at this point can proceed to Sections VI. 1 and VI. 2.)

Exercises

1. Let f be an entire function and suppose there is a constant M, an $R > 0$, and an integer $n \geq 1$ such that $|f(z)| \leq M|z|^n$ for $|z| > R$. Show that f is a polynomial of degree $\leq n$.

2. Give an example to show that G must be assumed to be connected in Theorem 3.7.

3. Find all entire functions f such that $f(x) = e^x$ for x in $\mathbb{R}$.

4. Prove that $e^{z+a} = e^z e^a$ by applying Corollary 3.8.

5. Prove that $\cos(a+b) = \cos a \cos b - \sin a \sin b$ by applying Corollary 3.8.

6. Let G be a region and suppose that $f: G \rightarrow \mathbb{C}$ is analytic and $a \in G$ such that $|f(a)| \leq |f(z)|$ for all z in G. Show that either $f(a) = 0$ or f is constant.

7. Give an elementary proof of the Maximum Modulus Theorem for polynomials.

8. Let G be a region and let f and g be analytic functions on G such that $f(z)g(z) = 0$ for all z in G. Show that either $f \equiv 0$ or $g \equiv 0$.

9. Let $U: \mathbb{C} \rightarrow \mathbb{R}$ be a harmonic function such that $U(z) \geq 0$ for all z in $\mathbb{C}$; prove that U is constant.

10. Show that if f and g are analytic functions on a region G such that $\bar{f}g$ is analytic then either f is constant or $g \equiv 0$.

§4. The index of a closed curve

We have already shown that $\int_\gamma (z-a)^{-1} \, dz = 2\pi i n$ if $\gamma(t) = a + e^{2\pi i n t}$. The following result shows that this is not peculiar to the path γ.

4.1 Proposition. *If $\gamma : [0, 1] \rightarrow \mathbb{C}$ is a closed rectifiable curve and $a \notin \{\gamma\}$ then*

$$\frac{1}{2\pi i} \int_\gamma \frac{dz}{z - a}$$

is an integer.

Proof. This is only proved under the hypothesis that γ is smooth. In this case define $g : [0, 1] \rightarrow \mathbb{C}$ by

$$g(t) = \int_0^t \frac{\gamma'(s)}{\gamma(s) - a} \, ds$$

Hence, $g(0) = 0$ and $g(1) = \int_\gamma (z - a)^{-1} dz$. We also have that

$$g'(t) = \frac{\gamma'(t)}{\gamma(t) - a} \quad \text{for} \quad 0 \le t \le 1.$$

But this gives

$$\frac{d}{dt} e^{-g} (\gamma - a) = e^{-g} \gamma' - g' e^{-g} (\gamma - a)$$

$$= e^{-g} \left[\gamma' - \gamma'(\gamma - a)^{-1} (\gamma - a) \right]$$

$$= 0$$

So $e^{-g}(\gamma - a)$ is the constant function $e^{-g(0)}(\gamma(0) - a) = \gamma(0) - a = e^{-g(1)}(\gamma(1) - a)$. Since $\gamma(0) = \gamma(1)$ we have that $e^{-g(1)} = 1$ or that $g(1) = 2\pi i k$ for some integer k. ∎

4.2 Definition. If γ is a closed rectifiable curve in $\mathbb{C}$ then for $a \notin \{\gamma\}$

$$n(\gamma; a) = \frac{1}{2\pi i} \int_\gamma (z - a)^{-1} dz$$

is called the *index of* γ with respect to the point a. It is also sometimes called the *winding number* of γ around a.

Recall that if $\gamma : [0, 1] \rightarrow \mathbb{C}$ is a curve, $-\gamma$ or γ^{-1} is the curve defined by $(-\gamma)(t) = \gamma(1 - t)$ (this is actually a reparametrization of the original definition). Also if γ and σ are curves defined on $[0, 1]$ with $\gamma(1) = \sigma(0)$ then $\gamma + \sigma$ is the curve $(\gamma + \sigma)(t) = \gamma(2t)$ if $0 \le t \le \frac{1}{2}$ and $(\gamma + \sigma)(t) = \sigma(2t - 1)$ if $\frac{1}{2} \le t \le 1$. The proof of the following proposition is left to the reader.

4.3 Proposition. *If γ and σ are closed rectifiable curves having the same initial points then*
(a) $n(\gamma; a) = -n(-\gamma; a)$ *for every* $a \notin \{\gamma\}$;
(b) $n(\gamma + \sigma; a) = n(\gamma; a) + n(\sigma; a)$ *for every* $a \notin \{\gamma\} \cup \{\sigma\}$.

Why is $n(\gamma; a)$ called the winding number of γ about a? As was said before if $\gamma(t) = a + e^{2\pi i n t}$ for $0 \le t \le 1$ then $n(\gamma; a) = n$. In fact if $|b - a| < 1$ then $n(\gamma; b) = n$ and if $|b - a| > 1$ then $n(\gamma; b) = 0$. This can be shown directly or one can invoke Theorem 4.4 below. So at least in this case $n(\gamma; b)$ measures the number of times γ wraps around $b -$ with the minus sign indicating that the curve goes in the clockwise direction.

The following discussion is intuitive and mathematically imprecise. Actually, with a little more sophistication this discussion can be corrected and gives insight into the Argument Principle (V.3).

If γ is smooth then

$$\int_\gamma (z-a)^{-1}\,dz = \int_0^1 \frac{\gamma'(t)}{\gamma(t)-a}\,dt.$$

Taking inspiration from calculus one is tempted to write $\int_\gamma (z-a)^{-1}\,dx = \log[\gamma(t)-a]|_{t=0}^{t=1}$. Since $\gamma(1)=\gamma(0)$, this would always give zero. The difficulty lies in the fact that $\gamma(t)-a$ is complex valued and unless $\gamma(t)-a$ lies in a region on which a branch of the logarithm can be defined, the above inspiration turns out to be only so much hot air. In fact if γ wraps around the point a then we cannot define $\log(\gamma(t)-a)$ since there is no analytic branch of the logarithm defined on $\mathbb{C}-\{a\}$.

Nevertheless there is a correct interpretation of the preceding discussion. If we think of $\log z = \log|z| + i\arg z$ as defined then

$$\int_\gamma (z-a)^{-1}\,dz = \log[\gamma(1)-a] - \log[\gamma(0)-a] =$$

$$\{\log|\gamma(1)-a| - \log|\gamma(0)-a|\} + i\{\arg[\gamma(1)-a] - \arg[\gamma(0)-a]\}.$$

Since the difficulty in defining $\log z$ is in choosing the correct value of $\arg z$, we can think of the real part of the last expression as equal to zero. Since $\gamma(1)=\gamma(0)$ it must be that even with the ambiguity in defining $\arg z$, $\arg[\gamma(1)-a] - \arg[\gamma(0)-a]$ must equal an integral multiple of 2π, and furthermore this integer counts the number of times γ wraps around a.

Let γ be a closed rectifiable curve and consider the open set $G = \mathbb{C} - \{\gamma\}$. Since $\{\gamma\}$ is compact $\{z:|z|>R\} \subset G$ for some sufficiently large R. This says that G has one, and only one, unbounded component.

4.4 Theorem. *Let γ be a closed rectifiable curve in $\mathbb{C}$. Then $n(\gamma;a)$ is constant for a belonging to a component of $G = \mathbb{C} - \{\gamma\}$. Also, $n(\gamma;a) = 0$ for a belonging to the unbounded component of G.*

Proof. Define $f: G \to \mathbb{C}$ by $f(a) = n(\gamma;a)$. It will be shown that f is continuous. If this is done then it follows that $f(D)$ is connected for each component D of G. But since $f(G)$ is contained in the set of integers it follows that $f(D)$ reduces to a single point. That is, f is constant on D.

To show that f is continuous recall that the components of G are open (Theorem II. 2.9). Fix a in G and let $r = d(a, \{\gamma\})$. If $|a-b| < \delta < \frac{1}{2}r$ then

$$|f(a)-f(b)| = \frac{1}{2\pi}\left|\int_\gamma \left[(z-a)^{-1} - (z-b)^{-1}\right]dz\right|$$

$$= \frac{1}{2\pi}\left|\int_\gamma \frac{(a-b)}{(z-a)(z-b)}\,dz\right|$$

$$\leq \frac{|a-b|}{2\pi}\int_\gamma \frac{|dz|}{|z-a||z-b|}$$

(margin notes:)

γ cpt., hence bounded. (cont. image of cpt. set)

this comp contains $B(a,N)$ for some N lg.

But for $|a - b| < \frac{1}{2}r$ and z on $\{\gamma\}$ we have that $|z - a| \geq r > \frac{1}{2}r$ and $|z - b| > \frac{1}{2}r$. It follows that $|f(a) - f(b)| < \frac{2\delta}{\pi r^2} V(\gamma)$. So if $\epsilon > 0$ is given then, by choosing δ to be smaller than $\frac{1}{2}r$ and $(\pi r^2 \epsilon)/2V(\gamma)$, we see that f must be continuous. (Also, see Exercise 2.3.)

Now let U be the unbounded component of G. As was mentioned before the theorem there is an $R > 0$ such that $U \supset \{z : |z| > R\}$. If $\epsilon > 0$ choose a with $|a| > R$ and $|z - a| > (2\pi\epsilon)^{-1} V(\gamma)$ uniformly for z on $\{\gamma\}$; then $|n(\gamma; a)| < \epsilon$. That is, $n(\gamma; a) \to 0$ as $a \to \infty$. Since $n(\gamma; a)$ is constant on U, it must be zero. ∎

Exercises

1. Prove Proposition 4.3.
2. Give an example of a closed rectifiable curve γ in $\mathbb{C}$ such that for any integer k there is a point $a \notin \{\gamma\}$ with $n(\gamma; a) = k$.
3. Let $p(z)$ be a polynomial of degree n and let $R > 0$ be sufficiently large so that p never vanishes in $\{z : z \geq R\}$. If $\gamma(t) = Re^{it}$, $0 \leq t \leq 2\pi$, show that $\int_\gamma \frac{p'(z)}{p(z)} dz = 2\pi i n$.
4. Fix $w = re^{i\theta} \neq 0$ and let γ be a rectifiable path in $\mathbb{C} - \{0\}$ from 1 to w. Show that there is an integer k such that $\int_\gamma z^{-1} dz = \log r + i\theta + 2\pi i k$.

§5. *Cauchy's Theorem and Integral Formula*

We have already proved Cauchy's Theorem for functions analytic in a disk: if G is an open disk then $\int_\gamma f = 0$ for any analytic function f on G and any closed rectifiable curve γ in G (Proposition 2.15). For which regions G does this result remain valid? There are regions for which the result is false. For example, if $G = \mathbb{C} - \{0\}$ and $f(z) = z^{-1}$ then $\gamma(t) = e^{it}$ for $0 \leq t \leq 2\pi$ gives that $\int_\gamma f = 2\pi i$. The difficulty with $\mathbb{C} - \{0\}$ is the presence of a hole (namely $\{0\}$). In the next section it will be shown that $\int_\gamma f = 0$ for every analytic function f and every closed rectifiable curve γ in regions G that have no "holes."

In the present section we adopt a different approach. Fix a region G and an analytic function f on G. Is there a condition on a closed rectifiable curve γ such that $\int_\gamma f = 0$? The answer is furnished by the index of γ with respect to points outside G. Before presenting this result we need the following lemma. (This has already been seen in Exercise 2.3.)

5.1 Lemma. *Let γ be a rectifiable curve and suppose φ is a function defined and continuous on $\{\gamma\}$. For each $m \geq 1$ let $F_m(z) = \int_\gamma \varphi(w)(w - z)^{-m} dw$ for $z \notin \{\gamma\}$. Then each F_m is analytic on $\mathbb{C} - \{\gamma\}$ and $F_m'(z) = mF_{m+1}(z)$.*

Proof. We first claim that each F_m is continuous. In fact, this follows in the same way that we showed that the index was continuous (see the proof of Theorem 4.4). One need only observe that, since $\{\gamma\}$ is compact, φ is bounded there; and use the factorization.

$$\frac{1}{(w-z)^m} - \frac{1}{(w-a)^m} = \left[\frac{1}{w-z} - \frac{1}{w-a}\right] \sum_{k=1}^{m} \frac{1}{(w-z)^{m-k}} \frac{1}{(w-a)^{k-1}}$$

5.2
$$= (z-a)\left[\frac{1}{(w-z)^m(w-a)} + \frac{1}{(w-z)^{m-1}(w-a)^2} + \cdots\right.$$

$$\left. + \frac{1}{(w-z)(w-a)^m}\right]$$

The details are left to the reader.

Now fix a in $G = \mathbb{C} - \{\gamma\}$ and let $z \in G$, $z \neq a$. It follows from (5.2) that

5.3
$$\frac{F_m(z) - F_m(a)}{z-a} = \int_\gamma \frac{\varphi(w)(w-a)^{-1}}{(w-z)^m} dw + \cdots + \int_\gamma \frac{\varphi(w)(w-a)^{-m}}{w-z} dw$$

Since $a \notin \{\gamma\}$, $\varphi(w)(w-a)^{-k}$ is continuous on $\{\gamma\}$ for each k. By the first part of this proof (the part left to the reader), each integral on the right hand side of (5.3) defines a continuous function of z, z in G. Hence letting $z \to a$, (5.3) gives that the limit exists and

$$F_m'(a) = \int_\gamma \frac{\varphi(w)}{(w-a)^{m+1}} dw + \cdots + \int_\gamma \frac{\varphi(w)}{(w-a)^{m+1}} dw$$

$$= m F_{m+1}(a). \ \blacksquare$$

5.4 Cauchy's Integral Formula (First Version). *Let G be an open subset of the plane and $f: G \to \mathbb{C}$ an analytic function. If γ is a closed rectifiable curve in G such that $n(\gamma; w) = 0$ for all w in $\mathbb{C} - G$, then for a in $G - \{\gamma\}$*

$$n(\gamma; a) f(a) = \frac{1}{2\pi i} \int_\gamma \frac{f(z)}{z-a} dz.$$

Proof. Define $\varphi: G \times G \to \mathbb{C}$ by $\varphi(z, w) = [f(z) - f(w)]/(z-w)$ if $z \neq w$ and $\varphi(z, z) = f'(z)$. It follows that φ is continuous; and for each w in G, $z \to \varphi(z, w)$ is analytic (Exercise 1). Let $H = \{w \in \mathbb{C} : n(\gamma; w) = 0\}$. Since $n(\gamma; w)$ is a continuous integer-valued function of w, H is open. Moreover $H \cup G = \mathbb{C}$ by the hypothesis.

Define $g: \mathbb{C} \to \mathbb{C}$ by $g(z) = \int_\gamma \varphi(z, w) dw$ if $z \in G$ and $g(z) = \int_\gamma (w - z)^{-1} f(w) dw$ if $z \in H$. If $z \in G \cap H$ then

$$\int_\gamma \varphi(z, w) dw = \int_\gamma \frac{f(w) - f(z)}{w - z} dw$$

$$= \int_\gamma \frac{f(w)}{w - z} dw - f(z) n(\gamma; z) 2\pi i$$

$$= \int_\gamma \frac{f(w)}{w - z} dw.$$

Hence g is a well-defined function.

By Lemma 5.1 g is analytic on $\mathbb{C}$; that is, g is an entire function. But Theorem 4.4 implies that H contains a neighborhood of ∞ in $\mathbb{C}_\infty$. Since f is bounded on $\{\gamma\}$ and $\lim_{z\to\infty}(w-z)^{-1}=0$ uniformly for w in $\{\gamma\}$,

5.5
$$\lim_{z\to\infty} g(z) = \lim_{z\to\infty} \int_\gamma \frac{f(w)}{w-z}\, dw = 0.$$

In particular (5.5) implies there is an $R>0$ such that $|g(z)|\le 1$ for $|z|\ge R$. Since g is bounded on $\bar{B}(0;R)$ it follows that g is a bounded entire function. Hence g is constant by Liouville's Theorem. But then (5.5) says that $g\equiv 0$. That is, if $a\in G-\{\gamma\}$ then

$$0=\int_\gamma \frac{f(z)-f(a)}{z-a}\, dz$$

$$=\int_\gamma \frac{f(z)}{z-a}\, dz - f(a)\int_\gamma \frac{dz}{z-a}.$$

This proves the theorem. ∎

Often there is a need for a more general version of Cauchy's Integral Formula that involves more than one curve. For example in dealing with an annulus one needs a formula involving two curves.

5.6 Cauchy's Integral Formula (Second Version). *Let G be an open subset of the plane and $f: G \to \mathbb{C}$ an analytic function. If $\gamma_1,\ldots,\gamma_m$ are closed rectifiable curves in G such that $n(\gamma_1; w) + \cdots + n(\gamma_m; w) = 0$ for all w in $\mathbb{C} - G$, then for a in $G - \bigcup_{k=1}^m \{\gamma_k\}$*

$$f(a) \sum_{k=1}^m n(\gamma_k; a) = \sum_{k=1}^m \frac{1}{2\pi i} \int_{\gamma_k} \frac{f(z)}{z-a}\, dz.$$

Proof. The proof follows the lines of Theorem 5.4. Define $g(z,w)$ as it was done there and let $H = \{w : n(\gamma_1; w) + \cdots + n(\gamma_m; w) = 0\}$. Now $g(z)$ is defined as in the proof of (5.4) except that the sum of the integrals over $\gamma_1,\ldots,\gamma_m$ is used. The remaining details of the proof are left to the reader. ∎

Though an easy corollary of the preceding theorem, the next result is very important in the development of the theory of analytic functions.

5.7 Cauchy's Theorem (First Version). *Let G be an open subset of the plane and $f: G\to\mathbb{C}$ an analytic function. If $\gamma_1,\ldots,\gamma_m$ are closed rectifiable curves in G such that $n(\gamma_1; w) + \cdots + n(\gamma_m; w)=0$ for all w in $\mathbb{C}-G$ then*

$$\sum_{k=1}^m \int_{\gamma_k} f = 0.$$

Proof. Substitute $f(z)(z-a)$ for f in Theorem 5.6. ∎

Let $G=\{z : R_1 < |z| < R_2\}$ and define curves γ_1 and γ_2 in G by $\gamma_1(t)= r_1 e^{it}$, $\gamma_2(t)=r_2 e^{-it}$ for $0\le t\le 2\pi$, where $R_1 < r_1 < r_2 < R_2$. If $|w|\le R_1$,

$n(\gamma_1; w) = 1 = -n(\gamma_2; w)$; if $|w| \geq R_2$ then $n(\gamma_1; w) = n(\gamma_2; w) = 0$. So $n(\gamma_1; w) + n(\gamma_2; w) = 0$ for all w in $\mathbb{C} - G$.

5.8 Theorem. *Let G be an open subset of the plane and $f: G \to \mathbb{C}$ an analytic function. If $\gamma_1, \ldots, \gamma_m$ are closed rectifiable curves in G such that $n(\gamma_1; w) + \cdots + n(\gamma_m; w) = 0$ for all w in $\mathbb{C} - G$ then for a in $G - \{\gamma\}$ and $k \geq 1$*

$$f^{(k)}(a) \sum_{j=1}^{m} n(\gamma_j; a) = k! \sum_{j=1}^{m} \frac{1}{2\pi i} \int_{\gamma_j} \frac{f(z)}{(z-a)^{k+1}} \, dz.$$

Proof. This follows immediately by differentiating both sides of the formula in Theorem 5.6 and applying Lemma 5.1. ∎

5.9 Corollary. *Let G be an open set and $f: G \to \mathbb{C}$ an analytic function. If γ is a closed rectifiable curve in G such that $n(\gamma; w) = 0$ for all w in $\mathbb{C} - G$ then for a in $G - \{\gamma\}$*

$$f^{(k)}(a) n(\gamma; a) = \frac{k!}{2\pi i} \int_{\gamma} \frac{f(z)}{(z-a)^{k+1}} \, dz.$$

Cauchy's Theorem and Integral Formula is the basic result of complex analysis. With a result that is so fundamental to an entire theory it is usual in mathematics to seek the outer limits of the theorem's validity. Are there other functions that satisfy $\int_\gamma f = 0$ for all closed curves γ? The answer is no as the following converse to Cauchy's Theorem shows.

A closed path T is said to be *triangular* if it is polygonal and has three sides.

5.10 Morera's Theorem. *Let G be a region and let $f: G \to \mathbb{C}$ be a continuous function such that $\int_T f = 0$ for every triangular path T in G; then f is analytic in G.*

Proof. First observe that f will be shown to be analytic if it can be proved that f is analytic on each open disk contained in G. Hence, without loss of generality, we may assume G to be an open disk; suppose $G = B(a; R)$.

Use the hypothesis to prove that f has a primitive. For z in G define $F(z) = \int_{[a,z]} f$. Fix z_0 in G; then for any point z in G the hypothesis gives that $F(z) = \int_{[a,z_0]} f + \int_{[z_0,z]} f$. Hence

$$\frac{F(z) - F(z_0)}{z - z_0} = \frac{1}{z - z_0} \int_{[z_0,z]} f.$$

This gives

$$\frac{F(z) - F(z_0)}{z - z_0} - f(z_0) = \frac{1}{(z - z_0)} \int_{[z_0,z]} f - f(z_0)$$

$$= \frac{1}{(z - z_0)} \int_{[z_0,z]} [f(w) - f(z_0)] \, dw.$$

But by taking absolute values

$$\left| \frac{F(z) - F(z_0)}{z - z_0} - f(z_0) \right| \leq \sup\{ |f(w) - f(z_0)| : w \in [z, z_0] \}$$

which shows that

$$\lim_{z \to z_0} \frac{F(z) - F(z_0)}{z - z_0} = f(z_0). \blacksquare$$

Exercises

1. Suppose $f : G \to \mathbb{C}$ is analytic and define $\varphi : G \times G \to \mathbb{C}$ by $\varphi(z, w) = [f(z) - f(w)](z - w)^{-1}$ if $z \neq w$ and $\varphi(z, z) = f'(z)$. Prove that φ is continuous and for each fixed w, $z \to \varphi(z, w)$ is analytic.

2. Give the details of the proof of Theorem 5.6.

3. Let $B_{\pm} = B(\pm 1; \frac{1}{2})$, $G = B(0; 3) - (B_+ \cup B_-)$. Let $\gamma_1, \gamma_2, \gamma_3$ be curves whose traces are $|z - 1| = 1$, $|z + 1| = 1$, and $|z| = 2$, respectively. Give $\gamma_1, \gamma_2,$ and γ_3 orientations such that $n(\gamma_1; w) + n(\gamma_2; w) + n(\gamma_3; w) = 0$ for all w in $\mathbb{C} - G$.

4. Show that the Integral Formula follows from Cauchy's Theorem.

5. Let γ be a closed rectifiable curve in $\mathbb{C}$ and $a \notin \{\gamma\}$. Show that for $n \geq 2$ $\int_\gamma (z - a)^{-n} dz = 0$.

6. Let f be analytic on $D = B(0; 1)$ and suppose $|f(z)| \leq 1$ for $|z| < 1$. Show $|f'(0)| \leq 1$.

7. Let $\gamma(t) = 1 + e^{it}$ for $0 \leq t \leq 2\pi$. Find $\int_\gamma \left(\frac{z}{z - 1} \right)^n dz$ for all positive integers n.

8. Let G be a region and suppose $f_n : G \to \mathbb{C}$ is analytic for each $n \geq 1$. Suppose that $\{f_n\}$ converges uniformly to a function $f : G \to \mathbb{C}$ Show that f is analytic.

9. Show that if $f : \mathbb{C} \to \mathbb{C}$ is a continuous function such that f is analytic off $[-1, 1]$ then f is an entire function.

10. Use Cauchy's Integral Formula to prove the Cayley–Hamilton Theorem: If A is an $n \times n$ matrix over $\mathbb{C}$ and $f(z) = \det(z - A)$ is the characteristic polynomial of A then $f(A) = 0$. (This exercise was taken from a paper by C. A. McCarthy, *Amer. Math. Monthly*, **82** (1975), 390–391). QA1. A515

§6 The homotopic version of Cauchy's Theorem and simple connectivity

This section presents a condition on a closed curve γ such that $\int_\gamma f = 0$ for an analytic function. This condition is less general but more geometric than the winding number condition of Theorem 5.7. This condition is also used to introduce the concept of a simply connected region; in a simply connected region Cauchy's Theorem is valid for every analytic function and every closed rectifiable curve. Let us illustrate this condition by considering a closed rectifiable curve in a disk, a region where Cauchy's Theorem is always valid (Proposition 2.15).

Let $G = B(a; R)$ and let $\gamma : [0, 1] \to G$ be a closed rectifiable curve. If $0 \le t \le 1$ and $0 \le s \le 1$, and we put $z = ta + (1-t)\gamma(s)$; then z lies on the straight line segment from a to $\gamma(s)$. Hence, z must lie in G. Let $\gamma_t(s) = ta + (1-t)\gamma(s)$ for $0 \le s \le 1$ and $0 \le t \le 1$. So, $\gamma_0 = \gamma$ and γ_1 is the curve constantly equal to a; the curves γ_t are somewhere in between. We were able to draw γ down to a because there were no holes. If a point inside γ were missing from G (imagine a stick protruding up from the disk with its base inside γ), then as γ shrinks it would get caught on the hole and could not go to the constant curve.

6.1 Definition. Let $\gamma_0, \gamma_1 : [0, 1] \to G$ be two closed rectifiable curves in a region G; then γ_0 is *homotopic* to γ_1 in G if there is a continuous function $\Gamma : [0, 1] \times [0, 1] \to G$ such that

6.2
$$\begin{cases} \Gamma(s,0) = \gamma_0(s) \quad \text{and} \quad \Gamma(s,1) = \gamma_1(s) \quad (0 \le s \le 1) \\ \Gamma(0,t) = \Gamma(1,t) \quad (0 \le t \le 1) \end{cases}$$

So if we define $\gamma_t : [0, 1] \to G$ by $\gamma_t(s) = \Gamma(s,t)$ then each γ_t is a closed curve and they form a continuous family of curves which start at γ_0 and go to γ_1. Notice however that there is no requirement that each γ_t be rectifiable. In practice when γ_0 and γ_1 are rectifiable (or smooth) each of the γ_t will also be rectifiable (or smooth).

If γ_0 is homotopic to γ_1 in G write $\gamma_0 \sim \gamma_1$. Actually a notation such as $\gamma_0 \sim \gamma_1(G)$ should be used because of the role of G. If the range of Γ is not required to be in G then, as we shall see shortly, all curves would be homotopic. However, unless there is the possibility of confusion, we will only write $\gamma_0 \sim \gamma_1$.

It is easy to show that "$\sim$" is an equivalence relation. Clearly any curve is homotopic to itself. If $\gamma_0 \sim \gamma_1$ and $\Gamma : [0, 1] \times [0, 1] \to G$ satisfies (6.2) then define $\Lambda(s,t) = \Gamma(s, 1-t)$ to see that $\gamma_1 \sim \gamma_0$. Finally, if $\gamma_0 \sim \gamma_1$ and $\gamma_1 \sim \gamma_2$ with Γ satisfying (6.2) and $\Lambda : [0, 1] \times [0, 1] \to G$ satisfying $\Lambda(s,0) =$

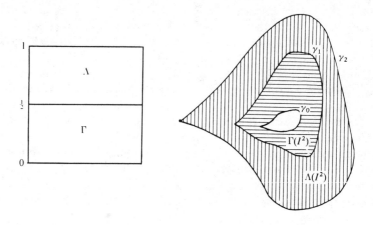

$\gamma_1(s)$, $\Lambda(s,1) = \gamma_2(s)$, and $\Lambda(0,t) = \Lambda(1,t)$ for all s and t; define $\Phi : [0,1] \times [0,1] \to G$ by

$$\Phi(s,t) = \begin{cases} \Gamma(s,2t) & \text{if } 0 \le t \le \frac{1}{2} \\ \Lambda(s,2t-1) & \text{if } \frac{1}{2} \le t \le 1 \end{cases}$$

Then Φ is continuous and shows that $\gamma_0 \sim \gamma_2$.

6.3 Definition. A set G is *convex* if given any two points a and b in G the line segment joining a and b, $[a,b]$, lies entirely in G. The set G is *star shaped* if there is a point a in G such that for each z in G, the line segment $[a,z]$ lies entirely in G. Clearly each convex set is star shaped but the converse is just as clearly false.

We will say that G is $a-$ star shaped if $[a,z] \subset G$ whenever $z \in G$. If G is $a-$ star shaped and z and w are points in G then $[z,a,w]$ is a polygon in G connecting z and w. Hence, each star shaped set is connected.

6.4 Proposition. *Let G be an open set which is $a-$ star shaped. If γ_0 is the curve which is constantly equal to a then every closed rectifiable curve in G is homotopic to γ_0.*

Proof. Let γ_1 be a closed rectifiable curve in G and put $\Gamma(s,t) = t\gamma_1(s) + (1-t)a$. Because G is $a-$ star shaped, $\Gamma(s,t) \in G$ for $0 \le s,t \le 1$. It is easy to see that Γ satisfies (6.2). ∎

The situation in which a curve is homotopic to a constant curve is one that we will often encounter. Hence it is convenient to introduce some new terminology.

6.5 Definition. If γ is a closed rectifiable curve in G then γ is *homotopic to zero* $(\gamma \sim 0)$ if γ is homotopic to a constant curve.

6.6 Cauchy's Theorem (Second Version). *If $f : G \to \mathbb{C}$ is an analytic function and γ is a closed rectifiable curve in G such that $\gamma \sim 0$, then*

$$\int_\gamma f = 0.$$

This version of Cauchy's Theorem would follow immediately from the first version if it could be shown that $n(\gamma;w) = 0$ for all w in $\mathbb{C} - G$ whenever $\gamma \sim 0$. This can be done. A plausible argument proceeds as follows.

Let $\gamma_1 = \gamma$ and let γ_0 be a constant curve such that $\gamma_1 \sim \gamma_0$. Let Γ satisfy (6.2) and define $h(t) = n(\gamma_t;w)$, where $\gamma_t(s) = \Gamma(s,t)$ for $0 \le s, t \le 1$ and w is fixed in $\mathbb{C} - G$. Now show that h is continuous on $[0,1]$. Since h is integer valued and $h(0) = 0$ it must be that $h(t) \equiv 0$. In particular, $n(\gamma;w) = 0$ for all w in $\mathbb{C} - G$.

The only point of difficulty with this argument is that for $0 < t < 1$ it may be that γ_t is not rectifiable.

As was stated after Definition 6.1, in practice each of the curves γ_t will not only be rectifiable but also smooth. So there is justification in making this assumption and providing the details to transform the preceding paragraph into a legitimate proof (Exercise 9). Indeed, in a course designed for physicists and engineers this is probably preferable. But this is not desirable for the training of mathematicians.

The statement of a theorem is not as important as its proof. Proofs are important in mathematics for several reasons, not the least of which is that a proof deepens our insight into the meaning of the theorems and gives a natural delineation of the extent of the theorem's validity. Most important for the education of a mathematician, it is essential to examine other proofs because they reveal methods.

A good method is worth a thousand theorems. Not only is this statement valid as a value judgement, but also in a literal sense. An important method can be reused in other situations to obtain further results.

With this in mind a complete proof of Theorem 6.6 will be presented. In fact, we will prove a somewhat more general fact since the proof of this new result necessitates only a little more effort than the proof that $n(\gamma; w) = 0$ for w in $\mathbb{C} - G$ whenever $\gamma \sim 0$. In fact, the proof of the next result more clearly reveals the usefulness of the method.

6.7 Cauchy's Theorem (Third Version). *If γ_0 and γ_1 are two closed rectifiable curves in G and $\gamma_0 \sim \gamma_1$ then*

$$\int_{\gamma_0} f = \int_{\gamma_1} f$$

for every function f analytic on G.

Before proceeding let us consider a special case. Suppose Γ satisfies (6.2) and also suppose Γ has continuous second partial derivatives. Hence

$$\frac{\partial^2 \Gamma}{\partial s \partial t} = \frac{\partial^2 \Gamma}{\partial t \partial s}$$

throughout the square $I^2 = [0, 1] \times [0, 1]$. Define

$$g(t) = \int_0^1 f(\Gamma(s, t)) \frac{\partial \Gamma}{\partial s}(s, t)\, ds;$$

then $g(0) = \int_{\gamma_0} f$ and $g(1) = \int_{\gamma_1} f$. By Leibniz's rule g has a continuous derivative,

$$g'(t) = \int_0^1 \left[f'(\Gamma(s, t)) \frac{\partial \Gamma}{\partial s} \frac{\partial \Gamma}{\partial t} + f(\Gamma(s, t)) \frac{\partial^2 \Gamma}{\partial t \partial s} \right] ds$$

But

$$\frac{\partial}{\partial s}\left[(f \circ \Gamma) \frac{\partial \Gamma}{\partial t} \right] = (f' \circ \Gamma) \frac{\partial \Gamma}{\partial s} \frac{\partial \Gamma}{\partial t} + f \circ \Gamma \frac{\partial^2 \Gamma}{\partial s \partial t};$$

hence

$$g'(t) = f(\Gamma(1, t)) \frac{\partial \Gamma}{\partial t} (1, t) - f(\Gamma(0, t)) \frac{\partial \Gamma}{\partial t} (0, t).$$

Since $\Gamma(1, t) = \Gamma(0, t)$ for all t we get $g'(t) = 0$ for all t. So g is a constant; in particular $\int_{\gamma_0} f = \int_{\gamma_1} f$.

Proof of Theorem 6.7. Let $\Gamma : I^2 \to G$ satisfy (6.2). Since Γ is continuous and I^2 is compact, Γ is uniformly continuous and $\Gamma(I^2)$ is a compact subset of G. Thus

$$r = d(\Gamma(I^2), \mathbb{C} - G) > 0$$

and there is an integer n such that if $(s - s')^2 + (t - t')^2 < 4/n^2$ then

$$|\Gamma(s, t) - \Gamma(s', t')| < r.$$

Let

$$Z_{jk} = \Gamma\left(\frac{j}{n}, \frac{k}{n}\right), 0 \leq j, k \leq n$$

and put

$$J_{jk} = \left[\frac{j}{n}, \frac{j+1}{n}\right] \times \left[\frac{k}{n}, \frac{k+1}{n}\right]$$

for $0 \leq j, k \leq n - 1$. Since the diameter of the square J_{jk} is $\frac{\sqrt{2}}{n}$, it follows that $\Gamma(J_{jk}) \subset B(Z_{jk}; r)$. So if we let P_{jk} be the closed polygon $[Z_{jk}, Z_{j+1, k}, Z_{j+1, k+1}, Z_{j, k+1}, Z_{jk}]$; then, because disks are convex, $P_{jk} \subset B(Z_{jk}; r)$. But from Proposition 2.15 it is known that

6.8
$$\int_{P_{jk}} f = 0$$

for any function f analytic in G.

It can now be shown that $\int_{\gamma_0} f = \int_{\gamma_1} f$ by going up the ladder we have constructed, one rung at a time. That is, let Q_k be the closed polygon $[Z_{0, k}, Z_{1, k} \ldots, Z_{nk}]$. We will show that $\int_{\gamma_0} f = \int_{Q_0} f = \int_{Q_1} f = \cdots = \int_{Q_n} f = \int_{\gamma_1} f$ (one rung at a time!). To see that $\int_{\gamma_0} f = \int_{Q_0} f$ observe that if $\sigma_j(t) = \gamma_0(t)$ for

$$\frac{j}{n} \leq t \leq \frac{j+1}{n}$$

then $\sigma_j + [Z_{j+1, 0}, Z_{j0}]$ (the + indicating that σ_j is to be followed by the polygon) is a closed rectifiable curve in the disk $B(Z_{j0}; r) \subset G$. Hence

$$\int_{\sigma_j} f = - \int_{[Z_{j+1, 0}, Z_{j0}]} f = \int_{[Z_{j0}, Z_{j+1, 0}]} f.$$

Adding both sides of this equation for $0 \leq j < n$ yields $\int_{\gamma_0} f = \int_{Q_0} f$. Similarly $\int_{\gamma_1} f = \int_{Q_n} f$.

To see that $\int_{Q_k} f = \int_{Q_{k+1}} f$ use equation (6.8); this gives

6.9
$$0 = \sum_{j=0}^{n-1} \int_{P_{jk}} f$$

However, notice that the integral $\int_{P_{jk}} f$ includes the integral over $[Z_{j+1,k},$ $Z_{j+1,k+1}]$, which is the negative of the integral over $[Z_{j+1,k+1}, Z_{j+1,k}]$, which is part of the integral $\int_{P_{j+1,k}} f$. Also,

$$Z_{0,k} = \Gamma\left(0, \frac{k}{n}\right) = \Gamma\left(1, \frac{k}{n}\right) = Z_{n,k}$$

so that $[Z_{0,k+1}, Z_{0,k}] = -[Z_{1k}, Z_{1,k+1}]$. Hence, taking these cancellations into consideration, equation (6.9) becomes

$$0 = \int_{Q_k} f - \int_{Q_{k+1}} f$$

This completes the proof of the theorem. ∎

The second version of Cauchy's Theorem immediately follows by letting γ_1 be a constant path in (6.7).

6.10 Corollary. *If γ is a closed rectifiable curve in G such that $\gamma \sim 0$ then* $n(\gamma; w) = 0$ *for all w in $\mathbb{C} - G$.*

The converse of the above corollary is not valid. That is, there is a closed rectifiable curve γ in a region G such that $n(\gamma; w) = 0$ for all w in $\mathbb{C} - G$ but γ is not homotopic to a constant curve (Exercise 8).

If G is an open set and γ_0 and γ_1 are closed rectifiable curves in G then $n(\gamma_0; a) = n(\gamma_1; a)$ for each a in $\mathbb{C} - G$ provided $\gamma_0 \sim \gamma_1(G)$. Let $\gamma_0(t) = e^{2\pi i t}$ and $\gamma_1(t) = e^{-2\pi i t}$ for $0 \le t \le 1$. Then $n(\gamma_0; 0) = 1$ and $n(\gamma_1; 0) = -1$ so that γ_0 and γ_1 are not homotopic in $\mathbb{C} - \{0\}$.

6.11 Definition. If $\gamma_0, \gamma_1 : [0, 1] \to G$ are two rectifiable curves in G such that $\gamma_0(0) = \gamma_1(0) = a$ and $\gamma_0(1) = \gamma_1(1) = b$ then γ_0 and γ_1 are *fixed-end-point* *homotopic* (*FEP homotopic*) if there is a continuous map $\Gamma: I^2 \to G$ such that

6.12
$$\Gamma(s, 0) = \gamma_0(s) \qquad \Gamma(s, 1) = \gamma_1(s)$$
$$\Gamma(0, t) = a \qquad \Gamma(1, t) = b$$

for $0 \le s, t \le 1$.

Again the relation of FEP homotopic is an equivalence relationship on curves from one given point to another (Exercise 3).

Notice that if γ_0 and γ_1 are rectifiable curves from a to b then $\gamma_0 - \gamma_1$ is a closed rectifiable curve. Suppose Γ satisfies (6.12) and define $\gamma: [0, 1] \to G$ by $\gamma(s) = \gamma_0(3s)$ for $0 \le s \le \frac{1}{3}$; $\gamma(s) = b$ for $\frac{1}{3} \le s \le \frac{2}{3}$; and $\gamma(s) = \gamma_1 (3 - 3s)$ for $\frac{2}{3} \le s \le 1$. We will show that $\gamma \sim 0$. In fact, define $\Lambda: I^2 \to G$ by

$$\Lambda(s, t) = \begin{cases} \Gamma(3s(1-t), t) & \text{if } 0 \le s \le \frac{1}{3} \\ \Gamma(1-t, 3s-1+2t-3st) & \text{if } \frac{1}{3} \le s \le \frac{2}{3} \\ \gamma_1((3-3s)(1-t)) & \text{if } \frac{2}{3} \le s \le 1. \end{cases}$$

Although this formula may appear mysterious it can easily be understood by seeing that for a given value of t, Λ is the restriction of Γ to the boundary of the square $[0, 1-t] \times [t, 1]$ (see the figure). It is left to the reader to check that Λ shows $\gamma \sim 0$.

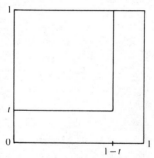

Hence, for f analytic on G the second version of Cauchy's Theorem gives

$$0 = \int_\gamma f = \int_{\gamma_0} f - \int_{\gamma_1} f.$$

This is summarized in the following.

6.13 Independence of Path Theorem. *If γ_0 and γ_1 are two rectifiable curves in G from a to b and γ_0 and γ_1 are FEP homotopic then*

$$\int_{\gamma_0} f = \int_{\gamma_1} f$$

for any function f analytic in G.

Those regions G for which the integral of an analytic function around a closed curve is always zero can be characterized.

6.14 Definition. An open set G is *simply connected* if G is connected and every closed curve in G is homotopic to zero.

6.15 Cauchy's Theorem (Fourth Version). *If G is simply connected then $\int_\gamma f = 0$ for every closed rectifiable curve and every analytic function f.*

Let us now take a few moments to digest the concept of simple connectedness. Clearly every star shaped region is simply connected. Also, examine the complement of the spiral $r = \theta$. That is, let $G = \mathbb{C} - \{\theta e^{i\theta} : 0 \le \theta < \infty\}$; then G is simply connected. In fact, it is easily seen that G is open and connected. If one argues in an intuitive way it is not difficult to become convinced that every curve in G is homotopic to zero. A rigorous proof will be postponed until we have proved the following: *A region G is simply connected iff $\mathbb{C}_\infty - G$, its complement in the extended plane, is connected in $\mathbb{C}_\infty$.* This will not be proved until Chapter VIII. If this criterion is applied to the region G above then G is simply connected since $\mathbb{C}_\infty - G$ consists of the spiral $r = \theta$ and the point at infinity.

Notice that for $G = \mathbb{C} - \{0\}$, $\mathbb{C} - G = \{0\}$ is connected but $\mathbb{C}_\infty - G = \{0, \infty\}$ is not. Also, the domain of the principal branch of the logarithm is simply connected.

It was shown earlier in this chapter (Corollary 1.22) that if an analytic function f has a primitive in a region G then the integral of f around every closed rectifiable curve in G is zero. The next result should not be too surprising in light of this.

6.16 Corollary. *If G is simply connected and $f : G \to \mathbb{C}$ is analytic in G then f has a primitive in G.*

Proof. Fix a point a in G and let γ_1, γ_2 be any two rectifiable curves in G from a to a point z in G. (Since G is open and connected there is always a path from a to any other point of G.) Then, by Theorem 6.15, $0 = \int_{\gamma_1 - \gamma_2} f = \int_{\gamma_1} f - \int_{\gamma_2} f$ (where $\gamma_1 - \gamma_2$ is the curve which goes from a to z along γ_1 and then back to a along $-\gamma_2$). Hence we can get a well defined function $F : G \to \mathbb{C}$ by setting $F(z) = \int_\gamma f$ where γ is any rectifiable curve from a to z. We claim that F is a primitive of f.

If $z_0 \in G$ and $r > 0$ is such that $B(z_0 ; r) \subset G$, then let γ be a path from a to z_0. For z in $B(z_0 ; r)$ let $\gamma_z = \gamma + [z_0, z]$; that is, γ_z is the path γ followed by the straight line segment from z_0 to z. Hence

$$\frac{F(z) - F(z_0)}{z - z_0} = \frac{1}{(z - z_0)} \int_{[z_0, z]} f.$$

Now proceed as in the proof of Morera's Theorem to show that $F'(z_0) = f(z_0)$. ∎

Perhaps a somewhat less expected consequence of simple connectedness is the fact that a branch of $\log f(z)$ where f is analytic and never vanishes, can be defined on a simply connected region. Nevertheless this is a direct consequence of the preceding corollary.

6.17 Corollary. *Let G be simply connected and let $f : G \to \mathbb{C}$ be an analytic function such that $f(z) \ne 0$ for any z in G. Then there is an analytic function*

$g: G \rightarrow \mathbb{C}$ such that $f(z) = \exp g(z)$. If $z_0 \in G$ and $e^{w_0} = f(z_0)$, we may choose g such that $g(z_0) = w_0$.

Proof. Since f never vanishes, $\dfrac{f'}{f}$ is analytic on G; so, by the preceding corollary, it must have a primitive g_1. If $h(z) = \exp g_1(z)$ then h is analytic and never vanishes. So, $\dfrac{f}{h}$ is analytic and its derivative is

$$\frac{h(z)f'(z) - h'(z)f(z)}{h(z)^2}$$

But $h' = g_1'h$ so that $hf' - fh' = 0$. Hence f/h is a constant c for all z in G. That is $f(z) = c \exp g_1(z) = \exp[g_1(z) + c']$ for some c'. By letting $g(z) = g_1(z) + c' + 2\pi i k$ for an appropriate k, $g(z_0) = w_0$ and the theorem is proved. ∎

Let us emphasize that the hypothesis of simple connectedness is a topological one and this was used to obtain some basic results of analysis. Not only are these last three theorems (6.15, 6.16, and 6.17) consequences of simple connectivity, but they are equivalent to it. It will be shown in Chapter VIII that if a region G has the conclusion of each of these theorems satisfied for every function analytic on G, then G must be simply connected.

We close this section with a definition.

6.18 Definition. If G is an open set then γ is *homologous to zero*, in symbols $\gamma \approx 0$, if $n(\gamma; w) = 0$ for all w in $\mathbb{C} - G$.

Using this notation, Corollary 6.10 says that $\gamma \sim 0$ implies $\gamma \approx 0$. This result appears in Algebraic Topology when it is shown that the first homology group of a space is isomorphic to the abelianization of the fundamental group. In fact, those familiar with homology theory will recognize in the proof of Theorem 6.7 the elements of simplicial approximation.

Exercises

1. Let G be a region and let $\sigma_1, \sigma_2: [0, 1] \rightarrow G$ be the constant curves $\sigma_1(t) \equiv a, \sigma_2(t) \equiv b$. Show that if γ is closed rectifiable curve in G and $\gamma \sim \sigma_1$ then $\gamma \sim \sigma_2$. (Hint: connect a and b by a curve.)

2. Show that if we remove the requirement "$\Gamma(0, t) = \Gamma(1, t)$ for all t" from Definition 6.1 then the curve $\gamma_0(t) = e^{2\pi i t}$, $0 \le t \le 1$, is homotopic to the constant curve $\gamma_1(t) \equiv 1$ in the region $G = \mathbb{C} - \{0\}$.

3. Let $\mathscr{C}$ = all rectifiable curves in G joining a to b and show that Definition 6.11 gives an equivalence relation on $\mathscr{C}$.

4. Let $G = \mathbb{C} - \{0\}$ and show that every closed curve in G is homotopic to a closed curve whose trace is contained in $\{z: |z| = 1\}$.

5. Evaluate the integral $\int_\gamma \dfrac{dz}{z^2+1}$ where $\gamma(\theta) = 2|\cos 2\theta|\, e^{i\theta}$ for $0 \le \theta \le 2\pi$.

6. Let $\gamma(\theta) = \theta e^{i\theta}$ for $0 \le \theta \le 2\pi$ and $\gamma(\theta) = 4\pi - \theta$ for $2\pi \le \theta \le 4\pi$.
Evaluate $\int_\gamma \dfrac{dz}{z^2+\pi^2}$.

7. Let $f(z) = [(z-\frac{1}{2}-i)\cdot(z-1-\frac{3}{2}i)\cdot(z-1-\frac{i}{2})\cdot(z-\frac{3}{2}-i)]^{-1}$ and let γ be the polygon $[0, 2, 2+2i, 2i, 0]$. Find $\int_\gamma f$.

8. Let $G = \mathbb{C} - \{a, b\}$, $a \ne b$, and let γ be the curve in the figure below.

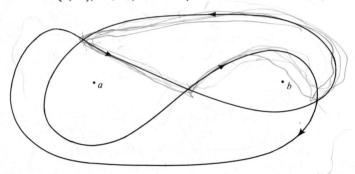

(a) Show that $n(\gamma; a) = n(\gamma; b) = 0$.

(b) Convince yourself that γ is not homotopic to zero. (Notice that the word is "convince" and not "prove". Can you prove it?) Notice that this example shows that it is possible to have a closed curve γ in a region such that $n(\gamma; z) = 0$ for all z not in G without γ being homotopic to zero. That is, the converse to Corollary 6.10 is false.

9. Let G be a region and let γ_0 and γ_1 be two closed smooth curves in G. Suppose $\gamma_0 \sim \gamma_1$ and Γ satisfies (6.2). Also suppose that $\gamma_t(s) = \Gamma(s,t)$ is smooth for each t. If $w \in \mathbb{C} - G$ define $h(t) = n(\gamma_t; w)$ and show that $h: [0, 1] \to \mathbb{Z}$ is continuous.

10. Find all possible values of $\int_\gamma \dfrac{dz}{1+z^2}$ where γ is any closed rectifiable curve in $\mathbb{C}$ not passing through $\pm i$.

11. Evaluate $\int_\gamma \dfrac{e^z - e^{-z}}{z^4}\, dz$ where γ is one of the curves depicted below. (Justify your answer.)

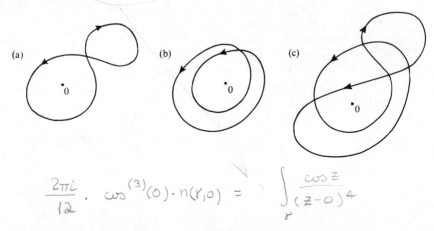

(a)   (b)   (c)

$\dfrac{2\pi i}{12} \cdot \cos^{(3)}(0) \cdot n(\gamma,0) = \int_\gamma \dfrac{\cos z}{(z-0)^4}$

§7. *Counting zeros; the Open Mapping Theorem*

In this section some applications of Cauchy's Integral Theorem are given. It is shown how to count the number of zeros inside a curve; also, using some information on the existence of roots of an analytic equation, it will be proved that a non-constant analytic function on a region maps open sets onto open sets.

In section 3 it was shown that if an analytic function f had a zero at $z = a$ we could write $f(z) = (z-a)^m g(z)$ where g is analytic and $g(a) \neq 0$. Suppose G is a region and let f be analytic in G with zeros at $a_1, \ldots, a_m$. (Where some of the a_k may be repeated according to the multiplicity of the zero.) So we can write $f(z) = (z-a_1)(z-a_2) \ldots (z-a_m)g(z)$ where g is analytic on G and $g(z) \neq 0$ for any z in G. Applying the formula for differentiating a product gives

7.1
$$\frac{f'(z)}{f(z)} = \frac{1}{z-a_1} + \frac{1}{z-a_2} + \cdots + \frac{1}{z-a_m} + \frac{g'(z)}{g(z)}$$

for $z \neq a_1, \ldots, a_m$. Now that this is done, the proof of the following theorem is straightforward.

7.2 Theorem. *Let G be a region and let f be an analytic function on G with zeros $a_1, \ldots, a_m$ (repeated according to multiplicity). If γ is a closed rectifiable curve in G which does not pass through any point a_k and if $\gamma \approx 0$ then*

$$\frac{1}{2\pi i} \int_\gamma \frac{f'(z)}{f(z)} \, dz = \sum_{k=1}^m n(\gamma; a_k)$$

Proof. If $g(z) \neq 0$ for any z in G then $g'(z)/g(z)$ is analytic in G; since $\gamma \approx 0$, Cauchy's Theorem gives $\int_\gamma \frac{g'(z)}{g(z)} \, dz = 0$. So, using (7.1) and the definition of the index, the proof of the theorem is finished. ∎

7.3 Corollary. *Let f, G, and γ be as in the preceding theorem except that $a_1, \ldots, a_m$ are the points in G that satisfy the equation $f(z) = \alpha$; then*

$$\frac{1}{2\pi i} \int_\gamma \frac{f'(z)}{f(z)-\alpha} \, dz = \sum_{k=1}^m n(\gamma; a_k)$$

As an illustration, let us calculate $\int_\gamma \frac{(2z+1)}{z^2+z+1} \, dz$ where γ is the circle $|z| = 2$. Since the denominator of the integrand factors into $(z-\omega_1)(z-\omega_2)$, where ω_1 and ω_2 are the non-real cubic roots of 1, Theorem 7.2 gives

$$\int_\gamma \frac{2z+1}{z^2+z+1} \, dz = 4\pi i.$$

As another illustration, let $\gamma : [0, 1] \to G$ be a closed rectifiable curve in $\mathbb{C}$, $\gamma \approx 0$. Suppose that f is analytic in G. Then $f \circ \gamma = \sigma$ is a closed rectifiable curve in $\mathbb{C}$ (Exercise 1). Suppose that α is some complex number with $\alpha \notin \{\sigma\} = f(\{\gamma\})$, and let us calculate $n(\sigma; \alpha)$. We get

$$n(\sigma; \alpha) = \frac{1}{2\pi i} \int_\sigma \frac{dw}{w - \alpha}$$

$$= \frac{1}{2\pi i} \int_\gamma \frac{f'(z)}{f(z) - \alpha} \, dz$$

$$= \sum_{k=1}^m n(\gamma; a_k)$$

where a_k are the points in G with $f(a_k) = \alpha$. (To show the second equality above takes a little effort, although for γ smooth it is easy. The details are left to the reader.)

Note. It may be that there are infinitely many points in G that satisfy the equation $f(z) = \alpha$. However, from what we have proved, this sequence must converge to the boundary of G. It follows that $n(\gamma; z) \neq 0$ for at most a finite number of solutions of $f(z) = \alpha$. (See Exercise 2.)

Now if β in $\mathbb{C} - \{\sigma\}$ belongs to the same component of $\mathbb{C} - \{\sigma\}$ as does α, then $n(\sigma; \alpha) = n(\sigma; \beta)$; or,

$$\sum_k n(\gamma; z_k(\alpha)) = \sum_j n(\gamma; z_j(\beta))$$

where $z_k(\alpha)$ and $z_j(\beta)$ are the points in G that satisfy $f(z) = \alpha$ and $f(z) = \beta$ respectively. If we had chosen γ so that $n(\gamma; z_k(\alpha)) = 1$ for each k, we would have that $f(G)$ contains the component of $\mathbb{C} - f(\{\gamma\})$ that contains α. We would also have some information about the number of solutions of $f(z) = \beta$. This procedure is used to prove the following result which, in addition to being of interest in itself, will yield the Open Mapping Theorem as a consequence.

7.4 Theorem. *Suppose f is analytic in $B(a; R)$ and let $\alpha = f(a)$. If $f(z) - \alpha$ has a zero of order m at $z = a$ then there is an $\epsilon > 0$ and $\delta > 0$ such that for $|\zeta - \alpha| < \delta$, the equation $f(z) = \zeta$ has exactly m simple roots in $B(a; \epsilon)$.*

A *simple root* of $f(z) = \zeta$ is a zero of $f(z) - \zeta$ of multiplicity 1. Notice that this theorem says that $f(B(a; \epsilon)) \supset B(\alpha; \delta)$. Also, the condition that $f(z) - \alpha$ have a zero of finite multiplicity guarantees that f is not constant.

Proof of Theorem. Since the zeros of an analytic function are isolated we can choose $\epsilon > 0$ such that $\epsilon < \frac{1}{2}R$, $f(z) = \alpha$ has no solutions with $0 < |z-a| < 2\epsilon$, and $f'(z) \neq 0$ if $0 < |z-a| < 2\epsilon$. (If $m \geq 2$ then $f'(a) = 0$.) Let $\gamma(t) = a + \epsilon \exp(2\pi it)$, $0 \leq t \leq 1$, and put $\sigma = f \circ \gamma$. Now $\alpha \notin \{\sigma\}$; so there is a $\delta > 0$ such that $B(\alpha; \delta) \cap \{\sigma\} = \square$. Thus, $B(\alpha; \delta)$ is contained in the same component of $\mathbb{C} - \{\sigma\}$; that is, $|\alpha - \zeta| < \delta$ implies $n(\sigma; \alpha) = n(\sigma;$
$$\zeta) = \sum_{k=1}^{p} n(\gamma; z_k(\zeta)).$$ But since $n(\gamma; z)$ must be either zero or one, we have that there are exactly m solutions to the equation $f(z) = \zeta$ inside $B(a; \epsilon)$. Since $f'(z) \neq 0$ for $0 < |z-a| < \epsilon$, each of these roots (for $\zeta \neq \alpha$) must be simple (Exercise 3). ∎

7.5 Open Mapping Theorem. *Let G be a region and suppose that f is a non constant analytic function on G. Then for any open set U in G, $f(U)$ is open.*

Proof. If $U \subset G$ is open, then we will have shown that $f(U)$ is open if we can show that for each a in U there is a $\delta > 0$ such that $B(\alpha; \delta) \subset f(U)$, where $\alpha = f(a)$. But only part of the strength of the preceding theorem is needed to find an $\epsilon > 0$ and a $\delta > 0$ such that $B(a; \epsilon) \subset U$ and $f(B(a; \epsilon)) \supset B(\alpha; \delta)$. ∎

If X and Ω are metric spaces and $f: X \to \Omega$ has the property that $f(U)$ is open in Ω whenever U is open in X, then f is called an *open map*. If f is a one-one and onto map then we can define the inverse map $f^{-1}: \Omega \to X$ by $f^{-1}(\omega) = x$ where $f(x) = \omega$. It follows that f^{-1} is continuous exactly when f is open; in fact, for $U \subset X$, $(f^{-1})^{-1}(U) = f(U)$.

7.6 Corollary. *Suppose $f: G \to \mathbb{C}$ is one-one, analytic and $f(G) = \Omega$. Then $f^{-1}: \Omega \to \mathbb{C}$ is analytic and $(f^{-1})'(\omega) = [f'(z)]^{-1}$ where $\omega = f(z)$.*

Proof. By the Open Mapping Theorem, f^{-1} is continuous and Ω is open. Since $z = f^{-1}(f(z))$ for each $z \in \Omega$, the result follows from Proposition III. 2.20. ∎

Exercises

1. Show that if $f: G \to \mathbb{C}$ is analytic and γ is a rectifiable curve in G then $f \circ \gamma$ is also a rectifiable curve. (First assume G is a disk.)

2. Let G be open and suppose that γ is a closed rectifiable curve in G such that $\gamma \approx 0$. Set $r = d(\{\gamma\}, \partial G)$ and $H = \{z \in \mathbb{C} : n(\gamma; z) = 0\}$. (a) Show that $\{z : d(z, \partial G) < \frac{1}{2}r\} \subset H$. (b) Use part (a) to show that if $f: G \to \mathbb{C}$ is analytic then $f(z) = \alpha$ has at most a finite number of solutions z such that $n(\gamma; z) \neq 0$.

3. Let f be analytic in $B(a; R)$ and suppose that $f(a) = 0$. Show that a is a zero of multiplicity m iff $f^{(m-1)}(a) = \ldots = f(a) = 0$ and $f^{(m)}(a) \neq 0$.

4. Suppose that $f\colon G \to \mathbb{C}$ is analytic and one-one; show that $f'(z) \neq 0$ for any z in G.

5. Let X and Ω be metric spaces and suppose that $f\colon X \to \Omega$ is one-one and onto. Show that f is an open map iff f is a closed map. (A function f is a closed map if it takes closed sets onto closed sets.)

6. Let $P\colon \mathbb{C} \to \mathbb{R}$ be defined by $P(z) = \operatorname{Re} z$; show that P is an open map but is not a closed map. (Hint: Consider the set $F = \{z\colon \operatorname{Im} z = (\operatorname{Re} z)^{-1}$ and $\operatorname{Re} z \neq 0\}$.)

7. Use Theorem 7.2 to give another proof of the Fundamental Theorem of Algebra.

§8. Goursat's Theorem

Most modern books define an analytic function as one which is differentiable on an open set (not assuming the continuity of the derivative). In this section it is shown that this definition is the same as ours.

Goursat's Theorem. *Let G be an open set and let $f\colon G \to \mathbb{C}$ be a differentiable function; then f is analytic on G.*

Proof. We need only show that f' is continuous on each open disk contained in G; so, we may assume that G is itself an open disk. It will be shown that f is analytic by an application of Morera's Theorem (5.10). That is, we must show that $\int_T f = 0$ for each triangular path T in G.

Let $T = [a, b, c, a]$ and let Δ be the closed set formed by T and its inside. Notice that $T = \partial\Delta$. Now using the midpoints of the sides of Δ form four triangles $\Delta_1, \Delta_2, \Delta_3, \Delta_4$ inside Δ and, by giving the boundaries appropriate

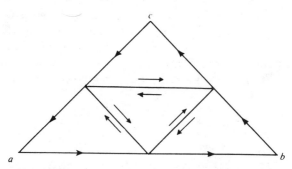

directions, we have that each $T_j = \partial\Delta_j$ is a triangle path and

8.1
$$\int_T f = \sum_{j=1}^{4} \int_{T_j} f$$

Among these four paths there is one, call it $T^{(1)}$, such that $\left|\int_{T^{(1)}} f\right| \geq \left|\int_{T_j} f\right|$ for $j = 1, 2, 3, 4$. Note that the length of each T_j (perimeter of Δ_j)—denoted by $\ell(T_j)$—is $\frac{1}{2}\ell(T)$. Also diam $T_j = \frac{1}{2}$ diam T; finally, using (8.1)

$$\left|\int_T f\right| \leq 4 \left|\int_{T^{(1)}} f\right|.$$

Now perform the same process on $T^{(1)}$, getting a triangle $T^{(2)}$ with the analogous properties. By induction we get a sequence $\{T^{(n)}\}$ of closed triangular paths such that if $\Delta^{(n)}$ is the inside of $T^{(n)}$ along with $T^{(n)}$ then;

8.2 $$\Delta^{(1)} \supset \Delta^{(2)} \supset \ldots ;$$

8.3 $$\left| \int_{T^{(n)}} f \right| \leq 4 \left| \int_{T^{(n+1)}} f \right| ;$$

8.4 $$\ell(T^{(n+1)}) = \tfrac{1}{2}\ell(T^{(n)});$$

8.5 $$\operatorname{diam} \Delta^{(n+1)} = \tfrac{1}{2} \operatorname{diam} \Delta^{(n)}.$$

These equations imply:

8.6 $$\left| \int_{T} f \right| \leq 4^n \left| \int_{T^{(n)}} f \right| ;$$

8.7 $$\ell(T^{(n)}) = (\tfrac{1}{2})^n \ell \quad \text{where} \quad \ell = \ell(T);$$

8.8 $$\operatorname{diam} \Delta^{(n)} = (\tfrac{1}{2})^n d \quad \text{where} \quad d = \operatorname{diam} \Delta.$$

Since each $\Delta^{(n)}$ is closed, (8.2) and (8.8) allow us to apply Cantor's Theorem (II. 3.6), and get that $\bigcap_{n=1}^{\infty} \Delta^{(n)}$ consists of a single point z_0.

Let $\epsilon > 0$; since f has a derivative at z_0 we can find a $\delta > 0$ such that $B(z_0; \delta) \subset G$ and

$$\left| \frac{f(z) - f(z_0)}{z - z_0} - f'(z_0) \right| < \epsilon$$

whenever $0 < |z - z_0| < \delta$. Alternately,

8.9 $$|f(z) - f(z_0) - f'(z_0)(z - z_0)| < \epsilon |z - z_0|$$

whenever $|z - z_0| < \delta$. Choose n such that $\operatorname{diam} \Delta^{(n)} = (\tfrac{1}{2})^n d < \delta$. Since $z_0 \in \Delta^{(n)}$ this gives $\Delta^{(n)} \subset B(z_0; \delta)$. Now Cauchy's Theorem implies that $0 = \int_{T^{(n)}} dz = \int_{T^{(n)}} z \, dz$. Hence

$$\left| \int_{T^{(n)}} f \right| = \left| \int_{T^{(n)}} [f(z) - f(z_0) - f'(z_0)(z - z_0)] \, dz \right|$$

$$\leq \epsilon \int_{T^{(n)}} |z - z_0| \, |dz|$$

$$\leq \epsilon [\operatorname{diam} \Delta^{(n)}] [\ell(T^{(n)})]$$

$$= \epsilon d \ell (\tfrac{1}{4})^n$$

But using (8.6) this gives

$$\left|\int_T f\right| \le 4^n \epsilon d\ell(\tfrac{1}{4})^n = \epsilon d\ell.$$

Since ϵ was arbitrary and d and ℓ are fixed, $\int_T f = 0.$ ∎

Chapter V

Singularities

In this chapter functions which are analytic in a punctured disk (an open disk with the center removed) are examined. From information about the behavior of the function near the center of the disk, a number of interesting and useful results will be derived. In particular, we will use these results to evaluate certain definite integrals over the real line which cannot be evaluated by the methods of calculus.

§1. Classification of singularities

This section begins by studying the best behaved singularity—the removable kind.

1.1 Definition. A function f has an *isolated singularity* at $z = a$ if there is an $R > 0$ such that f is defined and analytic in $B(a; R) - \{a\}$ but not in $B(a; R)$. The point a is called a *removable singularity* if there is an analytic function $g: B(a; R) \to \mathbb{C}$ such that $g(z) = f(z)$ for $0 < |z - a| < R$.

The functions $\dfrac{\sin z}{z}$, $\dfrac{1}{z}$, and $\exp \dfrac{1}{z}$ all have isolated singularities at $z = 0$. However, only $\dfrac{\sin z}{z}$ has a removable singularity (see Exercise 1). It is left to the reader to see that the other two functions do not have removable singularities.

How can we determine when a singularity is removable? Since the function has an analytic extension to $B(a; R)$, $\int_{\gamma} f = 0$ for any closed curve in the punctured disk; but this may be difficult to apply. Also it must happen that $\lim_{z \to a} f(z)$ exists. This is easier to verify, but a much weaker criterion is available.

1.2 Theorem. *If f has an isolated singularity at a then the point $z = a$ is a removable singularity iff*

$$\lim_{z \to a} (z - a)f(z) = 0$$

Proof. Suppose f is analytic in $\{z: 0 < |z - a| < R\}$, and define $g(z) = (z - a)f(z)$ for $z \neq a$ and $g(a) = 0$. Suppose $\lim_{z \to a} (z - a)f(z) = 0$; then g is clearly a continuous function. If we can show that g is analytic then it follows that a is a removable singularity. In fact, if g is analytic we have $g(z) = (z - a)h(z)$ for some analytic function defined on $B(a; R)$ because $g(a) = 0$ (IV. 3.9). But then $h(z)$ and $f(z)$ must agree for $0 < |z - a| < R$, so that a is, by definition, a removable singularity.

To show that g is analytic we apply Morera's Theorem. Let T be a triangle in $B(a; R)$ and let Δ be the inside of T together with T. If $a \notin \Delta$ then $T \sim 0$ in $\{z: 0 < |z-a| < R\}$ and so, $\int_T g = 0$ by Cauchy's Theorem. If a is a vertex of T then we have $T = [a, b, c, a]$. Let $x \in [a, b]$ and $y \in [c, a]$ and

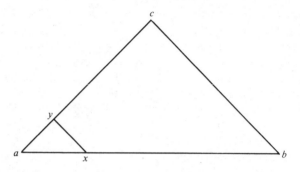

form the triangle $T_1 = [a, x, y, a]$. If P is the polygon $[x, b, c, y, x]$ then $\int_T g = \int_{T_1} g + \int_P g = \int_{T_1} g$ since $P \sim 0$ in the punctured disk. Since g is continuous and $g(a) = 0$, for any $\epsilon > 0$ x and y can be chosen such that $|g(z)| \le \epsilon/\ell$ for any z on T_1, where ℓ is the length of T. Hence $|\int_T g| = |\int_{T_1} g| \le \epsilon$; since ϵ was arbitrary we have $\int_T g = 0$.

If $a \in \Delta$ and $T = [x, y, z, x]$ then consider the triangles $T_1 = [x, y, a, x]$, $T_2 = [y, z, a, y]$, $T_3 = [z, x, a, z]$. From the preceding paragraph $\int_{T_j} g = 0$

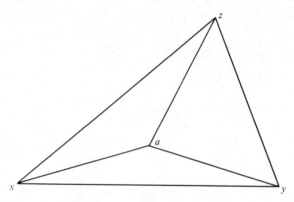

for $j = 1, 2, 3$ and so, $\int_T g = \int_{T_1} g + \int_{T_2} g + \int_{T_3} g = 0$. Since this exhausts all possibilities, g must be analytic by Morera's Theorem. Since the converse is obvious, the proof of the theorem is complete. ∎

The preceding theorem points out another stark difference between functions of a real variable and functions of a complex variable. The function $f(x) = |x|$, $x \in \mathbb{R}$, is not differentiable because it has a "corner" at $x = 0$. Such a situation does not occur in the complex case. For a function to have an honest singularity (i.e., a non-removable one) the function must behave badly in the vicinity of the point. That is, either $|f(z)|$ becomes infinite as z nears the point (and does so at least as quickly as $(z-a)^{-1}$), or $|f(z)|$ doesn't have any limit as $z \to a$.

1.3 Definition. If $z = a$ is an isolated singularity of f then a is a *pole* of f if $\lim\limits_{z \to a} |f(z)| = \infty$. That is, for any $M > 0$ there is a number $\epsilon > 0$ such that $|f(z)| \geq M$ whenever $0 < |z-a| < \epsilon$. If an isolated singularity is neither a pole nor a removable singularity it is called an *essential singularity*.

It is easy to see that $(z-a)^{-m}$ has a pole at $z = a$ for $m \geq 1$. Also, it is not difficult to see that although $z = 0$ is an isolated singularity of $\exp(z^{-1})$, it is neither a pole nor a removable singularity; hence it is an essential singularity.

Suppose that f has a pole at $z = a$; it follows that $[f(z)]^{-1}$ has a removable singularity at $z = a$. Hence, $h(z) = [f(z)]^{-1}$ for $z \neq a$ and $h(a) = 0$ is analytic in $B(a; R)$ for some $R > 0$. However, since $h(a) = 0$ it follows by Corollary IV. 3.9 that $h(z) = (z-a)^m h_1(z)$ for some analytic function h_1 with $h_1(a) \neq 0$ and some integer $m \geq 1$. But this gives that $(z-a)^m f(z) = [h_1(z)]^{-1}$ has a removable singularity at $z = a$. This is summarized as follows.

1.4 Proposition. *If G is a region with a in G and if f is analytic on $G - \{a\}$ with a pole at $z = a$ then there is a positive integer m and an analytic function $g: G \to \mathbb{C}$ such that*

$$\textbf{1.5} \qquad\qquad f(z) = \frac{g(z)}{(z-a)^m}.$$

1.6 Definition. If f has a pole at $z = a$ and m is the smallest positive integer such that $f(z)(z-a)^m$ has a removable singularity at $z = a$ then f has a *pole of order m* at $z = a$.

Notice that if m is the order of the pole at $z = a$ and g is chosen to satisfy (1.5) then $g(a) \neq 0$. (Why?)

Let f have a pole of order m at $z = a$ and put $f(z) = g(z)(z-a)^{-m}$. Since g is analytic in a disk $B(a; R)$ it has a power series expansion about a. Let

$$g(z) = A_m + A_{m-1}(z-a) + \cdots + A_1(z-a)^{m-1} + (z-a)^m \sum_{k=0}^{\infty} a_k(z-a)^k.$$

Hence

$$\textbf{1.7} \qquad\qquad f(z) = \frac{A_m}{(z-a)^m} + \cdots + \frac{A_1}{(z-a)} + g_1(z)$$

where g_1 is analytic in $B(a; R)$ and $A_m \neq 0$.

1.8 Definition. If f has a pole of order m at $z = a$ and f satisfies (1.7) then $A_m(z-a)^{-m} + \cdots + A_1(z-a)^{-1}$ is called the *singular part* of f at $z = a$.

As an example consider a rational function $r(z) = p(z)/q(z)$, where $p(z)$ and $q(z)$ are polynomials without common factors. That is, they have no common zeros; and consequently the poles of $r(z)$ are exactly the zeros of $q(z)$. The order of each pole of $r(z)$ is the order of the zero of $q(z)$. Suppose $q(a) = 0$ and let $S(z)$ be the singular part of $r(z)$ at a. Then $r(z) - S(z) = r_1(z)$ and $r_1(z)$ is a rational function whose poles are also poles of $r(z)$. Moreover, it is not difficult to see that the singular part of $r_1(z)$ at any of its poles is also the singular part of $r(z)$ at that pole. Using induction we arrive at the following: if $a_1, \cdots, a_n$ are the poles of $r(z)$ and $S_j(z)$ is the singular part of $r(z)$ at $z = a_j$ then

1.9
$$r(z) = \sum_{j=1}^{n} S_j(z) + P(z)$$

where $P(z)$ is a rational function without poles. But, by the Fundamental Theorem of Algebra, a rational function without poles is a polynomial! So $P(z)$ is a polynomial and (1.9) is nothing else but the expansion of a rational function by *partial fractions*.

Is this expansion by partial fractions (1.9) peculiar only to rational functions? Certainly it is if we require $P(z)$ in (1.9) to be a polynomial. But if we allow $P(z)$ to be any analytic function in a region G, then (1.9) is valid for any function $r(z)$ analytic in G except for a finite number of poles. Suppose we have a function f analytic in G except for infinitely many poles (e.g., $f(z) = (\cos z)^{-1}$); can we get an analogue of (1.9) where we replace the finite sum by an infinite sum? The answer to this is yes and is contained in Mittag-Leffler's Theorem which will be proved in Chapter VII.

There is an analogue of the singular part which is valid for essential singularities. Actually we will do more than this as we will investigate functions which are analytic in an annulus. But first, a few definitions.

1.10 Definition. If $\{z_n : n = 0, \pm 1, \pm 2, \ldots\}$ is a doubly infinite sequence of complex numbers, $\sum_{n=-\infty}^{\infty} z_n$ is *absolutely convergent* if both $\sum_{n=0}^{\infty} z_n$ and $\sum_{n=1}^{\infty} z_{-n}$ are absolutely convergent. In this case $\sum_{n=-\infty}^{\infty} z_n = \sum_{n=1}^{\infty} z_{-n} + \sum_{n=0}^{\infty} z_n$. If u_n is a function on a set S for $n = 0, \pm 1, \ldots$ and $\sum_{-\infty}^{\infty} u_n(s)$ is absolutely convergent for each $s \in S$, then the convergence is *uniform* over S if both $\sum_{n=0}^{\infty} u_n$ and $\sum_{n=1}^{\infty} u_{-n}$ converge uniformly on S.

The reason we are limiting ourselves to absolute convergence is that this is the type of convergence we will be most concerned with. One can define convergence of $\sum_{-\infty}^{\infty} z_n$, but the definition is not that the partial sums $\sum_{n=-m}^{m} z_n$ converge. In fact, the series $\sum_{n \neq 0} \frac{1}{n}$ satisfies this criterion but it is clearly not a series we wish to have convergent. On the other hand, if $\sum_{-\infty}^{\infty} z_n$ is absolutely convergent with sum z then it readily follows that $z = \lim \sum_{n=-m}^{m} z_n$.

If $0 \leq R_1 < R_2 \leq \infty$ and a is any complex number, define ann $(a; R_1, R_2) = \{z : R_1 < |z-a| < R_2\}$. Notice that ann $(a; 0, R_2)$ is a punctured disk.

1.11 Laurent Series Development. *Let f be analytic in the annulus* ann $(a; R_1, R_2)$. *Then*

$$f(z) = \sum_{n=-\infty}^{\infty} a_n(z-a)^n$$

where the convergence is absolute and uniform over ann $(a; r_1, r_2)^-$ *if* $R_1 < r_1 < r_2 < R_2$. *Also the coefficients* a_n *are given by the formula*

1.12
$$a_n = \frac{1}{2\pi i} \int_\gamma \frac{f(z)}{(z-a)^{n+1}} \, dz$$

where γ *is the circle* $|z-a| = r$ *for any* r, $R_1 < r < R_2$. *Moreover, this series is unique.*

Proof. If $R_1 < r_1 < r_2 < R_2$ and γ_1, γ_2 are the circles $|z-a| = r_1$, $|z-a| = r_2$ respectively, then $\gamma_1 \sim \gamma_2$ in ann $(a; R_1, R_2)$. By Cauchy's Theorem we have that for any function g analytic in ann $(a; R_1, R_2)$, $\int_{\gamma_1} g = \int_{\gamma_2} g$. In particular the integral appearing in (1.12) is independent of r so that for each integer n, a_n is a constant. Moreover, $f_2 : B(a; R_2) \to \mathbb{C}$ given by the formula

1.13
$$f_2(z) = \frac{1}{2\pi i} \int_{|w-a|=r_2} \frac{f(w)}{w-z} \, dw,$$

where $|z - a| < r_2$, $R_1 < r_2 < R_2$, is a well defined function. Also, by Lemma IV.5.1 f_2 is analytic in $B(a; r_2)$. Similarly, if $G = \{z : |z - a| > r_1\}$ then $f_1 : G \to \mathbb{C}$ defined by

1.14
$$f_1(z) = -\frac{1}{2\pi i} \int_{|w-a|=r_1} \frac{f(w)}{w-z} \, dw,$$

where $|z-a| > r_1$ and $R_1 < r_1 < R_2$, is analytic in G.

If $R_1 < |z-a| < R_2$ let r_1 and r_2 be chosen so that $R_1 < r_1 < |z-a| < r_2 < R_2$. Let $\gamma_1(t) = a + r_1 e^{it}$ and $\gamma_2(t) = a + r_2 e^{it}$, $0 \le t \le 2\pi$. Also choose a straight line segment λ going from a point on γ_1 radially to γ_2 which misses z. Since $\gamma_1 \sim \gamma_2$ in ann $(a; R_1, R_2)$ we have that the closed curve

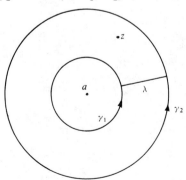

$\gamma = \gamma_2 - \lambda - \gamma_1 + \lambda$ is homotopic to zero. Also $n(\gamma_2; z) = 1$ and $n(\gamma_1, z) = 0$ gives, by Cauchy's Integral Formula, that

$$f(z) = \frac{1}{2\pi i}\int_{\gamma}\frac{f(w)}{w-z}\,dw$$

$$= \frac{1}{2\pi i}\int_{\gamma_2}\frac{f(w)}{w-z}\,dw - \frac{1}{2\pi i}\int_{\gamma_1}\frac{f(w)}{w-z}\,dw$$

$$= f_2(z) + f_1(z).$$

The plan now is to expand f_1 and f_2 in power series (f_1 having negative powers of $(z-a)$); then adding them together will give the Laurent series development of $f(z)$. Since f_2 is analytic in the disk $B(a; R_2)$ it has a power series expansion about a. Using Lemma IV. 5.1 to calculate $f_2^{(n)}(a)$,

1.15
$$f_2(z) = \sum_{n=0}^{\infty} a_n(z-a)^n$$

where the coefficients a_n are given by (1.12).

Now define $g(z)$ for

$$0 < |z| < \frac{1}{R_1} \text{ by } g(z) = f_1\left(a + \frac{1}{z}\right);$$

so $z = 0$ is an isolated singularity. We claim that $z = 0$ is a removable singularity. In fact, if $r > R_1$ then let $\rho(z) = d(z, C)$ where C is the circle $\{w: |w-a| = r\}$; also put $M = \max\{|f(w)|: w \in C\}$. Then for $|z-a| > r$

$$|f_1(z)| \le \frac{Mr}{\rho(z)}.$$

But $\lim_{z\to\infty} \rho(z) = \infty$; so that

$$\lim_{z\to 0} g(z) = \lim_{z\to 0} f_1\left(a + \frac{1}{z}\right) = 0.$$

Hence, if we define $g(0) = 0$ then g is analytic in $B(0; 1/R_1)$. Let

1.16
$$g(z) = \sum_{n=1}^{\infty} B_n z^n$$

be its power series expansion about 0. It is easy to show that this gives

1.17
$$f_1(z) = \sum_{n=1}^{\infty} a_{-n}(z-a)^{-n}$$

where a_{-n} is defined by (1.12) (the details are to be furnished by the reader in Exercise 3). Also, by the convergence properties of (1.15) and (1.17), $\sum_{-\infty}^{\infty} a_n(z-a)^n$ converges absolutely and uniformly on properly smaller annuli.

The uniqueness of this expansion can be demonstrated by showing that if $f(z) = \sum_{n=-\infty}^{\infty} a_n(z-a)^n$ converges absolutely and uniformly on proper annuli then the coefficients a_n must be given by the formula (1.12). ∎

We now use the Laurent Expansion to classify isolated singularities.

1.18 Corollary. *Let $z = a$ be an isolated singularity of f and let $f(z) = \sum\limits_{-\infty}^{\infty} a_n(z-a)^n$ be its Laurent Expansion in* ann $(a; 0, R)$. *Then*:

(a) $z = a$ *is a removable singularity iff* $a_n = 0$ *for* $n \leq -1$;
(b) $z = a$ *is a pole of order m iff* $a_{-m} \neq 0$ *and* $a_n = 0$ *for* $n \leq -(m+1)$;
(c) $z = a$ *is an essential singularity iff* $a_n \neq 0$ *for infinitely many negative integers n.*

Proof. (a) If $a_n = 0$ for $n \leq -1$ then let $g(z)$ be defined in $B(a; R)$ by $g(z) = \sum\limits_{n=0}^{\infty} a_n(z-a)^n$; thus, g must be analytic and agrees with f in the punctured disk. The converse is equally as easy.

(b) Suppose $a_n = 0$ for $n \leq -(m+1)$; then $(z-a)^m f(z)$ has a Laurent Expansion which has no negative powers of $(z-a)$. By part (a), $(z-a)^m f(z)$ has a removable singularity at $z-a$. Thus f has a pole of order m at $z = a$. The converse follows by retracing the steps in the preceding argument.

(c) Since f has an essential singularity at $z = a$ when it has neither a removable singularity nor a pole, part (c) follows from parts (a) and (b). ∎

One can also classify isolated singularities by examining the equations

1.19 $$\lim_{z \to a} |z-a|^s |f(z)| = 0$$

1.20 $$\lim_{z \to a} |z-a|^s |f(z)| = \infty$$

where s is some real number. This is outlined in Exercises 7, 8, and 9; the reader is strongly encouraged to work through these exercises.

The following gives the best information which can be proved at this time concerning essential singularities. We know that f has an essential singularity at $z = a$ when $\lim\limits_{z \to a} |f(z)|$ fails to exist ("existing" includes the possibility that the limit is infinity). This means that as z approaches a the values of $f(z)$ must wander through $\mathbb{C}$. The next theorem says that not only do they wander, but, as z approaches a, $f(z)$ comes arbitrarily close to every complex number. Actually, there is a result due to Picard that says that $f(z)$ assumes each complex value with at most one exception. However, this is not proved until Chapter XII.

1.21 Casorati–Weierstrass Theorem. *If f has an essential singularity at $z = a$ then for every $\delta > 0$, $\{f[\text{ann } (a; 0, \delta)]\}^- = \mathbb{C}$.*

Proof. Suppose that f is analytic in ann $(a; 0, R)$; it must be shown that if c and $\epsilon > 0$ are given then for each $\delta > 0$ we can find a z with $|z-a| < \delta$ and $|f(z)-c| < \epsilon$. Assume this to be false; that is, assume there is a c in $\mathbb{C}$ and $\epsilon > 0$ such that $|f(z)-c| \geq \epsilon$ for all z in $G = $ ann $(a; 0, \delta)$. Thus $\lim\limits_{z \to a} |z-a|^{-1}|f(z)-c| = \infty$, which implies that $(z-a)^{-1}(f(z)-c)$ has a pole at $z = a$. If m is the order of this pole then $\lim\limits_{z \to a} |z-a|^{m+1}|f(z)-c| = 0$. Hence $|z-a|^{m+1}|f(z)| \leq |z-a|^{m+1}|f(z)-c| + |z-a|^{m+1}|c|$ gives that

$\lim_{z \to a} |z - a|^{m+1}|f(z)| = 0$ since $m \geq 1$. But, according to Theorem 1.2, this gives that $f(z)(z-a)^m$ has a removable singularity at $z = a$. This contradicts the hypothesis and completes the proof of the theorem. ∎

Exercises

1. Each of the following functions f has an isolated singularity at $z = 0$. Determine its nature; if it is a removable singularity define $f(0)$ so that f is analytic at $z = 0$; if it is a pole find the singular part; if it is an essential singularity determine $f(\{z: 0 < |z| < \delta\})$ for arbitrarily small values of δ.

(a) $f(z) = \dfrac{\sin z}{z}$;

(b) $f(z) = \dfrac{\cos z}{z}$;

(c) $f(z) = \dfrac{\cos z - 1}{z}$;

(d) $f(z) = \exp(z^{-1})$;

(e) $f(z) = \dfrac{\log(z+1)}{z^2}$;

(f) $f(z) = \dfrac{\cos(z^{-1})}{z^{-1}}$;

(g) $f(z) = \dfrac{z^2+1}{z(z-1)}$;

(h) $f(z) = (1-e^z)^{-1}$;

(i) $f(z) = z \sin \dfrac{1}{z}$;

(j) $f(z) = z^n \sin \dfrac{1}{z}$.

2. Give the partial fraction expansion of $r(z) = \dfrac{z^2+1}{(z^2+z+1)(z-1)^2}$.

3. Give the details of the derivation of (1.17) from (1.16).

4. Let $f(z) = \dfrac{1}{z(z-1)(z-2)}$; give the Laurent Expansion of $f(z)$ in each of the following annuli: (a) ann $(0; 0, 1)$; (b) ann $(0; 1, 2)$; (c) ann $(0; 2, \infty)$.

5. Show that $f(z) = \tan z$ is analytic in $\mathbb{C}$ except for simple poles at $z = \dfrac{\pi}{2} + n\pi$, for each integer n. Determine the singular part of f at each of these poles.

6. If $f: G \to \mathbb{C}$ is analytic except for poles show that the poles of f cannot have a limit point in G.

7. Let f have an isolated singularity at $z = a$ and suppose $f \not\equiv 0$. Show that if either (1.19) or (1.20) holds for some s in $\mathbb{R}$ then there is an integer m such that (1.19) holds if $s > m$ and (1.20) holds if $s < m$.

8. Let f, a, and m be as in Exercise 7. Show: (a) $m = 0$ iff $z = a$ is a removable singularity and $f(a) \neq 0$; (b) $m < 0$ iff $z = a$ is a removable singularity and f has a zero at $z = a$ of order $-m$; (c) $m > 0$ iff $z = a$ is a pole of f of order m.

9. A function f has an essential singularity at $z = a$ iff neither (1.19) nor (1.20) holds for any real number s.

10. Suppose that f has an essential singularity at $z = a$. Prove the following strengthened version of the Casorati–Weierstrass Theorem. If $c \in \mathbb{C}$ and $\epsilon > 0$ are given then for each $\delta > 0$ there is a number α, $|c - \alpha| < \epsilon$, such that $f(z) = \alpha$ has infinitely many solutions in $B(a; \delta)$.

11. Give the Laurent series development of $f(z) = \exp\left(\dfrac{1}{z}\right)$. Can you generalize this result?

12. (a) Let $\lambda \in \mathbb{C}$ and show that

$$\exp\left\{\tfrac{1}{2}\lambda\left(z + \frac{1}{z}\right)\right\} = a_0 + \sum_{n=1}^{\infty} a_n\left(z^n + \frac{1}{z^n}\right)$$

for $0 < |z| < \infty$, where for $n \geq 0$

$$a_n = \frac{1}{\pi}\int_0^\pi e^{\lambda \cos t} \cos nt \, dt$$

(b) Similarly, show

$$\exp\left\{\tfrac{1}{2}\lambda\left(z - \frac{1}{z}\right)\right\} = b_0 + \sum_{n=1}^{\infty} b_n\left(z^n + \frac{(-1)^n}{z^n}\right)$$

for $0 < |z| < \infty$, where

$$b_n = \frac{1}{\pi}\int_0^\pi \cos(nt - \lambda \sin t) \, dt.$$

13. Let $R > 0$ and $G = \{z : |z| > R\}$; a function $f : G \to \mathbb{C}$ has a *removable singularity, a pole, or an essential singularity at infinity* if $f(z^{-1})$ has, respectively, a removable singularity, a pole, or an essential singularity at $z = 0$. If f has a pole at ∞ then the order of the pole is the order of the pole of $f(z^{-1})$ at $z = 0$.
(a) Prove that an entire function has a removable singularity at infinity iff it is a constant.
(b) Prove that an entire function has a pole at infinity of order m iff it is a polynomial of degree m.
(c) Characterize those rational functions which have a removable singularity at infinity.
(d) Characterize those rational functions which have a pole of order m at infinity.

14. Let $G = \{z : 0 < |z| < 1\}$ and let $f : G \to \mathbb{C}$ be analytic. Suppose that γ is a closed rectifiable curve in G such that $n(\gamma; a) = 0$ for all a in $\mathbb{C} - G$. What is $\int_\gamma f$? Why?

15. Let f be analytic in $G = \{z : 0 < |z - a| < r\}$ except that there is a sequence of poles $\{a_n\}$ in G with $a_n \to a$. Show that for any ω in $\mathbb{C}$ there is a sequence $\{z_n\}$ in G with $a = \lim z_n$ and $\omega = \lim f(z_n)$.

16. Determine the regions in which the functions $f(z) = (\sin\frac{1}{z})^{-1}$ and $g(z) = \int_0^1 (t-z)^{-1} dt$ are analytic. Do they have any isolated singularities? Do they have any singularities that are not isolated?

17. Let f be analytic in the region $G = \text{ann}(a; 0, R)$. Show that if $\iint_G |f(x + iy)|^2 dx\, dy < \infty$ then f has a removable singularity at $z = a$. Suppose that $p > 0$ and $\iint_G |f(x + iy)|^p dx\, dy < \infty$; what can be said about the nature of the singularity at $z = a$?

§2. Residues

The inspiration behind this section is the desire for an answer to the following question: If f has an isolated singularity at $z = a$ what are the possible values for $\int_\gamma f$ when γ is a closed curve homologous to zero and not passing through a? If the singularity is removable then clearly the integral will be zero. If $z = a$ is a pole or an essential singularity the answer is not always zero but can be found with little difficulty. In fact, for some curves γ, the answer is given by equation (1.12) with $n = -1$.

2.1 Definition. Let f have an isolated singularity at $z = a$ and let

$$f(z) = \sum_{n=-\infty}^{\infty} a_n(z-a)^n$$

be its Laurent Expansion about $z = a$. Then the *residue* of f at $z = a$ is the coefficient a_{-1}. Denote this by Res $(f; a) = a_{-1}$. The following is a generalization of formula (1.12) for $n = -1$.

2.2 Residue Theorem. *Let f be analytic in the region G except for the isolated singularities $a_1, a_2, \ldots, a_m$. If γ is a closed rectifiable curve in G which does not pass through any of the points a_k and if $\gamma \approx 0$ in G then*

$$\frac{1}{2\pi i} \int_\gamma f = \sum_{k=1}^{m} n(\gamma; a_k) \operatorname{Res}(f; a_k).$$

Proof. Let $m_k = n(\gamma; a_k)$ for $1 \le k \le m$, and choose positive numbers $r_1, \ldots, r_m$ such that no two disks $\bar{B}(a_k; r_k)$ intersect, none of them intersects $\{\gamma\}$, and each disk is contained in G. (This can be done by induction and by using the fact that γ does not pass through any of the singularities.) Let $\gamma_k(t) = a_k + r_k \exp(-2\pi i m_k t)$ for $0 \le t \le 1$. Then for $1 \le j \le m$

$$n(\gamma; a_j) + \sum_{k=1}^{m} n(\gamma_k; a_j) = 0.$$

Since $\gamma \approx 0(G)$ and $\bar{B}(a_k; r_k) \subset G$,

$$n(\gamma; a) + \sum_{k=1}^{m} n(\gamma_k; a) = 0$$

for all a not in $G-\{a_1,\dots,a_m\}$. Since f is analytic in $G-\{a_1,\dots,a_m\}$ Theorem IV.5.7 gives

2.3
$$0 = \int_\gamma f + \sum_{k=1}^{m} \int_{\gamma_k} f.$$

If $f(z) = \sum_{-\infty}^{\infty} b_n (z-a_k)^n$ is the Laurent expansion about $z = a_k$ then this series converges uniformly on $\partial B(a_k; r_k)$. Hence $\int_{\gamma k} f = \sum_{-\infty}^{\infty} b_n \int_{\gamma k} (z-a_k)^n$. But $\int_{\gamma_k} (z-a_k)^n = 0$ if $n \neq -1$ since $(z-a_k)^n$ has a primitive. Also $\int_{\gamma_k} (z-a_k)^{-1} = 2\pi i n(\gamma_k; a_k) \operatorname{Res}(f; a_k)$. Hence (2.3) implies the desired result. ∎

Remark. The condition in the Residue Theorem that f have only a finite number of isolated singularities was made to simplify the statement of the theorem and not because the theorem is invalid when f has infinitely many isolated singularities. In fact, if f has infinitely many singularities they can only accumulate on ∂G. (Why?) If $r = d(\{\gamma\}, \partial G)$ then the fact that $\gamma \approx 0$ gives that $n(\gamma; a) = 0$ whenever $d(a; \partial G) < \frac{1}{2} r$. (See Exercise IV.7.2.)

The Residue Theorem is a two edged sword; if you can calculate the residues of a function you can calculate certain line integrals and vice versa. Most often, however, it is used as a means to calculate line integrals. To use it in this way we will need a method of computing the residue of a function at a pole.

Suppose f has a pole of order $m \geq 1$ at $z = a$. Then $g(z) = (z-a)^m f(z)$ has a removable singularity at $z = a$ and $g(a) \neq 0$. Let $g(z) = b_0 + b_1(z-a) + \cdots$ be the power series expansion of g about $z = a$. It follows that for z near but not equal to a,

$$f(z) = \frac{b_0}{(z-a)^m} + \cdots + \frac{b_{m-1}}{(z-a)} + \sum_{k=0}^{\infty} b_{m+k}(z-a)^k.$$

This equation gives the Laurent Expansion of f in a punctured disk about $z = a$. But then $\operatorname{Res}(f; a) = b_{m-1}$; in particular, if $z = a$ is a simple pole $\operatorname{Res}(f; a) = g(a) = \lim_{z \to a} (z-a)f(z)$. This is summarized as follows.

2.4 Proposition. *Suppose f has a pole of order m at $z = a$ and put $g(z) = (z-a)^m f(z)$; then*

$$\operatorname{Res}(f; a) = \frac{1}{(m-1)!} g^{(m-1)}(a).$$

The remainder of this section will be devoted to calculating certain integrals by means of the Residue Theorem

2.5 Example. Show

$$\int_{-\infty}^{\infty} \frac{x^2}{1+x^4}\, dx = \frac{\pi}{\sqrt{2}}.$$

If $f(z) = \dfrac{z^2}{1+z^4}$ then f has as its poles the fourth roots of -1. These are exactly the numbers $e^{i\theta}$ where

$$\theta = \frac{\pi}{4}, \frac{3\pi}{4}, \frac{5\pi}{4}, \text{ and } \frac{7\pi}{4}.$$

Let

$$a_n = \exp\left(i\left[\frac{\pi}{4} + (n-1)\frac{\pi}{2}\right]\right)$$

for $n = 1, 2, 3, 4$; then it is easily seen that each a_n is a simple pole of f. Consequently,

$$\text{Res}(f; a_1) = \lim_{z \to a_1} (z-a_1)f(z) = a_1^2(a_1-a_2)^{-1}(a_1-a_3)^{-1}(a_1-a_4)^{-1}$$

$$= \frac{1-i}{4\sqrt{2}} = \tfrac{1}{4}\exp\left(-\frac{\pi i}{4}\right).$$

Similarly

$$\text{Res}(f; a_2) = \frac{-1-i}{4\sqrt{2}} = \tfrac{1}{4}\exp\left(\frac{-3\pi i}{4}\right).$$

Now let $R > 1$ and let γ be the closed path which is the boundary of the upper half of the disk of radius R with center zero, traversed in the counter-clockwise direction. The Residue Theorem gives

$$\frac{1}{2\pi i}\int_\gamma f = \text{Res}(f; a_1) + \text{Res}(f; a_2)$$

$$= \frac{-i}{2\sqrt{2}}$$

But, applying the definition of line integral,

$$\frac{1}{2\pi i}\int_\gamma f = \frac{1}{2\pi i}\int_{-R}^{R} \frac{x^2}{1+x^4}\,dx + \frac{1}{2\pi}\int_0^\pi \frac{R^3 e^{3it}}{1+R^4 e^{4it}}\,dt.$$

This gives

2.6 $$\int_{-R}^{R} \frac{x^2}{1+x^4}\,dx = \frac{\pi}{\sqrt{2}} - iR^3 \int_0^\pi \frac{e^{3it}}{1+R^4 e^{4it}}\,dt.$$

For $0 \le t \le \pi$, $1 + R^4 e^{4it}$ lies on the circle centered at 1 of radius R^4; hence $|1 + R^4 e^{4it}| \ge R^4 - 1$. Therefore

$$\left| iR^3 \int_0^\pi \frac{e^{3it}}{1 + R^4 e^{4it}} \, dt \right| \le \frac{\pi R^3}{R^4 - 1} \, ;$$

and since $\dfrac{x^2}{1 + x^4} \ge 0$ for all x in $\mathbb{R}$, it follows from (2.6) that

$$\int_{-\infty}^{\infty} \frac{x^2}{1 + x^4} \, dx = \lim_{R \to \infty} \int_{-R}^{R} \frac{x^2}{1 + x^4} \, dx$$

$$= \frac{\pi}{\sqrt{2}}$$

2.7 Example. Show

$$\int_0^\infty \frac{\sin x}{x} \, dx = \frac{\pi}{2} \, .$$

The function $f(z) = \dfrac{e^{iz}}{z}$ has a simple pole at $z = 0$. If $0 < r < R$ let γ be the closed curve that is depicted in the adjoining figure. It follows from Cauchy's

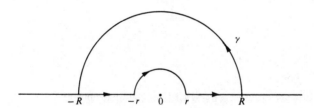

Theorem that $0 = \int_\gamma f$. Breaking γ into its pieces,

2.8
$$0 = \int_r^R \frac{e^{ix}}{x} \, dx + \int_{\gamma_R} \frac{e^{iz}}{z} \, dz + \int_{-R}^{-r} \frac{e^{ix}}{x} \, dx + \int_{\gamma_r} \frac{e^{iz}}{z} \, dz$$

where γ_R and γ_r are the semicircles from R to $-R$ and $-r$ to r respectively. But

$$\int_r^R \frac{\sin x}{x} \, dx = \frac{1}{2i} \int_r^R \frac{e^{ix} - e^{-ix}}{x} \, dx$$

$$= \frac{1}{2i} \int_r^R \frac{e^{ix}}{x} \, dx + \frac{1}{2i} \int_{-R}^{-r} \frac{e^{ix}}{x} \, dx$$

Also

$$\left| \int_{\gamma_R} \frac{e^{iz}}{z} \, dz \right| = \left| i \int_0^{\pi} \exp(i R e^{i\theta}) \, d\theta \right|$$

$$\leq \int_0^{\pi} \left| \exp(i R e^{i\theta}) \right| \, d\theta$$

$$= \int_0^{\pi} \exp(-R \sin \theta) \, d\theta$$

By the methods of calculus we see that, for $\delta > 0$ sufficiently small, the largest possible value of $\exp(-R \sin \theta)$, with $\delta \leq \theta \leq \pi-\delta$, is $\exp(-R \sin \delta)$. (Note that δ does not depend on R if R is larger than 1.) This gives that

$$\left| \int_{\gamma_R} \frac{e^{iz}}{z} \, dz \right| \leq 2\delta + \int_{\delta}^{\pi-\delta} \exp(-R \sin \theta) \, d\theta$$

$$< 2\delta + \pi \exp(-R \sin \delta).$$

If $\epsilon > 0$ is given then, choosing $\delta < \frac{1}{3} \epsilon$, there is an R_0 such that $\exp(-R \sin \delta) < \frac{\epsilon}{3\pi}$ for all $R > R_0$. Hence

$$\lim_{R \to \infty} \int_{\gamma_R} \frac{e^{iz}}{z} \, dz = 0.$$

Since $\frac{e^{iz}-1}{z}$ has a removable singularity at $z = 0$, there is a constant $M > 0$ such that $\left| \frac{e^{iz}-1}{z} \right| \leq M$ for $|z| \leq 1$. Hence,

$$\left| \int_{\gamma_r} \frac{e^{iz}-1}{z} \, dz \right| \leq \pi r M;$$

that is,

$$0 = \lim_{r \to 0} \int_{\gamma_r} \frac{e^{iz}-1}{z} \, dz.$$

But $\int_{\gamma_r} \frac{1}{z} \, dz = -\pi i$ for each r so that

$$-\pi i = \lim_{r \to 0} \int_{\gamma_r} \frac{e^{iz}}{z} \, dz$$

So, if we let $r \to 0$ and $R \to \infty$ in (2.8)

$$\int_0^{\infty} \frac{\sin x}{x} \, dx = \frac{\pi}{2}.$$

Notice that this example did not use the Residue Theorem. In fact, it could have been presented after Cauchy's Theorem. It was saved until now because the methods used to evaluate this integral are the same as the methods used in applying the Residue Theorem.

2.9 Example. Show that for $a > 1$,

$$\int_0^\pi \frac{d\theta}{a+\cos\theta} = \frac{\pi}{\sqrt{a^2-1}}.$$

If $z = e^{i\theta}$ then $\bar{z} = \dfrac{1}{z}$ and so

$$a+\cos\theta = a+\tfrac{1}{2}(z+\bar{z}) = a+\tfrac{1}{2}\left(z+\frac{1}{z}\right) = \frac{z^2+2az+1}{2z}.$$

Hence

$$\int_0^\pi \frac{d\theta}{a+\cos\theta} = \tfrac{1}{2}\int_0^{2\pi} \frac{d\theta}{a+\cos\theta}$$

$$= -i\int_\gamma \frac{dz}{z^2+2az+1}$$

where γ is the circle $|z| = 1$. But $z^2+2az+1 = (z-\alpha)(z-\beta)$ where $\alpha = -a+(a^2-1)^{\frac{1}{2}}$, $\beta = -a-(a^2-1)^{\frac{1}{2}}$. Since $a > 1$ it follows that $|\alpha| < 1$ and $|\beta| > 1$. By the Residue Theorem

$$\int_\gamma \frac{dz}{z^2+2az+1} = \frac{\pi i}{\sqrt{a^2-1}};$$

by combining this with the above equation we arrive at

$$\int_0^\pi \frac{d\theta}{a+\cos\theta} = \frac{\pi}{\sqrt{a^2-1}}.$$

2.10 Example. Show that

$$\int_0^\infty \frac{\log x}{1+x^2}\, dx = 0.$$

To solve this problem we *do not* use the principal branch of the logarithm. Instead define $\log z$ for z belonging to the region

$$G = \left\{ z \in \mathbb{C} : z \neq 0 \text{ and } -\frac{\pi}{2} < \arg z < \frac{3\pi}{2} \right\};$$

if $z = |z| e^{i\theta} \neq 0$ with $-\dfrac{\pi}{2} < \theta < \dfrac{3\pi}{2}$, let $\ell(z) = \log |z| + i\theta$. Let $0 < r < R$

and let γ be the same curve as in Example 2.7. Notice that $\ell(x) = \log x$ for $x > 0$, and $\ell(x) = \log |x| + \pi i$ for $x < 0$. Hence,

2.11
$$\int_\gamma \frac{\ell(z)}{1+z^2}\, dz = \int_r^R \frac{\log x}{1+x^2}\, dx + iR \int_0^\pi \frac{[\log R + i\theta]}{1+R^2 e^{2i\theta}}\, e^{i\theta}\, d\theta$$

$$+ \int_{-R}^{-r} \frac{\log |x| + \pi i}{1+x^2}\, dx + ir \int_\pi^0 \frac{[\log r + i\theta]}{1+r^2 e^{2i\theta}}\, e^{i\theta} d\theta$$

Now the only pole of $\ell(z)(1+z^2)^{-1}$ inside γ is at $z = i$; furthermore, this is a simple pole. By Proposition 2.4 the residue of $\ell(z)(1+z^2)^{-1}$ is $\dfrac{1}{2i}$

$[\log |i| + \tfrac{1}{2}\pi i] = \dfrac{\pi}{4}$. So,

$$\int_\gamma \frac{\ell(z)}{1+z^2}\, dz = \frac{\pi^2 i}{2}$$

Also,

$$\int_r^R \frac{\log x}{1+x^2}\, dx + \int_{-R}^{-r} \frac{\log |x| + \pi i}{1+x^2}\, dx = 2\int_r^R \frac{\log x}{1+x^2}\, dx + \pi i \int_r^R \frac{dx}{1+x^2}$$

Letting $r \to 0+$ and $R \to \infty$, and using the fact that

$$\int_0^\infty \frac{dx}{1+x^2} = \frac{\pi}{2}$$

(Exercise 2(f)), it follows from (2.11) that

$$\int_0^\infty \frac{\log x}{1+x^2}\, dx = \tfrac{1}{2} \lim_{r \to 0+} ir \int_0^\pi \frac{[\log r + i\theta]}{1+r^2 e^{2i\theta}}\, e^{i\theta} d\theta$$

$$- \tfrac{1}{2} \lim_{R \to \infty} iR \int_0^\pi \frac{[\log R + i\theta]}{1+R^2 e^{2i\theta}}\, e^{i\theta} d\theta.$$

We now show that both of these limits are zero. If $\rho > 0$ then

$$\left| \rho \int_0^\pi \frac{[\log \rho + i\theta]}{1+\rho^2 e^{i\theta}}\, e^{i\theta} d\theta \right| \leq \frac{\rho |\log \rho|}{|1-\rho^2|} \int_0^\pi d\theta + \frac{\rho}{|1-\rho^2|} \int_0^\pi \theta d\theta$$

$$= \frac{\pi\rho |\log \rho|}{|1-\rho^2|} + \frac{\rho\pi^2}{2|1-\rho^2|} ;$$

letting $\rho \to 0+$ or $\rho \to \infty$, the limit of this expression is zero.

2.12 Example. Show that $\displaystyle\int_0^\infty \frac{x^{-c}}{1+x}\,dx = \frac{\pi}{\sin \pi c}$ if $0 < c < 1$.

To evaluate this integral we must consider a branch of the function z^{-c}. The point $z = 0$ is called a branch point of z^{-c}, and the process used to

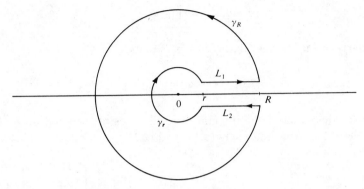

evaluate this integral is sometimes called integration around a branch point.

Let $G = \{z : z \neq 0 \text{ and } 0 < \arg z < 2\pi\}$; define a branch of the logarithm on G by putting $\ell(re^{i\theta}) = \log r + i\theta$ where $0 < \theta < 2\pi$. For z in G put $f(z) = \exp[-c\ell(z)]$; so f is a branch of z^{-c}. We now select an appropriate curve γ in G. Let $0 < r < 1 < R$ and let $\delta > 0$. Let L_1 be the line segment $[r+\delta i, R+\delta i]$; γ_R the part of the circle $|z| = R$ from $R+\delta i$ counterclockwise to $R-\delta i$; L_2 the line segment $[R-\delta i, r-\delta i]$; and γ_r the part of the circle $|z| = r$ from $r-\delta i$ clockwise to $r+\delta i$. Put $\gamma = L_1 + \gamma_R + L_2 + \gamma_r$.

Since $\gamma \sim 0$ in G and $\operatorname{Res}(f(z)(1+z)^{-1}; -1) = f(-1) = e^{-i\pi c}$, the Residue Theorem gives

2.13
$$\int_\gamma \frac{f(z)}{1+z}\,dz = 2\pi i e^{-i\pi c}.$$

Using the definition of a line integral

$$\int_{L_1} \frac{f(z)}{1+z}\,dz = \int_r^R \frac{f(t+i\delta)}{1+t+i\delta}\,dt.$$

Let $g(t,\delta)$ be defined on the compact set $[r,R]\times[0,\tfrac{1}{2}\pi]$ by

$$g(t,\delta) = \left| \frac{f(t+i\delta)}{1+t+i\delta} - \frac{t^{-c}}{1+t} \right|$$

when $\delta > 0$ and $g(t,0) \equiv 0$. Then g is continuous and hence uniformly continuous. If $\epsilon > 0$ then there is a δ_0 such that if $(t-t')^2+(\delta-\delta')^2 < \delta_0^2$ then $|g(t,\delta)-g(t',\delta')| < \epsilon/R$. In particular, $g(t,\delta) < \epsilon/R$ when $r \leq t \leq R$

and $\delta < \delta_0$. Thus

$$\int_r^R g(t, \delta) \, dt \le \epsilon$$

for $\delta < \delta_0$. This implies that

2.14
$$\int_r^R \frac{t^{-c}}{1+t} \, dt = \lim_{\delta \to 0+} \int_{L_1} \frac{f(z)}{1+z} \, dz.$$

Similarly, using the fact that $l(\bar{z}) = \overline{l(z)} + 2\pi i$

2.15
$$-e^{-2\pi i c} \int_r^R \frac{t^{-c}}{1+t} \, dt = \lim_{\delta \to 0+} \int_{L_2} \frac{f(z)}{1+z} \, dz.$$

Now the value of the integral in (2.13) does not depend on δ. Therefore, letting $\delta \to 0+$ and using (2.14) and (2.15) gives

2.16
$$2\pi i e^{-i\pi c} - (1 - e^{-2\pi i c}) \int_r^R \frac{t^{-c}}{1+t} \, dt = \lim_{\delta \to 0+} \left[\int_{\gamma_r} \frac{f(z)}{1+z} + \int_{\gamma_R} \frac{f(z)}{1+z} \right].$$

Now if $\rho > 0$ and $\rho \ne 1$ and if γ_ρ is the part of the circle $|z| = \rho$ from $\sqrt{\rho^2 - \delta^2} + i\delta$ to $\sqrt{\rho^2 - \delta^2} - i\delta$ then

$$\left| \int_{\gamma_\rho} \frac{f(z)}{1+z} \, dz \right| \le \frac{\rho^{-c}}{|1 - \rho|} 2\pi \rho.$$

Since this estimate is independent of δ, (2.16) implies

$$\left| 2\pi i e^{-i\pi c} - (1 - e^{-2\pi i c}) \int_r^R \frac{t^{-c}}{1+t} \, dt \right| \le \frac{r^{-c}}{|1 - r|} 2\pi r + \frac{R^{-c}}{|1 - R|} 2\pi R.$$

But as $r \to 0+$ and $R \to \infty$ the right-hand side of this last inequality converges to zero. Hence

$$2\pi i \, e^{-i\pi c} = (1 - e^{-2i\pi c}) \int_0^\infty \frac{t^{-c}}{1+t} \, dt;$$

or,

$$\int_0^\infty \frac{t^{-c}}{1+t} \, dt = \frac{2\pi i \, e^{-i\pi c}}{1 - e^{-2\pi i c}}$$

$$= \frac{2\pi i}{e^{\pi i c} - e^{-\pi i c}}$$

$$= \frac{\pi}{\sin \pi c}.$$

Exercises

1. Calculate the following integrals:

(a) $\displaystyle\int_0^\infty \frac{x^2 dx}{x^4+x^2+1}$
(b) $\displaystyle\int_0^\infty \frac{\cos x - 1}{x^2}\, dx$

(c) $\displaystyle\int_0^\pi \frac{\cos 2\theta d\theta}{1-2a\cos\theta + a^2}$ where $a^2 < 1$
(d) $\displaystyle\int_0^\pi \frac{d\theta}{(a+\cos\theta)^2}$ where $a > 1$.

2. Verify the following equations:

(a) $\displaystyle\int_0^\infty \frac{dx}{(x^2+a^2)^2} = \frac{\pi}{4a^3}$, $a > 0$;
(b) $\displaystyle\int_0^\infty \frac{(\log x)^3}{1+x^2}\, dx = 0$

(c) $\displaystyle\int_0^\infty \frac{\cos ax}{(1+x^2)^2}\, dx = \frac{\pi(a+1)\, e^{-a}}{4}$ if $a > 0$;

(d) $\displaystyle\int_0^{\pi/2} \frac{d\theta}{a+\sin^2\theta} = \frac{\pi}{2[a(a+1)]^{\frac{1}{2}}}$, if $a > 0$;

(e) $\displaystyle\int_0^\infty \frac{\log x}{(1+x^2)^2}\, dx = -\frac{\pi}{4}$;
(f) $\displaystyle\int_0^\infty \frac{dx}{1+x^2} = \frac{\pi}{2}$;

(g) $\displaystyle\int_{-\infty}^\infty \frac{e^{ax}}{1+e^x}\, dx = \frac{\pi}{\sin a\pi}$ if $0 < a < 1$;

(h) $\displaystyle\int_0^{2\pi} \log \sin^2 2\theta d\theta = 4\int_0^\pi \log \sin\theta d\theta = -4\pi \log 2.$ Ahlfors

3. Find all possible values of $\int_\gamma \exp z^{-1}\, dz$ where γ is any closed curve not passing through $z=0$.

4. Suppose that f has a simple pole at $z = a$ and let g be analytic in an open set containing a. Show that Res $(fg; a) = g(a)$ Res $(f; a)$.

5. Use Exercise 4 to show that if G is a region and f is analytic in G except for simple poles at $a_1, \ldots, a_n$; and if g is analytic in G then

$$\frac{1}{2\pi i}\int_\gamma fg = \sum_{k=1}^n n(\gamma; a_k)\, g(a_k)\, \text{Res}\, (f; a_k)$$

for any closed rectifiable curve γ not passing through $a_1, \ldots, a_n$ such that $\gamma \approx 0$ in G.

6. Let γ be the rectangular path $[n+\frac{1}{2}+ni, -n-\frac{1}{2}+ni, -n-\frac{1}{2}-ni, n+\frac{1}{2}-ni, n+\frac{1}{2}+ni]$ and evaluate the integral $\int_\gamma \pi(z+a)^{-2} \cot \pi z \, dz$ for $a \neq$ an integer. Show that $\lim\limits_{n \to \infty} \int_\gamma \pi(z+a)^{-2} \cot \pi z \, dz = 0$ and, by using the first part, deduce that

$$\frac{\pi^2}{\sin^2 \pi a} = \sum_{n=-\infty}^{\infty} \frac{1}{(a+n)^2}$$

(Hint: Use the fact that for $z = x+iy$, $|\cos z|^2 = \cos^2 x + \sinh^2 y$ and $|\sin z|^2 = \sin^2 x + \sinh^2 y$ to show that $|\cot \pi z| \leq 2$ for z on γ if n is sufficiently large.)

7. Use Exercise 6 to deduce that

$$\frac{\pi^2}{8} = \sum_{n=0}^{\infty} \frac{1}{(2n+1)^2}$$

8. Let γ be the polygonal path defined in Exercise 6 and evaluate $\int_\gamma \pi(z^2-a^2)^{-1} \cot \pi z \, dz$ for $a \neq$ an integer, Show that $\lim\limits_{n \to \infty} \int_\gamma \pi(z^2-a^2)^{-1} \cot \pi z \, dz = 0$, and consequently

$$\pi \cot \pi a = \frac{1}{a} + \sum_{n=1}^{\infty} \frac{2a}{a^2-n^2}$$

for $a \neq$ an integer.

9. Use methods similar to those of Exercises 6 and 8 to show that

$$\frac{\pi}{\sin \pi a} = \frac{1}{a} + \sum_{n=1}^{\infty} \frac{2(-1)^n a}{a^2-n^2}$$

for $a \neq$ an integer.

10. Let γ be the circle $|z| = 1$ and let m and n be non-negative integers. Show that

$$\frac{1}{2\pi i} \int_\gamma \frac{(z^2 \pm 1)^m dz}{z^{m+n+1}} = \begin{cases} \dfrac{(\pm 1)^p (n+2p)!}{p!\,(n+p)!}, & \text{if } m = 2p+n, \\ & \qquad p \geq 0 \\ \\ 0 & \text{otherwise} \end{cases}$$

11. In Exercise 1.12, consider a_n and b_n as functions of the parameter λ and use Exercise 10 to compute power series expansions for $a_n(\lambda)$ and $b_n(\lambda)$. ($b_n(\lambda)$ is called a Bessel function.)

12. Let f be analytic in the plane except for isolated singularities at $a_1, a_2, \ldots, a_m$. Show that

$$\text{Res}\,(f; \infty) = -\sum_{k=1}^{m} \text{Res}\,(f; a_k).$$

(Res(f; ∞)) is defined as the residue of $-z^{-2}f(z^{-1})$ at $z = 0$. Equivalently, Res(f; ∞)$= -\dfrac{1}{2\pi i} \int f$ when $\gamma(t) = Re^{it}$, $0 \le t \le 2\pi$, for sufficiently large R.) What can you say if f has infinitely many isolated singularities?

13. Let f be an entire function and let a, $b \in \mathbb{C}$ such that $|a| < R$ and $|b| < R$. If $\gamma(t) = Re^{it}$, $0 \le t \le 2\pi$, evaluate $\int_\gamma [(z-a)(z-b)]^{-1} f(z) dz$. Use this result to give another proof of Liouville's Theorem.

§3 The Argument Principle

Suppose that f is analytic and has a zero of order m at $z = a$. So $f(z) = (z-a)^m g(z)$ where $g(a) \ne 0$. Hence

3.1
$$\frac{f'(z)}{f(z)} = \frac{m}{z-a} + \frac{g'(z)}{g(z)}$$

and g'/g is analytic near $z = a$ since $g(a) \ne 0$. Now suppose that f has a pole of order m at $z = a$; that is, $f(z) = (z-a)^{-m} g(z)$ where g is analytic and $g(a) \ne 0$. This gives

3.2
$$\frac{f'(z)}{f(z)} = \frac{-m}{z-a} + \frac{g'(z)}{g(z)}$$

and again g'/g is analytic near $z = a$.

Also, to avoid the phrase "analytic except for poles" which may have already been used too frequently, we make the following standard definition.

3.3 Definition. If G is open and f is a function defined and analytic in G except for poles, then f is a *meromorphic function* on G.

Suppose that f is a meromorphic function on G and define $f: G \to \mathbb{C}_\infty$ by setting $f(z) = \infty$ whenever z is a pole of f. It easily follows that f is continuous from G into $\mathbb{C}_\infty$ (Exercise 4). This fact allows us to think of meromorphic functions as analytic functions with singularities for which we can remove the discontinuity of f, although we cannot remove the non-differentiability of f.

3.4 Argument Principle. *Let f be meromorphic in G with poles $p_1, p_2, \ldots, p_m$ and zeros $z_1, z_2, \ldots, z_n$ counted according to multiplicity. If γ is a closed rectifiable curve in G with $\gamma \approx 0$ and not passing through $p_1, \ldots, p_m$; $z_1, \ldots, z_n$; then*

3.5
$$\frac{1}{2\pi i} \int_\gamma \frac{f'(z)}{f(z)} dz = \sum_{k=1}^n n(\gamma; z_k) - \sum_{j=1}^m n(\gamma; p_j).$$

Proof. By a repeated application of (3.1) and (3.2)

$$\frac{f'(z)}{f(z)} = \sum_{k=1}^n \frac{1}{z - z_k} - \sum_{j=1}^m \frac{1}{z - p_j} + \frac{g'(z)}{g(z)}$$

where g is analytic and never vanishes in G. Since this gives that g'/g is analytic, Cauchy's Theorem gives the result. ∎

Why is this called the "Argument Principle"? The answer to this is not completely obvious, but it is suggested by the fact that if we could define $\log f(z)$ then it would be a primitive for f'/f. Thus Theorem 3.4 would give that as z goes around γ, $\log f(z)$ would change by $2\pi i K$ where K is the integer on the right hand side of (3.5). Since $2\pi i K$ is purely imaginary this would give that Im $\log f(z) = \arg f(z)$ changes by $2\pi K$.

Of course we can't define $\log f(z)$ (indeed, if we could then $\int_\gamma f'/f = 0$ since f'/f has a primitive). However, we can put the discussion in the above paragraph on a solid logical basis. Since no zero or pole of f lies on γ there is a disk $B(a; r)$, for each a in $\{\gamma\}$, such that a branch of $\log f(z)$ can be defined on $B(a; r)$ (simply select r sufficiently small that $f(z) \neq 0$ or ∞ in $B(a; r)$). The balls form an open cover of $\{\gamma\}$; and so, by Lebesgue's Covering Lemma, there is a positive number $\epsilon > 0$ such that for each a in $\{\gamma\}$ we can define a branch of $\log f(z)$ on $B(a; \epsilon)$. Using the uniform continuity of γ (suppose that γ is defined on $[0, 1]$), there is a partition $0 = t_0 < t_1 < \cdots < t_k = 1$ such that $\gamma(t) \in B(\gamma(t_{j-1}); \epsilon)$ for $t_{j-1} \le t \le t_j$ and $1 \le j \le k$. Let ℓ_j be a branch of $\log f$ defined on $B(\gamma(t_{j-1}); \epsilon)$ for $1 \le j \le k$. Also, since the j-th and $(j+1)$-st sphere both contain $\gamma(t_j)$ we can choose $\ell_1, \ldots, \ell_k$ so that $\ell_1(\gamma(t_1)) = \ell_2(\gamma(t_1))$; $\ell_2(\gamma(t_2)) = \ell_3(\gamma t_2))$; $\ldots$; $\ell_{k-1}(\gamma(t_{k-1})) = \ell_k(\gamma(t_{k-1}))$.

If γ_j is the path γ restricted to $[t_{j-1}, t_j]$ then, since $\ell_j' = f'/f$,

$$\int_{\gamma_j} \frac{f'}{f} = \ell_j[\gamma(t_j)] - \ell_j[\gamma(t_{j-1})]$$

for $1 \le j \le k$. Summing both sides of this equation the right hand side "telescopes" and we arrive at

$$\int_\gamma \frac{f'}{f} = \ell_k(a) - \ell_1(a)$$

where $a = \gamma(0) = \gamma(1)$. That is, $\ell_k(a) - \ell_1(a) = 2\pi i K$. Because $2\pi i K$ is purely imaginary we get Im $\ell_k(a) -$ Im $\ell_1(a) = 2\pi K$. This makes precise our contention that as z traces out γ, $\arg f(z)$ changes by $2\pi K$.

The proof of the following generalization is left to the reader (Exercise 1).

3.6 Theorem. *Let f be meromorphic in the region G with zeros $z_1, z_2, \ldots, z_n$ and poles $p_1, \ldots, p_m$ counted according to multiplicity. If g is analytic in G and γ is a closed rectifiable curve in G with $\gamma \approx 0$ and not passing through any z_i or p_j then*

$$\frac{1}{2\pi i} \int_\gamma g \frac{f'}{f} = \sum_{i=1}^n g(z_i) n(\gamma; z_i) - \sum_{j=1}^m g(p_j) n(\gamma; p_j).$$

We already know that a one-one analytic function f has an analytic inverse (IV. 7.6). It is a remarkable fact that Theorem 3.6 can be used to give a formula for calculating this inverse. Suppose $R > 0$ and that f is analytic and one-one on $\bar{B}(a; R)$; let $\Omega = f[B(a; R)]$. If $|z - a| < R$ and $\xi = f(z) \in \Omega$ then $f(w) - \xi$ has one, and only one, zero in $B(a; R)$. If we choose $g(w) \equiv w$, Theorem 3.6 gives

$$z = \frac{1}{2\pi i} \int_\gamma \frac{wf'(w)}{f(w) - \xi} \, dw$$

where γ is the circle $|w - a| = R$. But $z = f^{-1}(\xi)$; this gives the following

3.7 Proposition. *Let f be analytic on an open set containing $\bar{B}(a; R)$ and suppose that f is one-one on $B(a; R)$. If $\Omega = f[B(a; R)]$ and γ is the circle $|z - a| = R$ then $f^{-1}(\omega)$ is defined for each ω in Ω by the formula*

$$f^{-1}(\omega) = \frac{1}{2\pi i} \int_\gamma \frac{zf'(z)}{f(z) - \omega} \, dz.$$

This section closes with Rouché's Theorem.

3.8 Rouché's Theorem. *Suppose f and g are meromorphic in a neighborhood of $\bar{B}(a; R)$ with no zeros or poles on the circle $\gamma = \{z : |z - a| = R\}$. If Z_f, Z_g (P_f, P_g) are the number of zeros (poles) of f and g inside γ counted according to their multiplicities and if*

$$|f(z) + g(z)| < |f(z)| + |g(z)|$$

on γ, then

$$Z_f - P_f = Z_g - P_g.$$

Proof. From the hypothesis

$$\left| \frac{f(z)}{g(z)} + 1 \right| < \left| \frac{f(z)}{g(z)} \right| + 1$$

on γ. If $\lambda = f(z)/g(z)$ and if λ is a positive real number then this inequality becomes $\lambda + 1 < \lambda + 1$, a contradiction. Hence the meromorphic function f/g maps γ onto $\Omega = \mathbb{C} - [0, \infty)$. If l is a branch of the logarithm on Ω then $l(f(z)/g(z))$ is a well-defined primitive for $(f/g)'(f/g)^{-1}$ in a neighborhood of γ. Thus

$$0 = \frac{1}{2\pi i} \int_\gamma (f/g)'(f/g)^{-1}$$

$$= \frac{1}{2\pi i} \int_\gamma \left[\frac{f'}{f} - \frac{g'}{g} \right]$$

$$= (Z_f - P_f) - (Z_g - P_g). \quad \blacksquare$$

This statement of Rouché's Theorem was discovered by Irving Glicksberg (*Amer. Math. Monthly*, **83** (1976), 186–187). In the more classical statements of the theorem, f and g are assumed to satisfy the inequality

$|f+g| < |g|$ on γ. This weaker version often suffices in the applications as can be seen in the next paragraph.

Rouché's Theorem can be used to give another proof of the Fundamental Theorem of Algebra. If $p(z) = z^n + a_1 z^{n-1} + \cdots + a_n$ then

$$\frac{p(z)}{z^n} = 1 + \frac{a_1}{z} + \cdots + \frac{a_n}{z^n}$$

and this approaches 1 as z goes to infinity. So there is a sufficiently large number R with

$$\left| \frac{p(z)}{z^n} - 1 \right| < 1$$

for $|z| = R$; that is, $|p(z) - z^n| < |z|^n$ for $|z| = R$. Rouché's Theorem says that $p(z)$ must have n zeros inside $|z| = R$.

We also mention that the use of a circle in Rouché's Theorem was a convenience and not a necessity. Any closed rectifiable curve γ with $\gamma \sim 0$ in G could have been used, although the conclusion would have been modified by the introduction of winding numbers.

Exercises

1. Prove Theorem 3.6.
2. Suppose f is analytic on $\bar{B}(0; 1)$ and satisfies $|f(z)| < 1$ for $|z| = 1$. Find the number of solutions (counting multiplicities) of the equation $f(z) = z^n$ where n is an integer larger than or equal to 1.
3. Let f be analytic in $B(0; R)$ with $f(0) = 0$, $f'(0) \neq 0$ and $f(z) \neq 0$ for $0 < |z| \le R$. Put $\rho = \min\{|f(z)| : |z| = R\} > 0$. Define $g : B(0; \rho) \to \mathbb{C}$ by

$$g(\omega) = \frac{1}{2\pi i} \int_\gamma \frac{zf'(z)}{f(z) - \omega} \, dz$$

where γ is the circle $|z| = R$. Show that g is analytic and discuss the properties of g.
4. If f is meromorphic on G and $\tilde{f} : G \to \mathbb{C}_\infty$ is defined by $\tilde{f}(z) = \infty$ when z is a pole of f and $\tilde{f}(z) = f(z)$ otherwise, show that $\tilde{f}$ is continuous.
5. Let f be meromorphic on G; show that neither the poles nor the zeros of f have a limit point in G.
6. Let G be a region and let $H(G)$ denote the set of all analytic functions on G. (The letter "H" stands for holomorphic. Some authors call a differentiable function holomorphic and call functions analytic if they have a power series expansion about each point of their domain. Others reserve the term "analytic" for what many call the complete analytic function, which we will not describe here.) Show that $H(G)$ is an integral domain; that is, $H(G)$ is a commutative ring with no zero divisors. Show that $M(G)$, the meromorphic functions on G, is a field.

We have said that analytic functions are like polynomials; similarly, meromorphic functions are analogues of rational functions. The question arises, is every meromorphic function on G the quotient of two analytic functions on G? Alternately, is $M(G)$ the quotient field of $H(G)$? The answer is yes but some additional theory will be required before this answer can be proved.

7. State and prove a more general version of Rouché's Theorem for curves other than circles in G.

8. Is a non-constant meromorphic function on a region G an open mapping of G into $\mathbb{C}$? Is it an open mapping of G into $\mathbb{C}_\infty$?

9. Let $\lambda > 1$ and show that the equation $\lambda - z - e^{-z} = 0$ has exactly one solution in the half plane $\{z: \operatorname{Re} z > 0\}$. Show that this solution must be real. What happens to the solution as $\lambda \to 1$?

10. Let f be analytic in a neighborhood of $D = \bar{B}(0; 1)$. If $|f(z)| < 1$ for $|z| = 1$, show that there is a unique z with $|z| < 1$ and $f(z) = z$. If $|f(z)| \le 1$ for $|z| = 1$, what can you say?

Chapter VI

The Maximum Modulus Theorem

This chapter continues the study of a property of analytic functions first seen in Theorem IV. 3.11. In the first section this theorem is presented again with a second proof, and other versions of it are also given. The remainder of the chapter is devoted to various extensions and applications of this maximum principle.

§1. The Maximum Principle

Let Ω be any subset of $\mathbb{C}$ and suppose α is in the interior of Ω. We can, therefore, choose a positive number ρ such that $B(\alpha; \rho) \subset \Omega$; it readily follows that there is a point ξ in Ω with $|\xi| > |\alpha|$. To state this another way, if α is a point in Ω with $|\alpha| \geq |\xi|$ for each ξ in the set Ω then α belongs to $\partial\Omega$.

1.1 Maximum Modulus Theorem—First Version. *If f is analytic in a region G and a is a point in G with $|f(a)| \geq |f(z)|$ for all z in G then f must be a constant function.*

Proof. Let $\Omega = f(G)$ and put $\alpha = f(a)$. From the hypothesis we have that $|\alpha| \geq |\xi|$ for each ξ in Ω; as in the discussion preceding the theorem α is in $\partial\Omega \cap \Omega$. In particular, the set Ω cannot be open (because then $\Omega \cap \partial\Omega = \square$). Hence the Open Mapping Theorem (IV. 7.5) says that f must be constant. ∎

1.2 Maximum Modulus Theorem—Second Version. *Let G be a bounded open set in $\mathbb{C}$ and suppose f is a continuous function on G^- which is analytic in G. Then*

$$\max \{|f(z)| : z \in G^-\} = \max \{|f(z)| : z \in \partial G\}.$$

Proof. Since G is bounded there is a point $a \in G^-$ such that $|f(a)| \geq |f(z)|$ for all z in G^-. If f is a constant function the conclusion is trivial; if f is not constant then the result follows from Theorem 1.1. ∎

Note that in Theorem 1.2 we did not assume that G is connected as in Theorem 1.1. Do you understand how Theorem 1.1 puts the finishing touches on the proof of 1.2? Or, could the assumption of connectedness in Theorem 1.1 be dropped?

Let $G = \{z = x+iy : -\frac{1}{2}\pi < y < \frac{1}{2}\pi\}$ and put $f(z) = \exp [\exp z]$. Then f is continuous on G^- and analytic on G. If $z \in \partial G$ then $z = x \pm \frac{1}{2}\pi i$ so $|f(z)| = |\exp (\pm ie^x)| = 1$. However, as x goes to infinity through the real numbers, $f(x) \to \infty$. This does not contradict the Maximum Modulus Theorem because G is not bounded.

In light of the above example it is impossible to drop the assumption of the boundedness of G in Theorem 1.2; however, it can be replaced. The substitute is a growth condition on $|f(z)|$ as z approaches infinity. In fact, it is also possible to omit the condition that f be defined and continuous on G^-. To do this, the following definitions are needed.

1.3 Definition. If $f: G \to \mathbb{R}$ and $a \in G^-$ or $a = \infty$, then the *limit superior of* $f(z)$ *as* z *approaches* a, denoted by $\limsup\limits_{z \to a} f(z)$, is defined by

$$\limsup_{z \to a} f(z) = \lim_{r \to 0+} \sup \{f(z): z \in G \cap B(a; r)\}$$

(If $a = \infty$, $B(a; r)$ is the ball in the metric of $\mathbb{C}_\infty$.) Similarly, *the limit inferior* *of* $f(z)$ *as* z *approaches* a, denoted by $\liminf\limits_{z \to a} f(z)$, is defined by

$$\liminf_{z \to a} f(z) = \lim_{r \to 0+} \inf \{f(z): z \in G \cap B(a; r)\}.$$

It is easy to see that $\lim\limits_{z \to a} f(z)$ exists and equals α iff $\alpha = \limsup\limits_{z \to a} f(z) = \liminf\limits_{z \to a} f(z)$.

If $G \subset \mathbb{C}$ then let $\partial_\infty G$ denote the boundary of G in $\mathbb{C}_\infty$ and call it the *extended boundary* of G. Clearly $\partial_\infty G = \partial G$ if G is bounded and $\partial_\infty G = \partial G \cup \{\infty\}$ if G is unbounded.

After these preliminaries the final version of the Maximum Modulus Theorem can be stated.

1.4 Maximum Modulus Theorem—Third Version. *Let G be a region in* $\mathbb{C}$ *and f an analytic function on G. Suppose there is a constant M such that* $\limsup\limits_{z \to a} |f(z)| \le M$ *for all a in* $\partial_\infty G$. *Then* $|f(z)| \le M$ *for all z in G.*

Proof. Let $\delta > 0$ be arbitrary and put $H = \{z \in G: |f(z)| > M + \delta\}$. The theorem will be demonstrated if H is proved to be empty.

Since $|f|$ is continuous, H is open. Since $\limsup\limits_{z \to a} |f(z)| \le M$ for each a in $\partial_\infty G$, there is a ball $B(a; r)$ such that $|f(z)| < M + \delta$ for all z in $G \cap B(a; r)$. Hence $H^- \subset G$. Since this condition also holds if G is unbounded and $a = \infty$, H must be bounded. Thus, H^- is compact. So the second version of the Maximum Modulus Theorem applies. But for z in ∂H, $|f(z)| = M + \delta$ since $H^- \subset \{z: |f(z)| \ge M + \delta\}$; therefore, $H = \square$ or f is a constant. But the hypothesis implies that $H = \square$ if f is a constant. $\blacksquare$

Notice that in the example $G = \{z: |\operatorname{Im} z| < \frac{1}{2}\pi\}$, $f(z) = \exp(e^z)$, f satisfies the condition $\limsup\limits_{z \to a} |f(z)| \le 1$ for all a in ∂G but not for $a = \infty$.

Exercises

1. Prove the following Minimum Principle. If f is a non-constant analytic function on a bounded open set G and is continuous on G^-, then either f has a zero in G or $|f|$ assumes its minimum value on ∂G. (See Exercise IV. 3.6.)

2. Let G be a bounded region and suppose f is continuous on G^- and

analytic on G. Show that if there is a constant $c \geq 0$ such that $|f(z)| = c$ for all z on the boundary of G then either f is a constant function or f has a zero in G.

3. (a) Let f be entire and non-constant. For any positive real number c show that the closure of $\{z: |f(z)| < c\}$ is the set $\{z: |f(z)| \leq c\}$.

(b) Let p be a polynomial and show that each component of $\{z: |p(z)| < c\}$ contains a zero of p. (Hint: Use Exercise 2.)

(c) If p is a polynomial and $c > 0$ show that $\{z: |p(z)| = c\}$ is the union of a finite number of closed paths. Discuss the behavior of these paths as $c \rightarrow \infty$.

4. Let $0 < r < R$ and put $A = \{z: r \leq |z| \leq R\}$. Show that there is a positive number $\epsilon > 0$ such that for each polynomial p,

$$\sup \{|p(z) - z^{-1}|: z \in A\} \geq \epsilon$$

This says that z^{-1} is not the uniform limit of polynomials on A.

5. Let f be analytic on $\bar{B}(0; R)$ with $|f(z)| \leq M$ for $|z| \leq R$ and $|f(0)| = a > 0$. Show that the number of zeros of f in $B(0; \frac{1}{3}R)$ is less than or equal to $\dfrac{1}{\log 2} \log \left(\dfrac{M}{a} \right)$. Hint: If $z_1, \ldots, z_n$ are the zeros of f in $B(0; \frac{1}{3}R)$, consider the function

$$g(z) = f(z) \left[\prod_{k=1}^{n} \left(1 - \frac{z}{z_k} \right) \right]^{-1},$$

and note that $g(0) = f(0)$. $\left(\text{Notation: } \prod_{k=1}^{n} a_k = a_1 a_2 \ldots a_n. \right)$

6. Suppose that both f and g are analytic on $\bar{B}(0; R)$ with $|f(z)| = |g(z)|$ for $|z| = R$. Show that if neither f nor g vanishes in $B(0; R)$ then there is a constant λ, $|\lambda| = 1$, such that $f = \lambda g$.

7. Let f be analytic in the disk $B(0; R)$ and for $0 \leq r < R$ define $A(r) = \max \{\text{Re } f(z): |z| = r\}$. Show that unless f is a constant, $A(r)$ is a strictly increasing function of r.

8. Suppose G is a region, $f: G \rightarrow \mathbb{C}$ is analytic, and M is a constant such that whenever z is on $\partial_\infty G$ and $\{z_n\}$ is a sequence in G with $z = \lim z_n$ we have $\lim \sup |f(z_n)| \leq M$. Show that $|f(z)| \leq M$, for each z in G.

§2. Schwarz's Lemma

2.1 Schwarz's Lemma. *Let* $D = \{z: |z| < 1\}$ *and suppose f is analytic on D with*

(a) $|f(z)| \leq 1$ *for z in D,*
(b) $f(0) = 0$.

Then $|f'(0)| \leq 1$ *and* $|f(z)| \leq |z|$ *for all z in the disk D. Moreover if* $|f'(0)| = 1$ *or if* $|f(z)| = |z|$ *for some $z \neq 0$ then there is a constant c, $|c| = 1$, such that* $f(w) = cw$ *for all w in D.*

Proof. Define $g: D \to \mathbb{C}$ by $g(z) = \dfrac{f(z)}{z}$ for $z \neq 0$ and $g(0) = f'(0)$; then g is analytic in D. Using the Maximum Modulus Theorem, $|g(z)| \leq r^{-1}$ for $|z| \leq r$ and $0 < r < 1$. Letting r approach 1 gives $|g(z)| \leq 1$ for all z in D. That is, $|f(z)| \leq |z|$ and $|f'(0)| = |g(0)| \leq 1$. If $|f(z)| = |z|$ for some z in D, $z \neq 0$, or $|f'(0)| = 1$ then $|g|$ assumes its maximum value inside D. Thus, again applying the Maximum Modulus Theorem, $g(z) \equiv c$ for some constant c with $|c| = 1$. This yields $f(z) = cz$ and completes the proof of the theorem. ∎

We will apply Schwarz's Lemma to characterize the conformal maps of the open unit disk onto itself. First we introduce a class of such maps. If $|a| < 1$ define the Mobius transformation:

$$\varphi_a(z) = \frac{z - a}{1 - \bar{a}z}$$

Notice that φ_a is analytic for $|z| < |a|^{-1}$ so that it is analytic in an open disk containing the closure of $D = \{z: |z| < 1\}$. Also, it is an easy matter to check that

$$\varphi_a(\varphi_{-a}(z)) = z = \varphi_{-a}(\varphi_a(z)).$$

for $|z| < 1$. Hence φ_a maps D onto itself in a one-one fashion.

Let θ be a real number; then

$$|\varphi_a(e^{i\theta})| = \left| \frac{e^{i\theta} - a}{1 - \bar{a}e^{i\theta}} \right|$$

$$= \left| \frac{e^{i\theta} - a}{\bar{e}^{i\theta} - \bar{a}} \right|$$

$$= 1$$

This says that $\varphi_a(\partial D) = \partial D$.

These facts, and other pertinent information which can be easily checked, are summarized as follows.

2.2 Proposition. *If $|a| < 1$ then φ_a is a one-one map of $D = \{z: |z| < 1\}$ onto itself; the inverse of φ_a is φ_{-a}. Furthermore, φ_a maps ∂D onto ∂D, $\varphi_a(a)$ $= 0$, $\varphi_a'(0) = 1 - |a|^2$, and $\varphi_a'(a) = (1 - |a|^2)^{-1}$.*

Let us see how these functions φ_a can be used in applying Schwarz's Lemma. Suppose f is analytic on D with $|f(z)| \leq 1$. Also, suppose $|a| < 1$ and $f(a) = \alpha$ (so $|\alpha| < 1$ unless f is constant). Among all functions f having these properties what is the maximum possible value of $|f'(a)|$? To solve this problem let $g = \varphi_\alpha \circ f \circ \varphi_{-a}$. Then g maps D into D and also satisfies $g(0) = \varphi_\alpha(f(a)) = \varphi_\alpha(\alpha) = 0$. Thus we can apply Schwarz's Lemma to obtain that $|g'(0)| \leq 1$. Now obtain an explicit formula for $g'(0)$. Applying the chain rule

$$g'(0) = (\varphi_\alpha \circ f)'(\varphi_a(0))\varphi'_{-a}(0)$$

$$= (\varphi_\alpha \circ f)'(a)(1-|a|^2)$$

$$= \varphi'_\alpha(\alpha)f'(a)(1-|a|^2)$$

$$= \frac{1-|a|^2}{1-|\alpha|^2}f'(a).$$

Thus,

2.3
$$|f'(a)| \le \frac{1-|\alpha|^2}{1-|a|^2}.$$

Moreover equality will occur exactly when $|g'(0)| = 1$, or, by virtue of Schwarz's Lemma, when there is a constant c with $|c| = 1$ and

2.4
$$f(z) = \varphi_{-\alpha}(c\varphi_a(z))$$

for $|z| < 1$.

We are now ready to state and prove one of the main consequences of Schwarz's Lemma. Note that if $|c| = 1$ and $|a| < 1$ then $f = c\varphi_\alpha$ defines a one-one analytic map of the open unit disk D onto itself. The next result says that the converse is also true.

2.5 Theorem. *Let* $f: D \to D$ *be a one-one analytic map of* D *onto itself and suppose* $f(a) = 0$. *Then there is a complex number* c *with* $|c| = 1$ *such that* $f = c\varphi_\alpha$.

Proof. Since f is one-one and onto there is an analytic function $g: D \to D$ such that $g(f(z)) = z$ for $|z| < 1$. Applying inequality (2.3) to both f and g gives $|f'(a)| \le (1-|a|^2)^{-1}$ and $|g'(0)| \le 1-|a|^2$ (since $g(0) = a$). But since $1 = g'(0)f'(a)$, $|f'(a)| = (1-|a|^2)^{-1}$. Applying formula (2.4) we have that $f = c\varphi_\alpha$ for some c, $|c| = 1$. ∎

Exercises

1. Suppose $|f(z)| \le 1$ for $|z| < 1$ and f is analytic. By considering the function $g: D \to D$ defined by

$$g(z) = \frac{f(z)-a}{1-\bar{a}f(z)}$$

where $a = f(0)$, prove that

$$\frac{|f(0)|-|z|}{1+|f(0)|\,|z|} \le |f(z)| \le \frac{|f(0)|+|z|}{1-|f(0)|\,|z|}$$

for $|z| < 1$.

2. Does there exist an analytic function $f: D \to D$ with $f(\tfrac{1}{2}) = \tfrac{3}{4}$ and $f'(\tfrac{1}{2}) = \tfrac{2}{3}$?

3. Suppose $f: D \to \mathbb{C}$ satisfies $\operatorname{Re} f(z) \geq 0$ for all z in D and suppose that f is analytic and not constant.

(a) Show that $\operatorname{Re} f(z) > 0$ for all z in D.

(b) By using an appropriate Möbius transformation, apply Schwarz's Lemma to prove that if $f(0) = 1$ then

$$|f(z)| \leq \frac{1+|z|}{1-|z|}$$

for $|z| < 1$. What can be said if $f(0) \neq 1$?

(c) Show that if $f(0) = 1$, f also satisfies

$$|f(z)| \geq \frac{1-|z|}{1+|z|}.$$

(Hint: Use part (a)).

4. Prove Carathéodory's Inequality whose statement is as follows: Let f be analytic on $\bar{B}(0; R)$ and let $M(r) = \max\{|f(z)|: |z| = r\}$, $A(r) = \max\{\operatorname{Re} f(z): |z| = r\}$; then for $0 < r < R$, if $A(r) \geq 0$,

$$M(r) \leq \frac{R+r}{R-r}[A(R)+|f(0)|]$$

(Hint: First consider the case where $f(0) = 0$ and examine the function $g(z) = f(Rz)[2A(R)+f(Rz)]^{-1}$ for $|z| < 1$.)

5. Let f be analytic in $D = \{z: |z| < 1\}$ and suppose that $|f(z)| \leq M$ for all z in D. (a) If $f(z_k) = 0$ for $1 \leq k \leq n$ show that

$$|f(z)| \leq M \prod_{k=1}^{n} \frac{|z-z_k|}{|1-\bar{z}_k z|}$$

for $|z| < 1$. (b) If $f(z_k) = 0$ for $1 \leq k \leq n$, each $z_k \neq 0$, and $f(0) = Me^{iz}$ $(z_1 z_2 \ldots z_n)$, find a formula for f.

6. Suppose f is analytic in some region containing $\bar{B}(0; 1)$ and $|f(z)| = 1$ where $|z| = 1$. Find a formula for f. (Hint: First consider the case where f has no zeros in $B(0; 1)$.)

7. Suppose f is analytic in a region containing $\bar{B}(0; 1)$ and $|f(z)| = 1$ when $|z| = 1$. Suppose that f has a simple zero at $z = \frac{1}{4}(1+i)$ and a double zero at $z = \frac{1}{2}$. Can $f(0) = \frac{1}{2}$?

8. Is there an analytic function f on $B(0; 1)$ such that $|f(z)| < 1$ for $|z| < 1$, $f(0) = \frac{1}{2}$, and $f'(0) = \frac{3}{4}$? If so, find such an f. Is it unique?

§3. Convex functions and Hadamard's Three Circles Theorem

In this section we will study convex functions and logarithmically convex functions and show that such functions appear in connection with the study of analytic functions.

3.1 Definition. If $[a, b]$ is an interval in the real line, a function $f: [a, b] \to \mathbb{R}$ is *convex* if for any two points x_1 and x_2 in $[a, b]$

$$f(tx_2 + (1-t)x_1) \leq tf(x_2) + (1-t)f(x_1)$$

whenever $0 \leq t \leq 1$. A subset $A \subset \mathbb{C}$ is *convex* if whenever z and w are in A, $tz + (1-t)w$ is in A for $0 \leq t \leq 1$; that is, A is convex when for any two points in A the line segment joining the two points is also in A. (See IV. 4.3 and IV. 4.4.)

What is the relation between convex functions and convex sets? The answer is that a function is convex if and only if the portion of the plane lying above the graph of the function is a convex set.

3.2 Proposition. *A function $f: [a, b] \to \mathbb{R}$ is convex iff the set*

$$A = \{(x, y): a \leq x \leq b \text{ and } f(x) \leq y\}$$

is convex.

Proof. Suppose $f: [a, b] \to \mathbb{R}$ is a convex function and let (x_1, y_1) and (x_2, y_2) be points in A. If $0 \leq t \leq 1$ then, by the definition of convex function, $f(tx_2 + (1-t)x_1) \leq tf(x_2) + (1-t)f(x_1) \leq ty_2 + (1-t)y_1$. Thus $t(x_2, y_2) + (1-t)(x_1, y_1) = (tx_2 + (1-t)x_1, ty_2 + (1-t)y_1)$ is in A; so A is convex.

Suppose A is a convex set and let x_1, x_2 be two points in $[a, b]$. Then $(tx_2 + (1-t)x_1, tf(x_2) + (1-t)f(x_1))$ is in A if $0 \leq t \leq 1$ by virtue of its convexity. But the definition of A gives that $f(tx_2 + (1-t)x_1) \leq tf(x_2) + (1-t)f(x_1)$; that is, f is convex. ∎

The proof of the next proposition is left to the reader.

3.3 Proposition. (a) *A function $f: [a, b] \to \mathbb{R}$ is convex iff for any points $x_1, \ldots, x_n$ in $[a, b]$ and real numbers $t_1, \ldots, t_n \geq 0$ with $\sum\limits_{k=1}^{n} t_k = 1$,*

$$f\left(\sum_{k=1}^{n} t_k x_k\right) \leq \sum_{k=1}^{n} t_k f(x_k).$$

(b) *A set $A \subset \mathbb{C}$ is convex iff for any points $z_1, \ldots, z_n$ in A and real numbers $t_1, \ldots, t_n \geq 0$ with $\sum\limits_{k=1}^{n} t_k = 1$, $\sum\limits_{k=1}^{n} t_k z_k$ belongs to A.*

What are the virtues of convex functions and sets? We have already seen the convex sets used in connection with complex integration. Also, the fact that disks are convex sets has played a definite role, although this may not have been apparent since this fact is taken for granted. The use of convex functions may not be so familiar to the reader; however it should be. In the first course of calculus the fact (proved below) that f is convex when f'' is non-negative is used to obtain a local minimum at a point t_0 whenever $f'(t_0) = 0$. Moreover, convex functions (and concave functions) are used to obtain inequalities. If $f: [a, b] \to \mathbb{R}$ is convex then it follows from Proposition 3.2 that $f(x) \leq \max \{f(a), f(b)\}$ for all x in $[a, b]$. We now give a necessary condition for the convexity of a function.

3.4 Proposition. *A differentiable function f on* $[a, b]$ *is convex iff* f' *is increasing.*

Proof. First assume that f is convex; to show that f' is increasing let $a \leq x < y \leq b$ and suppose that $0 < t < 1$. Since $0 < (1-t)x+ty-x = t(y-x)$, the definition of convexity gives that

$$\frac{f((1-t)x+ty)-f(x)}{t(y-x)} \leq \frac{f(y)-f(x)}{y-x}.$$

Letting $t \to 0$ gives that

3.5 $$f'(x) \leq \frac{f(y)-f(x)}{y-x}.$$

Similarly, using the fact that $0 > (1-t)x+ty-y = (1-t)(x-y)$ and letting $t \to 1$ gives that

3.6 $$f'(y) \geq \frac{f(y)-f(x)}{y-x}.$$

So, combining (3.5) and (3.6), we have that f' is increasing.

Now supposing that f' is increasing and that $x < u < y$, apply the Mean Value Theorem for differentiation to find r and s with $x < r < u < s < y$ such that

$$f'(r) = \frac{f(u)-f(x)}{u-x}$$

and

$$f'(s) = \frac{f(y)-f(u)}{y-u}.$$

Since $f'(r) \leq f'(s)$ this gives that

$$\frac{f(u)-f(x)}{u-x} \leq \frac{f(y)-f(u)}{y-u}$$

whenever $x < u < y$. In particular by letting $u = (1-t)x+ty$ where $0 < t < 1$,

$$\frac{f(u)-f(x)}{t(y-x)} \leq \frac{f(y)-f(u)}{(1-t)(y-x)};$$

and hence

$$(1-t)[f(u)-f(x)] \leq t[f(y)-f(u)].$$

This shows that f must be convex. ∎

In actuality we will mostly be concerned with functions which are not only convex, but which are *logarithmically convex*; that is, $\log f(x)$ is convex. Of course this assumes that $f(x) > 0$ for each x. It is easy to see that a logarithmically convex function is convex, but not conversely.

3.7 Theorem. *Let* $a < b$ *and let* G *be the vertical strip* $\{x+iy: a < x < b\}$.

Suppose $f\colon G^- \to \mathbb{C}$ is continuous and f is analytic in G. If we define $M\colon [a, b] \to \mathbb{R}$ by

$$M(x) = \sup \{|f(x+iy)|\colon -\infty < y < \infty\},$$

and $|f(z)| < B$ for all z in G, then $\log M(x)$ is a convex function.

Before proving this theorem, note that to say that $\log M(x)$ is convex means (Exercise 3) that for $a \leq x < u < y \leq b$,

$$(y-x) \log M(u) \leq (y-u) \log M(x) + (u-x) \log M(y)$$

Taking the exponential of both sides gives

3.8 $M(u)^{(y-x)} \leq M(x)^{(y-u)} M(y)^{(u-x)}$

whenever $a \leq x < u < y \leq b$. Also, since $\log M(x)$ is convex we have that $\log M(x)$ is bounded by max $\{\log M(a), \log M(b)\}$. That is, for $a \leq x \leq b$

$$M(x) \leq \max \{M(a), M(b)\}.$$

This gives the following.

3.9 Corollary. *If f and G are as in Theorem 3.7 and f is not constant then*

$$|f(z)| < \sup \{|f(w)|\colon w \in \partial G\}$$

for all z in G.

To prove Theorem 3.7 the following lemma is used.

3.10 Lemma. *Let f and G be as in Theorem 3.7 and further suppose that $|f(z)| \leq 1$ for z on ∂G. Then $|f(z)| \leq 1$ for all z in G.*

Proof. For each $\epsilon > 0$ let $g_\epsilon(z) = [1+\epsilon(z-a)]^{-1}$ for each z in G^-. Then for $z = x+iy$ in G^-

$$|g_\epsilon(z)| \leq |\mathrm{Re}\,[1+\epsilon(z-a)]|^{-1}$$

$$= [1+\epsilon(x-a)]^{-1}$$

$$\leq 1.$$

So for z in ∂G $|f(z)g_\epsilon(z)| \leq 1$. Also, since f is bounded by B in G,

3.11
$$|f(z)g_\epsilon(z)| \leq B|1+\epsilon(z-a)|^{-1}$$

$$\leq B[\epsilon\,|\mathrm{Im}\,z|]^{-1}$$

So if $R = \{x+iy\colon a \leq x \leq b, |y| \leq B/\epsilon\}$, inequality (3.11) gives $|f(z)g_\epsilon(z)| \leq 1$ for z in ∂R. It follows from the Maximum Modulus Theorem that $|f(z)g_\epsilon(z)| \leq 1$ for z in R. But if $|\mathrm{Im}\,z| > B/\epsilon$ then (3.11) gives that $|f(z)g_\epsilon(z)| \leq 1$. Thus for all z in G.

$$|f(z)| \leq |1+\epsilon(z-a)|.$$

Letting ϵ approach zero the desired result follows. ∎

Proof of Theorem 3.7. First observe that to prove the theorem we need only establish

$$M(u)^{(b-a)} \leq M(a)^{(b-u)} M(b)^{(u-a)}$$

for $a < u < b$ (see (3.8)). To do this recall that for a constant $A > 0$, $A^z = \exp(z \log A)$ is an entire function of z with no zeros. So $g(z)$ defined by

$$g(z) = M(a)^{(b-z)/(b-a)} M(b)^{(z-a)/(b-a)}$$

is entire, never vanishes, and (because $|A^z| = A^{\text{Re } z}$) for $z = x + iy$

3.12 $$|g(z)| = M(a)^{(b-x)/(b-a)} M(b)^{(x-a)/(b-a)}.$$

(It is assumed here that $M(a)$ and $M(b) \neq 0$. However, if either $M(a)$ or $M(b)$ is zero then $f \equiv 0$.) Since the expression on the right hand side of (3.12) is continuous for x in $[a, b]$ and never vanishes, $|g|^{-1}$ must be bounded in G^-. Also, $|g(a+iy)| = M(a)$ and $|g(b+iy)| = M(b)$ so that $|f(z)/g(z)| \leq 1$ for z in ∂G; and f/g satisfies the hypothesis of Lemma 3.10. Thus

$$|f(z)| \leq |g(z)|, \, z \in G.$$

Using (3.12) this gives for $a < u < b$

$$M(u) \leq M(a)^{(b-u)/(b-a)} M(b)^{(u-a)/(b-a)}$$

which is the desired conclusion. ∎

Hadamard's Three Circles Theorem is an analogue of the preceding theorem for an annulus. Consider ann $(0; R_1, R_2) = A$ where $0 < R_1 < R_2 < \infty$. If G is the strip $\{x+iy: \log R_1 < x < \log R_2\}$ then the exponential function maps G onto A and ∂G onto ∂A. Using this fact one can prove the following from Theorem 3.7 (the details are left to the reader).

3.13 Hadamard's Three Circles Theorem. *Let* $0 < R_1 < R_2 < \infty$ *and suppose* f *is analytic on* ann $(0; R_1, R_2)$. *If* $R_1 < r < R_2$, *define* $M(r) = \max \{|f(re^{i\theta})|: 0 \leq \theta \leq 2\pi\}$. *Then for* $R_1 < r_1 \leq r \leq r_2 < R_2$,

$$\log M(r) \leq \frac{\log r_2 - \log r}{\log r_2 - \log r_1} \log M(r_1) + \frac{\log r - \log r_1}{\log r_2 - \log r_1} \log M(r_2).$$

Another way of expressing Hadamard's Theorem is to say that $\log M(r)$ is a convex function of $\log r$.

Exercises

1. Let $f: [a, b] \to \mathbb{R}$ and suppose that $f(x) > 0$ for all x and that f has a continuous second derivative. Show that f is logarithmically convex iff $f''(x)f(x) - [f'(x)]^2 \geq 0$ for all x.

2. Show that if $f: (a, b) \to \mathbb{R}$ is convex then f is continuous. Does this remain true if f is defined on the closed interval $[a, b]$?

3. Show that a function $f: [a, b] \to \mathbb{R}$ is convex iff any of the following equivalent conditions is satisfied:

(a) $a \leq x < u < y \leq b$ gives $\det \begin{pmatrix} f(u) & u & 1 \\ f(x) & x & 1 \\ f(y) & y & 1 \end{pmatrix} \geq 0$;

(b) $a \le x < u < y \le b$ gives $\dfrac{f(u)-f(x)}{u-x} \le \dfrac{f(y)-f(x)}{y-x}$;

(c) $a \le x < u < y \le b$ gives $\dfrac{f(u)-f(x)}{u-x} \le \dfrac{f(y)-f(u)}{y-u}$.

Interpret these conditions geometrically.

4. Supply the details in the proof of Hadamard's Three Circle Theorem.

5. Give necessary and sufficient conditions on the function f such that equality occurs in the conclusion of Hadamard's Three Circle Theorem.

6. Prove Hardy's Theorem: If f is analytic on $B(0; R)$ and not constant then

$$I(r) = \frac{1}{2\pi} \int_0^{2\pi} |f(re^{i\theta})| d\theta$$

is strictly increasing and log $I(r)$ is a convex function of log r. Hint: If $0 < r_1 < r < r_2$ find a continuous function $\varphi: [0, 2\pi] \to \mathbb{C}$ such that $\varphi(\theta)f(re^{i\theta}) = |f(re^{i\theta})|$ and consider the function $F(z) = \int_0^{2\pi} f(ze^{i\theta})\varphi(\theta)d\theta$. (Note that r is fixed, so φ may depend on r.)

7. Let f be analytic in ann $(0; R_1, R_2)$ and define

$$I_2(r) = \frac{1}{2\pi} \int_0^{2\pi} |f(re^{i\theta})|^2 d\theta.$$

Show that log $I_2(r)$ is a convex function of log r, $R_1 < r < R_2$.

§4. The Phragmén-Lindelöf Theorem

This section presents some results of E. Phragmén and E. Lindelöf (published in 1908) which extend the Maximum Principle by easing the requirement of boundedness on the boundary.

The Phragmén-Lindelöf Theorem bears a relation to the Maximum Modulus Theorem which is analogous to the relationship of the following result to Liouville's theorem. If f is entire and $|f(z)| \le 1+|z|^{\frac{1}{2}}$ then f is a constant function. (Prove it!) So it is not necessary to assume that an entire function is bounded in order to prove that it is constant; it is sufficient to assume that its growth as $z \to \infty$ is restricted by $1+|z|^{\frac{1}{2}}$. The Phragmén-Lindelöf Theorem places a growth restriction on an analytic function $f: G \to \mathbb{C}$ as z nears a point on the extended boundary. Nevertheless, the conclusion, like that of the Maximum Modulus Theorem, is that f is bounded.

4.1 Phragmén-Lindelöf Theorem. *Let G be a simply connected region and let f be an analytic function on G. Suppose there is an analytic function $\varphi: G \to \mathbb{C}$ which never vanishes and is bounded on G. If M is a constant and $\partial_\infty G = A \cup B$ such that:*

(a) *for every a in A,* $\limsup_{z \to a} |f(z)| \le M$;

(b) *for every b in B, and* $\eta > 0$, $\displaystyle\lim_{z \to b} \sup |f(z)| \, |\varphi(z)|^\eta \leq M$;

then $|f(z)| \leq M$ *for all z in G.*

Proof. Let $|\varphi(z)| \leq \kappa$ for all z in G. Also because G is simply connected there is an analytic branch of $\log \varphi(z)$ on G (Corollary IV. 4.16). Hence $g(z) = \exp (\eta \log \varphi(z))$ is an analytic branch of $\varphi(z)^\eta$ for $\eta > 0$; and $|g(z)| = |\varphi(z)|^\eta$. Define $F: G \to \mathbb{C}$ by $F(z) = f(z)g(z)\kappa^{-\eta}$; then F is analytic on G and $|F(z)| \leq |f(z)|$ since $|\varphi(z)| \leq \kappa$ for all z in G. But then, by conditions (a) and (b) on $\partial_\infty G$, F satisfies the hypothesis of Theorem 1.4. Thus $|F(z)| \leq \max (M, \kappa^{-\eta}M)$ for all z in G. This gives

$$|f(z)| \leq |\kappa/\varphi(z)|^\eta \max(M, \kappa^{-\eta}M)$$

for all z in G and for all $\eta > 0$. Letting $\eta \to 0+$ gives that $|f(z)| \leq M$ for all z in G. ∎

4.2 Corollary. *Let* $a \geq \frac{1}{2}$ *and put*

$$G = \left\{z \colon |\arg z| < \frac{\pi}{2a}\right\}.$$

Suppose that f is analytic on G and there is a constant M such that $\displaystyle\lim_{z \to w} \sup |f(z)| \leq M$ *for all w in* ∂G. *If there are positive constants P and* $b < a$ *such that*

4.3 $$|f(z)| \leq P \exp (|z|^b)$$

for all z with $|z|$ *sufficiently large, then* $|f(z)| \leq M$ *for all z in G.*

Proof. Let $b < c < a$ and put $\varphi(z) = \exp (-z^c)$ for z in G. If $z = re^{i\theta}$, $|\theta| < \pi/2a$, then $\operatorname{Re} z^c = r^c \cos c\theta$. So for z in G

$$|\varphi(z)| = \exp (-r^c \cos c\theta)$$

when $z = re^{i\theta}$. Since $c < a$, $\cos c\theta \geq \rho > 0$ for all z in G. This gives that φ is bounded on G. Also, if $\eta > 0$ and $z = re^{i\theta}$ is sufficiently large,

$$|f(z)| \, |\varphi(z)|^\eta \leq P \exp (r^b - \eta r^c \cos c\theta)$$

$$\leq P \exp (r^b - \eta r^c \rho)$$

But $r^b - \eta r^c \rho = r^c (r^{b-c} - \eta\rho)$. Since $b < c$, $r^{b-c} \to 0+$ as $r \to \infty$ so that $r^b - \eta r^c \rho \to -\infty$ as $r \to \infty$. Thus

$$\lim \sup |f(z)| \, |\varphi(z)|^\eta = 0$$

Hence, f and φ satisfy the hypothesis of the Phragmén-Lindelöf Theorem so that $|f(z)| \leq M$ for each z in G. ∎

Note that the size of the angle of the sector G is the only relevant fact in this corollary; its position is inconsequential. So if G is any sector of angle π/a the conclusion remains valid.

4.4 Corollary. *Let* $a \geq \frac{1}{2}$,

$$G = \left\{ z : |\arg z| < \frac{\pi}{2a} \right\},$$

and suppose that for every w in ∂G, $\limsup\limits_{z \to w} |f(z)| \leq M$. *Moreover, assume that for every* $\delta > 0$ *there is a constant P (which may depend on δ) such that*

4.5 $$|f(z)| \leq P \exp(\delta |z|^a)$$

for z in G and $|z|$ sufficiently large. Then $|f(z)| \leq M$ for all z in G.

Proof. Define $F: G \to \mathbb{C}$ by $F(z) = f(z) \exp(-\epsilon z^a)$ where $\epsilon > 0$ is arbitrary. If $x > 0$ and δ is chosen with $0 < \delta < \epsilon$ then there is a constant P with

$$|F(x)| \leq P \exp[(\delta - \epsilon)x^a].$$

But then $|F(x)| \to 0$ as $x \to \infty$ in $\mathbb{R}$; so $M_1 = \sup \{|F(x)|: 0 < x < \infty\} < \infty$. Define $M_2 = \max \{M_1, M\}$ and

$$H_+ = \{z \in G: 0 < \arg z < \pi/2a\},$$

$$H_- = \{z \in G: 0 > \arg z > -\pi/2a\};$$

then $\limsup\limits_{z \to w} |f(z)| \leq M_2$ for all z in ∂H_+ and ∂H_-. Using hypothesis (4.5), Corollary 4.2 gives $|F(z)| \leq M_2$ for all z in H_+ and H_- hence, $|F(z)| \leq M_2$ for all z in G.

We claim that $M_2 = M$. In fact, if $M_2 = M_1 > M$ then $|F|$ assumes its maximum value in G at some point x, $0 < x < \infty$ (because $|F(x)| \to 0$ as $x \to \infty$ and $\limsup\limits_{x \to 0} |f(x)| = \limsup\limits_{x \to 0} |F(x)| \leq M < M_1$). This would give that F is a constant by the Maximum Principle and so $M = M_1$. Thus, $M_2 = M$ and $|F(z)| \leq M$ for all z in G; that is,

$$|f(z)| \leq M \exp(\epsilon \, \text{Re} \, z^a)$$

for all z in G; since M is independent of ϵ, we can let $\epsilon \to 0$ and get $|f(z)| \leq M$ for all z in G. ∎

Let $G = \{z: z \neq 0 \text{ and } |\arg z| < \pi/2a\}$ and let $f(z) = \exp(z^a)$ for $z \in G$. Then $|f(z)| = \exp(|z|^a \cos a\theta)$ where $\theta = \arg z$. So for z in ∂G $|f(z)| = 1$; but $f(z)$ is clearly unbounded in G. In fact, on any ray in G we have that $|f(z)| \to \infty$. This shows that the growth condition (4.5) is very delicate and can't be improved.

Exercises

1. In the statement of the Phragmén-Lindelöf Theorem, the requirement that G be simply connected is not necessary. Extend Theorem 4.1 to regions G with the property that for each z in $\partial_\infty G$ there is a sphere V in $\mathbb{C}_\infty$ centered at z such that $V \cap G$ is simply connected. Give some examples of regions that are not simply connected but have this property and some which don't.

2. In Theorem 4.1 suppose there are bounded analytic functions $\varphi_1, \varphi_2, \ldots,$ φ_n on G that never vanish and $\partial_\infty G = A \cup B_1 \cup \ldots \cup B_n$ such that condition (a) is satisfied and condition (b) is also satisfied for each φ_k and B_k. Prove that $|f(z)| \leq M$ for all z in G.

3. Let $G = \{z : |\operatorname{Im} z| < \tfrac{1}{2}\pi\}$ and suppose $f : G \to \mathbb{C}$ is analytic and $\limsup_{z \to w} |f(z)| \leq M$ for w in ∂G. Also, suppose $A < \infty$ and $a < 1$ can be found such that

$$|f(z)| < \exp [A \exp (a \, |\operatorname{Re} z|)]$$

for all z in G. Show that $|f(z)| \leq M$ for all z in G. Examine $\exp (\exp z)$ to see that this is the best possible growth condition. Can we take $a = 1$ above?

4. Let $f : G \to \mathbb{C}$ be analytic and suppose M is a constant such that $\limsup |f(z_n)| \leq M$ for each sequence $\{z_n\}$ in G which converges to a point in $\partial_\infty G$. Show that $|f(z)| \leq M$. (See Exercise 1.8).

5. Let $f : G \to \mathbb{C}$ be analytic and suppose that G is bounded. Fix z_0 in ∂G and suppose that $\limsup_{z \to w} |f(z)| \leq M$ for w in ∂G, $w \neq z_0$. Show that if $\lim_{z \to z_0} |z - z_0|^\epsilon \, |f(z)| = 0$ for every $\epsilon > 0$ then $|f(z)| \leq M$ for every z in ∂G. (Hint: If $a \notin G$, consider $\varphi(z) = (z - z_0) \, (z - a)^{-1}$.)

6. Let $G = \{z : \operatorname{Re} z > 0\}$ and let $f : G \to \mathbb{C}$ be an analytic function with $\limsup_{z \to w} |f(z)| \leq M$ for w in ∂G, and also suppose that for every $\delta > 0$,

$$\limsup_{r \to 0} \{\exp (-\epsilon/r \, |f(re^{i\theta})| : |\theta| < \tfrac{1}{2}\pi\} = 0.$$

Show that $|f(z)| \leq M$ for all z in G.

7. Let $G = \{z : \operatorname{Re} z > 0\}$ and let $f : G \to \mathbb{C}$ be analytic such that $f(1) = 0$ and such that $\limsup_{z \to w} |f(z)| \leq M$ for w in ∂G. Also, suppose that for every δ, $0 < \delta < 1$, there is a constant P such that

$$|f(z)| \leq P \exp (|z|^{1-\delta}).$$

Prove that

$$|f(z)| \leq M \left[\frac{(1-x)^2 + y^2}{(1+x)^2 + y^2} \right]^{\frac{1}{2}}.$$

$\left(\text{Hint: Consider } f(z) \left(\dfrac{1+z}{1-z} \right). \right)$

Chapter VII

Compactness and Convergence in the Space of Analytic Functions

In this chapter a metric is put on the set of all analytic functions on a fixed region G, and compactness and convergence in this metric space is discussed. Among the applications obtained is a proof of the Riemann Mapping Theorem.

Actually some more general results are obtained which enable us to also study spaces of meromorphic functions.

§1. The space of continuous functions $C(G,\Omega)$

In this chapter (Ω, d) will always denote a complete metric space. Although much of what is said does not need the completeness of Ω, those results which hold the most interest are not true if (Ω, d) is not assumed to be complete.

1.1 Definition. If G is an open set in $\mathbb{C}$ and (Ω, d) is a complete metric space then designate by $C(G, \Omega)$ the set of all continuous functions from G to Ω.

The set $C(G, \Omega)$ is never empty since it always contains the constant functions. However, it is possible that $C(G, \Omega)$ contains only the constant functions. For example, suppose that G is connected and $\Omega = \mathbb{N} = \{1, 2, \ldots\}$. If f is in $C(G, \Omega)$ then $f(G)$ must be connected in Ω and, hence, must reduce to a point.

However, our principal concern will be when Ω is either $\mathbb{C}$ or $\mathbb{C}_\infty$. For these two choices of Ω, $C(G, \Omega)$ has many non constant elements. In fact, each analytic function on G is in $C(G, \mathbb{C})$ and each meromorphic function on G is in $C(G, \mathbb{C}_\infty)$ (see Exercise V. 3.4).

To put a metric on $C(G, \Omega)$ we must first prove a fact about open subsets of $\mathbb{C}$. The third part of the next proposition will not be used until Chapter VIII.

1.2 Proposition. *If G is open in $\mathbb{C}$ then there is a sequence $\{K_n\}$ of compact subsets of G such that $G = \bigcup_{n=1}^{\infty} K_n$. Moreover, the sets K_n can be chosen to satisfy the following conditions:*

(a) *$K_n \subset \text{int } K_{n+1}$;*
(b) *$K \subset G$ and K compact implies $K \subset K_n$ for some n;*
(c) *Every component of $\mathbb{C}_\infty - K_n$ contains a component of $\mathbb{C}_\infty - G$.*

Proof. For each positive integer n let

$$K_n = \{z: |z| \le n\} \cap \left\{z: d(z, \mathbb{C}-G) \ge \frac{1}{n}\right\};$$

since K_n is clearly bounded and it is the intersection of two closed subsets of $\mathbb{C}$, K_n is compact. Also, the set

$$\{z: |z| < n+1\} \cap \left\{z: d(z, \mathbb{C}-G) > \frac{1}{n+1}\right\}$$

is open, contains K_n, and is contained in K_{n+1}. This gives that (a) is satisfied. Since it easily follows that $G = \bigcup_{n=1}^{\infty} K_n$ we also get that $G = \bigcup_{n=1}^{\infty} \text{int } K_n$; so if K is a compact subset of G the sets $\{\text{int } K_n\}$ form an open cover of K. This gives part (b).

    To see part (c) note that the unbounded component of $\mathbb{C}_\infty - K_n$ ($\supset \mathbb{C}_\infty - G$) must contain ∞ and must therefore contain the component of $\mathbb{C}_\infty - G$ which contains ∞. Also the unbounded component contains $\{z: |z| > n\}$. So if D is a bounded component of $\mathbb{C}_\infty - K_n$ it contains a point z with $d(z, \mathbb{C}-G) < \frac{1}{n}$. But by definition this gives a point w in $\mathbb{C}-G$ with $|w-z| < \frac{1}{n}$. But then $z \in B\left(w; \frac{1}{n}\right) \subset \mathbb{C}_\infty - K_n$; since disks are connected and z is in the component D of $\mathbb{C}_\infty - K_n$, $B\left(w; \frac{1}{n}\right) \subset D$. If D_1 is the component of $\mathbb{C}_\infty - G$ that contains w it follows that $D_1 \subset D$. ∎

    If $G = \bigcup_{n=1}^{\infty} K_n$ where each K_n is compact and $K_n \subset \text{int } K_{n+1}$, define

1.3 $\rho_n(f, g) = \sup \{d(f(z), g(z)): z \in K_n\}$

for all functions f and g in $C(G, \Omega)$. Also define

1.4 $\rho(f, g) = \sum_{n=1}^{\infty} (\tfrac{1}{2})^n \frac{\rho_n(f, g)}{1 + \rho_n(f, g)};$

since $t(1+t)^{-1} \le 1$ for all $t \ge 0$, the series in (1.4) is dominated by $\sum (\tfrac{1}{2})^n$ and must converge. It will be shown that ρ is a metric for $C(G, \Omega)$. To do this the following lemma, whose proof is left as an exercise, is needed.

1.5 Lemma. *If (S, d) is a metric space then*

$$\mu(s, t) = \frac{d(s, t)}{1 + d(s, t)}$$

is also a metric on S. A set is open in (S, d) iff it is open in (S, μ); a sequence is a Cauchy sequence in (S, d) iff it is a Cauchy sequence in (S, μ).

1.6 Proposition. *$(C(G, \Omega), \rho)$ is a metric space.*

Proof. It is clear that $\rho(f, g) = \rho(g, f)$. Also, since each ρ_n satisfies the triangle inequality, the preceding lemma can be used to show that ρ satisfies the triangle inequality. Finally, the fact that $G = \bigcup_{n=1}^{\infty} K_n$ gives that $f = g$ whenever $\rho(f, g) = 0$. ∎

The next lemma concerns subsets of $C(G, \Omega) \times C(G, \Omega)$ and is very useful because it gives insight into the behavior of the metric ρ. Those who know the appropriate definitions will recognize that this lemma says that two uniformities are equivalent.

1.7 Lemma. *Let the metric ρ be defined as in (1.4). If $\epsilon > 0$ is given then there is a $\delta > 0$ and a compact set $K \subset G$ such that for f and g in $C(G, \Omega)$,*

1.8 $\sup \{d(f(z), g(z)): z \in K\} < \delta \Rightarrow \rho(f, g) < \epsilon.$

Conversely, if $\delta > 0$ and a compact set K are given, there is an $\epsilon > 0$ such that for f and g in $C(G, \Omega)$,

1.9 $\rho(f, g) < \epsilon \Rightarrow \sup \{d(f(z), g(z)): z \in K\} < \delta.$

Proof. If $\epsilon > 0$ is fixed let p be a positive integer such that $\sum_{n=p+1}^{\infty} (\tfrac{1}{2})^n < \tfrac{1}{2}\epsilon$ and put $K = K_p$. Choose $\delta > 0$ such that $0 \le t < \delta$ gives $\dfrac{t}{1+t} < \tfrac{1}{2}\epsilon$. Suppose f and g are functions in $C(G, \Omega)$ that satisfy $\sup \{d(f(z), g(z)): z \in K\} < \delta$. Since $K_n \subset K_p = K$ for $1 \le n \le p$, $\rho_n(f, g) < \delta$ for $1 \le n \le p$. This gives

$$\frac{\rho_n(f, g)}{1+\rho_n(f, g)} < \tfrac{1}{2}\epsilon$$

for $1 \le n \le p$. Therefore

$$\rho(f, g) < \sum_{n=1}^{p} (\tfrac{1}{2})^n (\tfrac{1}{2}\epsilon) + \sum_{n=p+1}^{\infty} (\tfrac{1}{2})^n$$

$$< \epsilon$$

That is, (1.8) is satisfied.

Now suppose K and δ are given. Since $G = \bigcup_{n=1}^{\infty} K_n = \bigcup_{n=1}^{\infty} \text{int } K_n$ and K is compact there is an integer $p \ge 1$ such that $K \subset K_p$; this gives

$$\rho_p(f, g) \ge \sup \{d(f(z), g(z)): z \in K\}$$

Let $\epsilon > 0$ be chosen so that $0 \le s < 2^p \epsilon$ implies $\dfrac{s}{1-s} < \delta$; then $\dfrac{t}{1+t} < 2^p \epsilon$ implies $t < \delta$. So if $\rho(f, g) < \epsilon$ then $\dfrac{\rho_p(f, g)}{1+\rho_p(f, g)} < 2^p \epsilon$ and this gives $\rho_p(f, g) < \delta$. But this is exactly the statement contained in (1.9). ∎

1.10 Proposition. (a) *A set $\mathcal{O} \subset (C(G, \Omega), \rho)$ is open iff for each f in $\mathcal{O}$ there is a compact set K and a $\delta > 0$ such that*

$$\mathcal{O} \supset \{g: d(f(z), g(z)) < \delta, z \in K\}$$

(b) *A sequence $\{f_n\}$ in $(C(G, \Omega), \rho)$ converges to f iff $\{f_n\}$ converges to f uniformly on all compact subsets of G.*

Proof. If $\mathcal{O}$ is open and $f \in \mathcal{O}$ then for some $\epsilon > 0$, $\mathcal{O} \supset \{g: \rho(f, g) < \epsilon\}$. But now the first part of the preceding lemma says that there is a $\delta > 0$ and a compact set K with the desired properties. Conversely, if $\mathcal{O}$ has the stated property and $f \in \mathcal{O}$ then the second part of the lemma gives an $\epsilon > 0$ such that $\mathcal{O} \supset \{g: \rho(f, g) < \epsilon\}$; this means that $\mathcal{O}$ is open.

The proof of part (b) will be left to the reader. ∎

1.11 Corollary. *The collection of open sets is independent of the choice of the sets $\{K_n\}$. That is, if $G = \bigcup_{n=1}^{\infty} K'_n$ where each K'_n is compact and $K'_n \subset$ int K'_{n+1} and if μ is the metric defined by the sets $\{K'_n\}$ then a set is open in $(C(G, \Omega), \mu)$ iff it is open in $(C(G, \Omega), \rho)$.*

Proof. This is a direct consequence of part (a) of the preceding proposition since the characterization of open sets does not depend on the choice of the sets $\{K_n\}$. ∎

Henceforward, whenever we consider $C(G, \Omega)$ as a metric space it will be assumed that the metric ρ is given by formula (1.4) for some sequence $\{K_n\}$ of compact sets such that $K_n \subset$ int K_{n+1} and $G = \bigcup_{n=1}^{\infty} K_n$. Actually, the requirement that $K_n \subset$ int K_{n+1} can be dropped and the above results will remain valid. However, to show this requires some extra effort (e.g., the Baire Category Theorem) which, though interesting, would be a detour.

Nothing done so far has used the assumption that Ω is complete. However, if Ω is not complete then $C(G, \Omega)$ is not complete. In fact, if $\{\omega_n\}$ is a non-convergent Cauchy sequence in Ω and $f_n(z) = \omega_n$ for all z in G, then $\{f_n\}$ is a non-convergent Cauchy sequence in $C(G, \Omega)$. However, we are assuming that Ω is complete and this gives the following.

1.12 Proposition. *$C(G, \Omega)$ is a complete metric space.*

Proof. Again utilize Lemma 1.7. Suppose $\{f_n\}$ is a Cauchy sequence in $C(G, \Omega)$. Then for each compact set $K \subset G$ the restrictions of the functions f_n to K gives a Cauchy sequence in $C(K, \Omega)$. That is, for every $\delta > 0$ there is an integer N such that

1.13 $$\sup \{d(f_n(z), f_m(z)): z \in K\} < \delta$$

for $n, m \geq N$. In particular $\{f_n(z)\}$ is a Cauchy sequence in Ω; so there is a point $f(z)$ in Ω such that $f(z) = \lim f_n(z)$. This gives a function $f: G \to \Omega$; it must be shown that f is continuous and $\rho(f_n, f) \to 0$.

Let K be compact and fix $\delta > 0$; choose N so that (1.13) holds for n, $m \geq N$. If z is arbitrary in K but fixed then there is an integer $m \geq N$ so that $d(f(z), f_m(z)) < \delta$. But then

$$d(f(z), f_n(z)) < 2\delta$$

for all $n \geq N$. Since N does not depend on z this gives

$$\sup \{d(f(z), f_n(z)): z \in K\} \to 0$$

as $n \to \infty$. That is, $\{f_n\}$ converges to f uniformly on every compact set in G. In particular, the convergence is uniform on all closed balls contained in G. This gives (Theorem II. 6.1) that f is continuous at each point of G. Also, Proposition 1.10 (b) gives that $\rho(f_n, f) \to 0$. ■

The next definition is derived from the classical origins of this subject. Actually it could have been omitted without interfering with the development of the chapter. However, even though there is virtue in maintaining a low ratio of definitions to theorems, the classical term is widely used and should be known by the reader.

1.14 Definition. A set $\mathscr{F} \subset C(G, \Omega)$ is *normal* if each sequence in $\mathscr{F}$ has a subsequence which converges to a function f in $C(G, \Omega)$.

This of course looks like the definition of sequentially compact subsets, but the limit of the subsequence is not required to be in the set $\mathscr{F}$. The next proof is left to the reader.

1.15 Proposition. *A set $\mathscr{F} \subset C(G, \Omega)$ is normal iff its closure is compact.*

1.16 Proposition. *A set $\mathscr{F} \subset C(G, \Omega)$ is normal iff for every compact set $K \subset G$ and $\delta > 0$ there are functions $f_1, \ldots, f_n$ in $\mathscr{F}$ such that for f in $\mathscr{F}$ there is at least one k, $1 \leq k \leq n$, with*

$$\sup \{d(f(z), f_k(z)): z \in K\} < \delta.$$

Proof. Suppose $\mathscr{F}$ is normal and let K and $\delta > 0$ be given. By Lemma 1.7 there is an $\epsilon > 0$ such that (1.9) holds. But since $\mathscr{F}^-$ is compact, $\mathscr{F}$ is totally bounded (actually there are a few details to fill in here). So there are $f_1, \ldots, f_n$ in $\mathscr{F}$ such that

$$\mathscr{F} \subset \bigcup_{k=1}^{n} \{f: \rho(f, f_k) < \epsilon\}$$

But from the choice of ϵ this gives

$$\mathscr{F} \subset \bigcup_{k=1}^{n} \{f: d(f(z), f_k(z)) < \delta, z \in K\};$$

that is, $\mathscr{F}$ satisfies the condition of the proposition.

For the converse, suppose $\mathscr{F}$ has the stated property. Since it readily follows that $\mathscr{F}^-$ also satisfies this condition, assume that $\mathscr{F}$ is closed. But since $C(G, \Omega)$ is complete $\mathscr{F}$ must be complete. And, again using Lemma 1.7, it readily follows that $\mathscr{F}$ is totally bounded. From Theorem II. 4.9 $\mathscr{F}$ is compact and therefore normal. ■

This section concludes by presenting the Arzela-Ascoli Theorem. Although its proof is not overly complicated it is a deep result which has proved extremely useful in many areas of analysis. Before stating the theorem a few results of a more general nature are needed.

Let (X_n, d_n) be a metric space for each $n \geq 1$ and let $X = \prod_{n=1}^{\infty} X_n$ be their cartesian product. That is, $X = \{\xi = \{x_n\}: x_n \in X_n \text{ for each } n \geq 1\}$. For $\xi = \{x_n\}$ and $\eta = \{y_n\}$ in X define

1.17
$$d(\xi, \eta) = \sum_{n=1}^{\infty} (\tfrac{1}{2})^n \frac{d_n(x_n, y_n)}{1 + d_n(x_n, y_n)}.$$

1.18 Proposition. $\left(\prod_{n=1}^{\infty} X_n, d\right)$, where d is defined by (1.17), is a metric space. If $\xi^k = \{x_n^k\}_{n=1}^{\infty}$ is in $X = \prod_{n=1}^{\infty} X_n$ then $\xi^k \to \xi = \{x_n\}$ iff $x_n^k \to x_n$ for each n. Also, if each (X_n, d_n) is compact then X is compact.

Proof. The proof that d is a metric is left to the reader. Suppose $d(\xi^k, \xi) \to 0$; since

$$\frac{d_n(x_n^k, x_n)}{1 + d_n(x_n^k, x_n)} \leq 2^n d(\xi^k, \xi)$$

we have that

$$\lim_{k \to \infty} \frac{d_n(x_n^k, x_n)}{1 + d_n(x_n^k, x_n)} = 0.$$

This gives that $x_n^k \to x_n$ for each $n \geq 1$. The proof of the converse is left to the reader.

Now suppose that each (X_n, d_n) is compact. To show that (X, d) is compact it suffices to show that every sequence in X has a convergent subsequence; this is accomplished by the Cantor diagonalization process. Let $\xi^k = \{x_n^k\} \in X$ for each $k \geq 1$ and consider the sequence of the first entries of the ξ^k; that is, consider $\{x_1^k\}_{k=1}^{\infty} \subset X_1$. Since X_1 is compact there is a point x_1 in X_1 and a subsequence of $\{x_1^k\}$ which converges to it. We are now faced with a problem in notation. If this subsequence of $\{x_1^k\}_{k=1}^{\infty}$ is denoted by $\{x_1^{k_j}\}_{j=1}^{\infty}$ there is little confusion at this stage. However, the next step in the proof is to consider the corresponding subsequence of second entries $\{x_2^{k_j}\}_{j=1}^{\infty}$ and take a subsequence of this. Furthermore, it is necessary to continue this process for all the entries. It is easy to see that this is opening up a notational Pandora's Box. However, there is an alternative. Denote the convergent subsequence of $\{x_1^k\}$ by $\{x_1^k: k \in \mathbb{N}_1\}$, where $\mathbb{N}_1$ is an infinite subset of the positive integers $\mathbb{N}$. Consider the sequence of second entries of $\{\xi^k: k \in \mathbb{N}_1\}$. Then there is a point x_2 in X_2 and an infinite subset $\mathbb{N}_2 \subset \mathbb{N}_1$ such that $\lim \{x_2^k: k \in \mathbb{N}_2\} = x_2$. (Notice that we still have $\lim \{x_1^k: k \in \mathbb{N}_2\} = x_1$.) Continuing this process gives a decreasing sequence of infinite subsets of $\mathbb{N}$, $\mathbb{N}_1 \supset \mathbb{N}_2 \ldots$; and points x_n in X_n such that

1.19
$$\lim \{x_n^k: k \in \mathbb{N}_n\} = x_n$$

Let k_j be the jth integer in $\mathbb{N}_j$ and consider $\{\xi^{k_j}\}$; we claim that $\xi^{k_j} \to \xi = \{x_n\}$ as $k \to \infty$. To show this it suffices to show that

1.20
$$x_n = \lim_{k_j \to \infty} x_n^{k_j}$$

for each $n \geq 1$. But since $\mathbb{N}_j \subset \mathbb{N}_n$ for $j \geq n$, $\{x_n^{k_j}: j \geq n\}$ is a subsequence of $\{x_n^k: k \in \mathbb{N}_n\}$. So (1.20) follws from (1.19). ∎

The following definition plays a central role in the Arzela-Ascoli Theorem.

1.21 Definition. A set $\mathcal{F} \subset C(G, \Omega)$ is *equicontinuous at a point* z_0 in G iff for every $\epsilon > 0$ there is a $\delta > 0$ such that for $|z - z_0| < \delta$,

$$d(f(z), f(z_0)) < \epsilon$$

for every f in $\mathcal{F}$. $\mathcal{F}$ *is equicontinuous over a set* $E \subset G$ if for every $\epsilon > 0$ there is a $\delta > 0$ such that for z and z' in E and $|z - z'| < \delta$,

$$d(f(z), f(z')) < \epsilon$$

for all f in $\mathcal{F}$.

Notice that if $\mathcal{F}$ consists of a single function f then the statement that $\mathcal{F}$ is equicontinuous at z_0 is only the statement that f is continuous at z_0. The important thing about equicontinuity is that the same δ will work for all the functions in $\mathcal{F}$. Also, for $\mathcal{F} = \{f\}$ to be equicontinuous over E is to require that f is uniformly continuous on E. For a larger family $\mathcal{F}$ to be equicontinuous there must be uniform uniform continuity.

Because of this analogy with continuity and uniform continuity the following proposition should not come as a surprise.

1.22 Proposition. *Suppose* $\mathcal{F} \subset C(G, \Omega)$ *is equicontinuous at each point of* G; *then* $\mathcal{F}$ *is equicontinuous over each compact subset of* G.

Proof. Let $K \subset G$ be compact and fix $\epsilon > 0$. Then for each w in K there is a $\delta_w > 0$ such that

$$d(f(w'), f(w)) < \tfrac{1}{2}\epsilon$$

for all f in $\mathcal{F}$ whenever $|w - w'| < \delta_w$. Now $\{B(w; \delta_w): w \in K\}$ forms an open cover of K; by Lebesgue's Covering Lemma (II. 4.8) there is a $\delta > 0$ such that for each z in K, $B(z; \delta)$ is contained in one of the sets of this cover. So if z and z' are in K and $|z - z'| < \delta$ there is a w in K with $z' \in B(z; \delta) \subset B(w; \delta_w)$. That is, $|z - w| < \delta_w$ and $|z' - w| < \delta_w$. This gives $d(f(z), f(w)) < \tfrac{1}{2}\epsilon$ and $d(f(z'), f(w)) < \tfrac{1}{2}\epsilon$; so that $d(f(z), f(z')) < \epsilon$ and $\mathcal{F}$ is equicontinuous over K. ∎

1.23 Arzela-Ascoli Theorem. *A set* $\mathcal{F} \subset C(G, \Omega)$ *is normal iff the following two conditions are satisfied*:

(a) *for each* z *in* G, $\{f(z): f \in \mathcal{F}\}$ *has compact closure in* Ω;
(b) $\mathcal{F}$ *is equicontinuous at each point of* G.

Proof. First assume that $\mathcal{F}$ is normal. Notice that for each z in G the map of $C(G, \Omega) \to \Omega$ defined by $f \rightsquigarrow f(z)$ is continuous; since $\mathcal{F}^-$ is compact its image is compact in Ω and (a) follows. To show (b) fix a point z_0 in G and let $\epsilon > 0$. If $R > 0$ is chosen so that $K = \bar{B}(z_0; R) \subset G$ then K is compact and Proposition 1.16 implies there are functions $f_1, \ldots, f_n$ in $\mathscr{F}$ such that for each f in $\mathscr{F}$ there is at least one f_k with

1.24
$$\sup \{d(f(z), f_k(z)): z \in K\} < \frac{\epsilon}{3}.$$

But since each f_k is continuous there is a δ, $0 < \delta < R$, such that $|z-z_0| < \delta$ implies that

$$d(f_k(z), f_k(z_0)) < \frac{\epsilon}{3}$$

for $1 \leq k \leq n$. Therefore, if $|z-z_0| < \delta$, $f \in \mathscr{F}$, and k is chosen so that (1.24) holds, then

$$d(f(z), f(z_0)) \leq d(f(z), f_k(z)) + d(f_k(z), f_k(z_0)) + d(f_k(z_0), f(z_0))$$

$$< \epsilon$$

That is, $\mathscr{F}$ is equicontinuous at z_0.

Now suppose $\mathscr{F}$ satisfies conditions (a) and (b); it must be shown that $\mathscr{F}$ is normal. Let $\{z_n\}$ be the sequence of all points in G with rational real and imaginary parts (so for z in G and $\delta > 0$ there is a z_n with $|z-z_n| < \delta$). For each $n \geq 1$ let

$$X_n = \{f(z_n): f \in \mathscr{F}\}^- \subset \Omega;$$

from part (a), (X_n, d) is a compact metric space. Thus, by Proposition 1.18, $X = \prod_{n=1}^{\infty} X_n$ is a compact metric space. For f in $\mathscr{F}$ define $\tilde{f}$ in X by

$$\tilde{f} = \{f(z_1), f(z_2), \ldots\}.$$

Let $\{f_k\}$ be a sequence in $\mathscr{F}$; so $\{\tilde{f}_k\}$ is a sequence in the compact metric space X. Thus there is a ξ in X and a subsequence of $\{\tilde{f}_k\}$ which converges to ξ. For the sake of convenient notation, assume that $\xi = \lim \tilde{f}_k$. Again from Proposition 1.18,

1.25
$$\lim_{k \to \infty} f_k(z_n) = \omega_n$$

where $\xi = \{\omega_n\}$.

It will be shown that $\{f_k\}$ converges to a function f in $C(G, \Omega)$. By (1.25) this function f will have to satisfy $f(z_n) = \omega_n$. The importance of (1.25) is that it imposes control over the behavior of $\{f_k\}$ on a dense subset of G. We will use the fact that $\{f_k\}$ is equicontinuous to spread this control to the rest of G.

To find the function f and show that $\{f_k\}$ converges to f it suffices to show that $\{f_k\}$ is a Cauchy sequence. So let K be compact set in G and let $\epsilon > 0$; by Lemma 1.10(b) it suffices to find an integer J such that for $k, j \geq J$,

1.26
$$\sup \{d(f_k(z), f_j(z)): z \in K\} < \epsilon.$$

Since K is compact $R = d(K, \partial G) > 0$. Let $K_1 = \{z: d(z, K) \leq \frac{1}{2}R\}$; then K_1 is compact and $K \subset \text{int } K_1 \subset K_1 \subset G$. Since $\mathscr{F}$ is equicontinuous at each point of G it is equicontinuous on K_1 by Proposition 1.22. So choose δ, $0 < \delta < \frac{1}{2}R$, such that

1.27
$$d(f(z), f(z')) < \frac{\epsilon}{3}$$

for all f in $\mathscr{F}$ whenever z and z' are in K_1 with $|z-z'| < \delta$. Now let D be the collection of points in $\{z_n\}$ which are also points in K_1; that is

$$D = \{z_n : z_n \in K_1\}$$

If $z \in K$ then there is a z_n with $|z-z_n| < \delta$; but $\delta < \frac{1}{2}R$ gives that $d(z_n, K) < \frac{1}{2}R$, or that $z_n \in K_1$. Hence $\{B(w; \delta) : w \in D\}$ is an open cover of K. Let $w_1, \ldots, w_n \in D$ such that

$$K \subset \bigcup_{i=1}^{n} B(w_i; \delta).$$

Since $\lim_{k \to \infty} f_k(w_i)$ exists for $1 \le i \le n$ (by (1.25)) there is an integer J such that for $j, k \ge J$

1.28
$$d(f_k(w_i), f_j(w_i)) < \frac{\epsilon}{3}$$

for $i = 1, \ldots, n$.

Let z be an arbitrary point in K and let w_i be such that $|w_i - z| < \delta$. If k and j are larger than J then (1.27) and (1.28) give

$$d(f_k(z), f_j(z)) \le d(f_k(z), f_k(w_i)) + d(f_k(w_i), f_j(w_i)) + d(f_j(w_i), f_j(z))$$
$$< \epsilon.$$

Since z was arbitrary this establishes (1.26). ∎

Exercises

1. Prove Lemma 1.5 (Hint: Study the function $f(t) = \dfrac{t}{1+t}$ for $t > -1$.)

2. Find the sets K_n obtained in Proposition 1.2 for each of the following choices of G: (a) G is an open disk; (b) G is an open annulus; (c) G is the plane with n pairwise disjoint closed disks removed; (d) G is an infinite strip; (e) $G = \mathbb{C} - \mathbb{Z}$.

3. Supply the omitted details in the proof of Proposition 1.18.

4. Let F be a subset of a metric space (X, d) such that F^- is compact. Show that F is totally bounded.

5. Suppose $\{f_n\}$ is a sequence in $C(G, \Omega)$ which converges to f and $\{z_n\}$ is a sequence in G which converges to a point z in G. Show $\lim f_n(z_n) = f(z)$.

6. (Dini's Theorem) Consider $C(G, \mathbb{R})$ and suppose that $\{f_n\}$ is a sequence in $C(G, \mathbb{R})$ which is monotonically increasing (i.e., $f_n(z) \le f_{n+1}(z)$ for all z in G) and $\lim f_n(z) = f(z)$ for all z in G where $f \in C(G, \mathbb{R})$. Show that $f_n \to f$.

7. Let $\{f_n\} \subset C(G, \Omega)$ and suppose that $\{f_n\}$ is equicontinuous. If $f \in C(G, \Omega)$ and $f(z) = \lim f_n(z)$ for each z then show that $f_n \to f$.

8. (a) Let f be analytic on $B(0; R)$ and let $f(z) = \sum_{n=0}^{\infty} a_n z^n$ for $|z| < R$. If $f_n(z) = \sum_{k=0}^{n} a_k z^k$, show that $f_n \to f$ in $C(G; \mathbb{C})$.

(b) Let $G = \text{ann}(0; 0, R)$ and let f be analytic on G with Laurent series development $f(z) = \sum_{n=-\infty}^{\infty} a_n z^n$. Put $f_n(z) = \sum_{k=-\infty}^{n} a_k z^k$ and show that $f_n \to f$ in $C(G; \mathbb{C})$.

§2. Spaces of analytic functions

Let G be an open subset of the complex plane. If $H(G)$ is the collection of analytic functions on G, we can consider $H(G)$ as a subset of $C(G,\mathbb{C})$. We use $H(G)$ to denote the analytic functions on G rather than $A(G)$ because it is a universal practice to let $A(G)$ denote the collection of continuous functions $f: G^- \to \mathbb{C}$ that are analytic in G. Thus $A(G) \neq H(G)$. The letter H is used in reference to "analytic" because the word holomorphic is commonly used for analytic. Another term used in place of analytic is regular.

The first question to ask about $H(G)$ is: Is $H(G)$ closed in $C(G,\mathbb{C})$? The next result answers this question positively and also says that the function $f \to f'$ is continuous from $H(G)$ into $H(G)$.

2.1 Theorem. *If $\{f_n\}$ is a sequence in $H(G)$ and f belongs to $C(G, \mathbb{C})$ such that $f_n \to f$ then f is analytic and $f_n^{(k)} \to f^{(k)}$ for each integer $k \geq 1$.*

Proof. We will show that f is analytic by applying Morera's Theorem (IV. 5.10). So let T be a triangle contained inside a disk $D \subset G$. Since T is compact, $\{f_n\}$ converges to f uniformly over T. Hence $\int_T f = \lim \int_T f_n = 0$ since each f_n is analytic. Thus f must be analytic in every disk $D \subset G$; but this gives that f is analytic in G.

To show that $f_n^{(k)} \to f^{(k)}$, let $D = \bar{B}(a; r) \subset G$; then there is a number $R > r$ such that $\bar{B}(a; R) \subset G$. If γ is the circle $|z-a| = R$ then Cauchy's Integral Formula gives

$$f_n^{(k)}(z) - f^{(k)}(z) = \frac{k!}{2\pi i} \int_\gamma \frac{f_n(w) - f(w)}{(w-z)^{k+1}} \, dw.$$

for z in D. Using Cauchy's Estimate,

2.2 $$|f_n^{(k)}(z) - f^{(k)}(z)| \leq \frac{k! M_n R}{(R-r)^{k+1}} \text{ for } |z-a| \leq r,$$

where $M_n = \sup\{|f_n(w) - f(w)|: |w-a| = R\}$. But since $f_n \to f$, $\lim M_n = 0$. Hence, it follows from (2.2) that $f_n^{(k)} \to f^{(k)}$ uniformly on $\bar{B}(a; r)$. Now if K is an arbitrary compact subset of G and $0 < r < d(K, \partial G)$ then there are $a_1, \ldots, a_n$ in K such that $K \subset \bigcup_{j=1}^{n} B(a_j; r)$. Since $f_n^{(k)} \to f^{(k)}$ uniformly on each $B(a_j; r)$, the convergence is uniform on K. ∎

We will always assume that the metric on $H(G)$ is the metric which it inherits as a subset of $C(G, \mathbb{C})$. The next result follows because $C(G, \mathbb{C})$ is complete.

2.3 Corollary. $H(G)$ *is a complete metric space.*

2.4 Corollary. *If* $f_n: G \to \mathbb{C}$ *is analytic and* $\sum\limits_{n=1}^{\infty} f_n(z)$ *converges uniformly on compact sets to* $f(z)$ *then*

$$f^{(k)}(z) = \sum_{n=1}^{\infty} f_n^{(k)}(z).$$

It should be pointed out that the above theorem has no analogue in the theory of functions of a real variable. For example it is easy to convince oneself by drawing pictures that the absolute value function can be obtained as the uniform limit of a sequence of differentiable functions. Also, it can be shown (using a Theorem of Weierstrass) that a continuous nowhere differentiable function on [0, 1] is the limit of a sequence of polynomials. Surely this is the most emphatic contradiction of the corresponding theorem for Real Variables. A contradiction in another direction is furnished by the following. Let $f_n(x) = \dfrac{1}{n} x^n$ for $0 \le x \le 1$. Then $0 = u-\lim f_n$; however the sequence of derivatives $\{f_n'\}$ does not converge uniformly on [0, 1].

To further illustrate how special analytic functions are, let us examine a result of A. Hurwitz. As a consequence it follows that if $f_n \to f$ and each f_n never vanishes then either $f \equiv 0$ or f never vanishes.

2.5 Hurwitz's Theorem. *Let G be a region and suppose the sequence $\{f_n\}$ in $H(G)$ converges to f. If $f \not\equiv 0$, $\bar{B}(a; R) \subset G$, and $f(z) \ne 0$ for $|z-a| = R$ then there is an integer N such that for $n \ge N$, f and f_n have the same number of zeros in $B(a; R)$.*

Proof. Since $f(z) \ne 0$ for $|z-a| = R$,

$$\delta = \inf \{|f(z)|: |z-a| = R\} > 0.$$

But $f_n \to f$ uniformly on $\{z : |z - a| = R\}$ so there is an integer N such that if $n \ge N$ and $|z - a| = R$ then

$$|f(z) - f_n(z)| < \frac{1}{2}\delta < |f(z)| \le |f(z)| + |f_n(z)|.$$

Hence Rouché's Theorem (V.3.8) implies that f and f_n have the same number of zeros in $B(a; R)$. ∎

2.6 Corollary. *If $\{f_n\} \subset H(G)$ converges to f in $H(G)$ and each f_n never vanishes on G then either $f \equiv 0$ or f never vanishes.*

In order to discuss normal families in $H(G)$ the following terminology is needed.

2.7 Definition. A set $\mathscr{F} \subset H(G)$ is *locally bounded* if for each point a in G there are constants M and $r > 0$ such that for all f in $\mathscr{F}$,

$$|f(z)| \leq M, \text{ for } |z-a| < r.$$

Alternately, $\mathscr{F}$ is locally bounded if there is an $r > 0$ such that

$$\sup \{|f(z)|: |z-a| < r, f \in \mathscr{F}\} < \infty.$$

That is, $\mathscr{F}$ is locally bounded if about each point a in G there is a disk on which $\mathscr{F}$ is uniformly bounded. This immediately extends to the requirement that $\mathscr{F}$ be uniformly bounded on compact sets in G.

2.8 Lemma. *A set $\mathscr{F}$ in $H(G)$ is locally bounded iff for each compact set $K \subset G$ there is a constant M such that*

$$|f(z)| \leq M$$

for all f in $\mathscr{F}$ and z in K.

The proof is left to the reader.

2.9 Montel's Theorem. *A family $\mathscr{F}$ in $H(G)$ is normal iff $\mathscr{F}$ is locally bounded.*

Proof. Suppose $\mathscr{F}$ is normal but fails to be locally bounded; then there is a compact set $K \subset G$ such that $\sup \{|f(z)|: z \in K, f \in \mathscr{F}\} = \infty$. That is, there is a sequence $\{f_n\}$ in $\mathscr{F}$ such that $\sup \{|f_n(z)|: z \in K\} \geq n$. Since $\mathscr{F}$ is normal there is a function f in $H(G)$ and a subsequence $\{f_{n_k}\}$ such that $f_{n_k} \to f$. But this gives that $\sup \{|f_{n_k}(z)-f(z)|: z \in K\} \to 0$ as $k \to \infty$. If $|f(z)| \leq M$ for z in K,

$$n_k \leq \sup \{|f_{n_k}(z)-f(z)|: z \in K\}+M;$$

since the right hand side converges to M, this is a contradiction.

Now suppose $\mathscr{F}$ is locally bounded; the Ascoli-Arzela Theorem (1.23) will be used to show that $\mathscr{F}$ is normal. Since condition (a) of Theorem 1.23 is clearly satisfied, we must show that $\mathscr{F}$ is equicontinuous at each point of G. Fix a point a in G and $\epsilon > 0$; from the hypothesis there is an $r > 0$ and $M > 0$ such that $\bar{B}(a; r) \subset G$ and $|f(z)| \leq M$ for all z in $\bar{B}(a; r)$ and for all f in $\mathscr{F}$. Let $|z - a| < \frac{1}{2}r$ and $f \in \mathscr{F}$; then using Cauchy's Formula with $\gamma(t) = a + re^{it}, 0 \leq t \leq 2\pi$,

$$|f(a)-f(z)| \leq \frac{1}{2\pi} \left| \int_\gamma \frac{f(w)(a-z)}{(w-a)(w-z)} dw \right|$$

$$\leq \frac{4M}{r} |a-z|$$

Letting $\delta < \min \left\{ \frac{1}{2}r, \frac{r}{4M} \epsilon \right\}$ it follows that $|a-z| < \delta$ gives $|f(a)-f(z)| < \epsilon$ for all f in $\mathscr{F}$. ∎

2.10 Corollary. *A set $\mathscr{F} \subset H(G)$ is compact iff it is closed and locally bounded.*

Exercises

1. Let $f, f_1, f_2, \ldots$ be elements of $H(G)$ and show that $f_n \to f$ iff for each closed rectifiable curve γ in G, $f_n(z) \to f(z)$ uniformly for z in $\{\gamma\}$.

2. Let G be a region, let $a \in \mathbb{R}$, and suppose that $f: [a, \infty] \times G \to \mathbb{C}$ is a continuous function. Define the integral $F(z) = \int_a^\infty f(t, z)dt$ to be *uniformly convergent on compact subsets* of G if $\lim_{b \to \infty} \int_a^b f(t, z)dt$ exists uniformly for z in any compact subset of G. Suppose that this integral does converge uniformly on compact subsets of G and that for each t in (a, ∞), $f(t, \cdot)$ is analytic on G. Prove that F is analytic and

$$F^{(k)}(z) = \int\limits_a^\infty \frac{\partial^k f(t, z)}{\partial z^k} \, dt$$

3. The proof of Montel's Theorem can be broken up into the following sequence of definitions and propositions: (a) Definition. A set $\mathscr{F} \subset C(G, \mathbb{C})$ is *locally Lipschitz* if for each a in G there are constants M and $r > 0$ such that $|f(z) - f(a)| \le M|z - a|$ for all f in $\mathscr{F}$ and $|z - a| < r$. (b) If $\mathscr{F} \subset C(G, \mathbb{C})$ is locally Lipschitz then $\mathscr{F}$ is equicontinuous at each point of G. (c) If $\mathscr{F} \subset H(G)$ is locally bounded then $\mathscr{F}$ is locally Lipschitz.

4. Prove Vitali's Theorem: If G is a region and $\{f_n\} \subset H(G)$ is locally bounded and $f \in H(G)$ that has the property that $A = \{z \in G : \lim f_n(z) = f(z)\}$ has a limit point in G then $f_n \to f$.

5. Show that for a set $\mathscr{F} \subset H(G)$ the following are equivalent conditions:
   (a) $\mathscr{F}$ is normal;
   (b) For every $\epsilon > 0$ there is a number $c > 0$ such that $\{cf: f \in \mathscr{F}\} \subset B(0; \epsilon)$ (here $B(0; \epsilon)$ is the ball in $H(G)$ with center at 0 and radius ϵ).

6. Show that if $\mathscr{F} \subset H(G)$ is normal then $\mathscr{F}' = \{f': f \in \mathscr{F}\}$ is also normal. Is the converse true? Can you add something to the hypothesis that $\mathscr{F}'$ is normal to insure that $\mathscr{F}$ is normal?

7. Suppose $\mathscr{F}$ is normal in $H(G)$ and Ω is open in $\mathbb{C}$ such that $f(G) \subset \Omega$ for every f in $\mathscr{F}$. Show that if g is analytic on Ω and is bounded on bounded sets then $\{g \circ f: f \in \mathscr{F}\}$ is normal.

8. Let $D = \{z: |z| < 1\}$ and show that $\mathscr{F} \subset H(D)$ is normal iff there is a sequence $\{M_n\}$ of positive constants such that $\limsup \sqrt[n]{M_n} \le 1$ and if $f(z) = \sum_0^\infty a_n z^n$ is in $\mathscr{F}$ then $|a_n| \le M_n$ for all n.

9. Let $D = B(0; 1)$ and for $0 < r < 1$ let $\gamma_r(t) = re^{2\pi i t}$, $0 \le t \le 1$. Show that a sequence $\{f_n\}$ in $H(D)$ converges to f iff $\int_{\gamma_r} |f(z) - f_n(z)| \, |dz| \to 0$ as $n \to \infty$ for each r, $0 < r < 1$.

10. Let $\{f_n\} \subset H(G)$ be a sequence of one-one functions which converge to f. Show that either f is one-one or f is a constant function.

11. Suppose that $\{f_n\}$ is a sequence in $H(G)$, f is a non-constant function, and $f_n \to f$ in $H(G)$. Let $a \in G$ and $\alpha = f(a)$; show that there is a sequence $\{a_n\}$ in G such that: (i) $a = \lim a_n$; (ii) $f_n(a_n) = \alpha$ for sufficiently large n.

12. Show that $\lim \tan nz = -i$ uniformly for z in any compact subset of $G = \{z: \operatorname{Im} z > 0\}$.

13. (a) Show that if f is analytic on an open set containing the disk $\bar{B}(a; R)$ then

$$|f(a)|^2 \leq \frac{1}{\pi R^2} \int_0^{2\pi} \int_0^R |f(a+re^{i\theta})|^2 \, r\,dr\,d\theta.$$

(b) Let G be a region and let M be a fixed positive constant. Let $\mathscr{F}$ be the family of all functions f in $H(G)$ such that $\int\int_G |f(z)|^2 \, dx\,dy \leq M$. Show that $\mathscr{F}$ is normal.

§3. Spaces of meromorphic functions

If G is a region and f is a meromorphic function on G, and if $f(z) = \infty$ whenever z is a pole of f then $f: G \to \mathbb{C}_\infty$ is a continuous function (Exercise V. 3.4). If $M(G)$ is the set of all meromorphic functions on G then consider $M(G)$ as a subset of $C(G, \mathbb{C}_\infty)$ and endow it with the metric of $C(G, \mathbb{C}_\infty)$. In this section this metric space will be discussed as $H(G)$ was discussed in the previous section.

Recall from Chapter I that the metric d is defined on $\mathbb{C}_\infty$ as follows: for z_1 and z_2 in $\mathbb{C}$

$$d(z_1, z_2) = \frac{2|z_1 - z_2|}{[(1+|z_1|^2)(1+|z_2|^2)]^{\frac{1}{2}}} \, ;$$

and for z in $\mathbb{C}$

$$d(z, \infty) = \frac{2}{(1+|z|^2)^{\frac{1}{2}}} \, .$$

Notice that for non zero complex numbers z_1 and z_2,

3.1 $$d(z_1, z_2) = d\left(\frac{1}{z_1}, \frac{1}{z_2}\right) ;$$

and for $z \neq 0$

3.2 $$d(z, 0) = d\left(\frac{1}{z}, \infty\right) .$$

Also recall that if $\{z_n\}$ is a sequence in $\mathbb{C}$ and $z \in \mathbb{C}$ that satisfies $d(z, z_n) \to 0$ then $|z - z_n| \to 0$.

Some facts about the relationship between the metric spaces $\mathbb{C}$ and $\mathbb{C}_\infty$ are summarized in the next proposition. In order to avoid confusion $B(a; r)$ will be used to designate a ball in $\mathbb{C}$ and $B_\infty(a; r)$ to designate a ball in $\mathbb{C}_\infty$.

3.3 Proposition. (a) *If a is in $\mathbb{C}$ and $r > 0$ then there is a number $\rho > 0$ such that $B_\infty(a; \rho) \subset B(a; r)$.*

(b) *Conversely, if $\rho > 0$ is given and $a \in \mathbb{C}$ then there is a number $r > 0$ such that $B(a; r) \subset B_\infty(a; \rho)$.*

(c) *If $\rho > 0$ is given then there is a compact set $K \subset \mathbb{C}$ such that $\mathbb{C}_\infty - K \subset B_\infty(\infty; \rho)$.*

(d) *Conversely, if a compact set $K \subset \mathbb{C}$ is given, there is a number $\rho > 0$ such that $B_\infty(\infty; \rho) \subset \mathbb{C}_\infty - K$.*

The proof is left to the reader.

The first observation is that $M(G)$ is not complete. In fact if $f_n(z) \equiv n$ then $\{f_n\}$ is a Cauchy sequence in $M(G)$. But $\{f_n\}$ converges to the function which is identically ∞ in $C(G, \mathbb{C}_\infty)$ and this is not meromorphic.

However this is the worst that can happen.

3.4 Theorem. *Let $\{f_n\}$ be a sequence in $M(G)$ and suppose $f_n \to f$ in $C(G, \mathbb{C}_\infty)$. Then either f is meromorphic or $f \equiv \infty$. If each f_n is analytic then either f is analytic or $f \equiv \infty$.*

Proof. Suppose there is a point a in G with $f(a) \neq \infty$ and put $M = |f(a)|$. Using part (a) of Proposition 3.3 we can find a number $\rho > 0$ such that $B_\infty(f(a); \rho) \subset B(f(a); M)$. But since $f_n \to f$ there is an integer n_0 such that $d(f_n(a), f(a)) < \frac{1}{2}\rho$ for all $n \geq n_0$. Also $\{f, f_1, f_2, \ldots\}$ is compact in $C(G, \mathbb{C}_\infty)$ so that it is equicontinuous. That is, there is an $r > 0$ such that $|z - a| < r$ implies $d(f_n(z), f_n(a)) < \frac{1}{2}\rho$. That gives that $d(f_n(z), f(a)) \leq \rho$ for $|z - a| \leq r$ and for $n \geq n_0$. But by the choice of ρ, $|f_n(z)| \leq |f_n(z) - f(a)| + |f(a)| \leq 2M$ for all z in $\bar{B}(a; r)$ and $n \geq n_0$. But then (from the formula for the metric d)

$$\frac{2}{(1 + 4M^2)} |f_n(z) - f(z)| \leq d(f_n(z), f(z))$$

for z in $\bar{B}(a; r)$ and $n \geq n_0$. Since $d(f_n(z), f(z)) \to 0$ uniformly for z in $\bar{B}(a; r)$, this gives that $|f_n(z) - f(z)| \to 0$ uniformly for z in $\bar{B}(a; r)$. Since the tail end of the sequence $\{f_n\}$ is bounded on $B(a; r)$, f_n has no poles and must be analytic near $z = a$ for $n \geq n_0$. It follows that f is analytic in a disk about a.

Now suppose there is a point a in G with $f(a) = \infty$. For a function g in $C(G, \mathbb{C}_\infty)$ define $\frac{1}{g}$ by $\left(\frac{1}{g}\right)(z) = \frac{1}{g(z)}$ if $g(z) \neq 0$ or ∞; $\left(\frac{1}{g}\right)(z) = 0$ if $g(z) = \infty$; and $\left(\frac{1}{g}\right)(z) = \infty$ if $g(z) = 0$. It follows that $\frac{1}{g} \in C(G, \mathbb{C}_\infty)$. Also, since $f_n \to f$ in $C(G, \mathbb{C}_\infty)$ it follows from formulas (3.1) and (3.2) that $\frac{1}{f_n} \to \frac{1}{f}$ in $C(G, \mathbb{C}_\infty)$. Now each function $\frac{1}{f_n}$ is meromorphic on G; so the preceding paragraph gives a number $r > 0$ and an integer n_0 such that $\frac{1}{f}$ and $\frac{1}{f_n}$ are analytic on $B(a; r)$ for $n \geq n_0$ and $\frac{1}{f_n} \to \frac{1}{f}$ uniformly on $B(a; r)$. From

Hurwitz's Theorem (2.5) either $\dfrac{1}{f} \equiv 0$ or $\dfrac{1}{f}$ has isolated zeros in $B(a; r)$. So

if $f \not\equiv \infty$ then $\dfrac{1}{f} \not\equiv 0$ and f must be meromorphic in $B(a; r)$. Combining this

with the first part of the proof we have that f is meromorphic in G if f is not identically infinite.

If each f_n is analytic then $\dfrac{1}{f_n}$ has no zeros in $B(a; r)$. It follows from

Corollary 2.6 to Hurwitz's Theorem that either $\dfrac{1}{f} \equiv 0$ or $\dfrac{1}{f}$ never vanishes.

But since $f(a) = \infty$ we have that $\dfrac{1}{f}$ has at least one zero; thus $f \equiv \infty$ in

$B(a; r)$. Combining this with the first part of the proof we see that $f \equiv \infty$ or f is analytic. ∎

3.5 Corollary. $M(G) \cup \{\infty\}$ *is a complete metric space.*

3.6 Corollary. $H(G) \cup \{\infty\}$ *is closed in* $C(G, \mathbb{C}_\infty)$.

To discuss normality in $M(G)$ one must introduce the quantity

$$\frac{2|f'(z)|}{1+|f(z)|^2},$$

for each meromorphic function f. However if z is a pole of f then the above expression is meaningless since $f'(z)$ has no meaning. To rectify this take the limit of the above expression as z approaches the pole. To show that the limit exists let a be a pole of f of order $m \geq 1$; then

$$f(z) = g(z) + \frac{A_m}{(z-a)^m} + \ldots + \frac{A_1}{(z-a)}$$

for z in some disk about a and g analytic in that disk. For $z \neq a$

$$f'(z) = g'(z) - \left[\frac{mA_m}{(z-a)^{m+1}} + \ldots + \frac{A_1}{(z-a)^2} \right]$$

Thus

$$\frac{2|f'(z)|}{1+|f(z)|^2} = \frac{2\left| \dfrac{mA_m}{(z-a)^{m+1}} + \ldots + \dfrac{A_1}{(z-a)^2} - g'(z) \right|}{1 + \left| \dfrac{A_m}{(z-a)^m} + \ldots + \dfrac{A_1}{(z-a)} + g(z) \right|^2}$$

$$= \frac{2|z-a|^{m-1}|mA_m + \cdots + A_1(z-a)^{m-1} - g'(z)(z-a)^{m+1}|}{|z-a|^{2m} + |A_m + \cdots + A_1(z-a)^{m-1} + g(z)(z-a)^m|^2}$$

So if $m \geq 2$

$$\lim_{z \to a} \frac{2|f'(z)|}{1+|f(z)|^2} = 0$$

If $m = 1$ then

$$\lim_{z \to a} \frac{2|f'(z)|}{1+|f(z)|^2} = \frac{2}{|A_1|}.$$

3.7 Definition. If f is a meromorphic function on the region G then define $\mu(f): G \to \mathbb{R}$ by

$$\mu(f)(z) = \frac{2|f'(z)|}{1+|f(z)|^2}$$

whenever z is not a pole of f, and

$$\mu(f)(a) = \lim_{z \to a} \frac{2|f'(z)|}{1+|f(z)|^2}$$

if a is a pole of f.

It follows that $\mu(f) \in C(G, \mathbb{C})$.

The reason for introducing $\mu(f)$ is as follows: If $f: G \to \mathbb{C}_\infty$ is meromorphic then for z close to z' we have that $d(f(z), f(z'))$ is approximated by $\mu(f)(z)|z-z'|$. So if a bound can be obtained for $\mu(f)$ then f is a Lipschitz function. If f belongs to a family of functions and $\mu(f)$ is uniformly bounded for f in this family, then the family is a uniformly Lipschitz set of functions. This is made precise in the following proof.

3.8 Theorem. *A family $\mathscr{F} \subset M(G)$ is normal in $C(G, \mathbb{C}_\infty)$ iff $\mu(\mathscr{F}) \equiv \{\mu(f): f \in \mathscr{F}\}$ is locally bounded.*

Note. If $f_n(z) = nz$ for $n \geq 1$ then $\mu(f_n)(z) = \dfrac{2n}{1+n^2|z|^2}$. Thus $\mathscr{F} = \{f_n\}$ is normal in $C(G, \mathbb{C}_\infty)$ and $\mu(\mathscr{F})$ is locally bounded. However, $\mathscr{F}$ is not normal in $M(G)$ since the sequence $\{f_n\}$ converges to the constantly infinite function which does not belong to $M(G)$.

Proof of Theorem 3.8. We will assume that $\mu(\mathscr{F})$ is locally bounded and prove that $\mathscr{F}$ is normal by applying the Arzela-Ascoli Theorem. Since $\mathbb{C}_\infty$ is compact it suffices to show that $\mathscr{F}$ is equicontinuous at each point of G. So let K be an arbitrary closed disk contained in G and let M be a constant with $\mu(f)(z) \leq M$ for all z in K and all f in $\mathscr{F}$. Let z and z' be arbitrary points in K.

Suppose neither z nor z' are poles of a fixed function f in $\mathscr{F}$ and let $\alpha > 0$ be an arbitrary number. Choose points $w_0 = z, w_1, \ldots, w_n = z'$ in K which satisfy the following conditions:

3.9 w in $[w_{k-1}, w_k]$ implies w is not a pole of f;

3.10 $\displaystyle\sum_{k=1}^{n} |w_k - w_{k-1}| \le 2|z - z'|;$

3.11 $\left| \dfrac{(1 + |f(w_{k-1})|^2)}{\left[(1 + |f(w_k)|^2)(1 + |f(w_{k-1})|^2)\right]^{\frac{1}{2}}} - 1 \right| < \alpha, \quad 1 \le k \le n;$

3.12 $\left| \dfrac{f(w_k) - f(w_{k-1})}{w_k - w_{k-1}} - f'(w_{k-1}) \right| < \alpha, \quad 1 \le k \le n.$

To see that such points can be found select a polygonal path P in K satisfying (3.9) and (3.10). Cover P by small disks in which conditions similar to (3.11) and (3.12) hold, choose a finite subcover, and then pick points $w_0, \dots, w_n$ on P such that each segment $[w_{k-1}, w_k]$ lies in one of these disks. Then $\{w_0, \dots, w_n\}$ will satisfy all of these conditions. If $\beta_k = [(1 + |f(w_{k-1})|^2)(1 + |f(w_k)|^2)]^{\ddagger}$ then

$$d(f(z), f(z')) \le \sum_{k=1}^{n} d(f(w_{k-1}), f(w_k))$$

$$= \sum_{k=1}^{n} \frac{2}{\beta_k} |f(w_k) - f(w_{k-1})|$$

$$\le \sum_{k=1}^{n} \frac{2}{\beta_k} \left| \frac{f(w_k) - f(w_{k-1})}{w_k - w_{k-1}} - f'(w_{k-1}) \right| |w_k - w_{k-1}|$$

$$+ \sum_{k=1}^{n} \frac{2}{\beta_k} |f'(w_{k-1})| |w_k - w_{k-1}|$$

Using the fact that $2|f'(w_k)| \le M(1 + |f(w_k)|^2)$ and the conditions on $w_0, \dots, w_n$ this becomes

$$d(f(z), f(z')) \le 2\alpha \sum_{k=1}^{n} \frac{1}{\beta_k} |w_k - w_{k-1}| + M \sum_{k=1}^{n} \left(\frac{1 + |f(w_{k-1})|^2}{\beta_k} \right) |w_k - w_{k-1}|$$

$$\le (4\alpha + 2\alpha M) |z - z'| + \sum_{k=1}^{n} M |w_k - w_{k-1}|$$

$$\le (4\alpha + 2\alpha M + 2M)|z - z'|$$

Since $\alpha > 0$ was arbitrary this gives that if z and z' are not poles of f then

3.13 $$d(f(z), f(z')) \le 2M|z - z'|.$$

Now suppose z' is a pole of f but z is not. If w is in K and is not a pole then it follows from (3.13) that

$$d(f(z), \infty) \le d(f(z), f(w)) + d(f(w), \infty)$$

$$\le 2M|z - w| + d(f(w), \infty).$$

Since it is possible to let w approach z' without w ever being a pole of f (poles are isolated!), this gives that $f(w) \to f(z') = \infty$ and $|z - w| \to |z - z'|$. Thus (3.13) holds if at most one of z and z' is a pole. But a similar procedure gives that (3.13) holds for all z and z' in K. So if $K = B(a; r)$ and $\epsilon > 0$ are given then for $\delta < \min \{r, \epsilon/2M\}$ we have that $|z - a| < \delta$ implies $d(f(z), f(a)) < \epsilon$, and δ is independent of f in $\mathscr{F}$. This gives that $\mathscr{F}$ is equicontinuous at each point a in G.

The proof of the converse is left to the reader. ∎

Exercises

1. Prove Proposition 3.3.
2. Show that if $\mathscr{F} \subset M(G)$ is a normal family in $C(G, \mathbb{C}_\infty)$ then $\mu(\mathscr{F})$ is locally bounded.

§4. The Riemann Mapping Theorem

We wish to define an equivalence relation between regions in $\mathbb{C}$. After doing this it will be shown that all proper simply connected regions in $\mathbb{C}$ are equivalent to the open disk $D = \{z: |z| < 1\}$, and hence are equivalent to one another.

4.1 Definition. A region G_1 is *conformally equivalent* to G_2 if there is analytic function $f: G_1 \to \mathbb{C}$ such that f is one-one and $f(G_1) = G_2$. Clearly, this is an equivalence relation.

It is immediate that $\mathbb{C}$ is not equivalent to any bounded region by Liouville's Theorem. Also it is easy to show from the definitions that if G_1 is simply connected and G_1 is equivalent to G_2 then G_2 must be simply connected. If f is the principal branch of the square root then f is one-one and shows that $\mathbb{C} - \{z: z \le 0\}$ is equivalent to the right half plane.

4.2 Riemann Mapping Theorem. *Let G be a simply connected region which is not the whole plane and let $a \in G$. Then there is a unique analytic function $f: G \to \mathbb{C}$ having the properties:*

(a) $f(a) = 0$ and $f'(a) > 0$;
(b) f is one-one;
(c) $f(G) = \{z: |z| < 1\}$.

The proof that the function f is unique is rather easy. In fact, if g also has the properties of f and $D = \{z: |z| < 1\}$ then $f \circ g^{-1}: D \to D$ is analytic, one-one, and onto. Also $f \circ g^{-1}(0) = f(a) = 0$ so Theorem VI. 2.5 implies there is a constant c with $|c| = 1$ and $f \circ g^{-1}(z) = cz$ for all z. But then $f(z) = cg(z)$ gives that $0 < f'(a) = cg'(a)$; since $g'(a) > 0$ it follows that $c = 1$, or $f = g$.

To motivate the proof of the existence of f, consider the family $\mathscr{F}$ of all analytic functions f having properties (a) and (b) and satisfying $|f(z)| < 1$ for z in G. The idea is to choose a member of $\mathscr{F}$ having property (c). Suppose

$\{K_n\}$ is a sequence of compact subsets of G such that $\bigcup\limits_{n=1}^{\infty} K_n = G$ and $a \in K_n$ for each n. Then $\{f(K_n)\}$ is a sequence of compact subsets of $D = \{z: |z| < 1\}$. Also, as n becomes larger $f(K_n)$ becomes larger and larger and tries to fill out the disk D. By choosing a function f in $\mathscr{F}$ with the largest possible derivative at a, we choose the function which "starts out the fastest" at $z = a$. It thus has the best possible chance of finishing first; that is, of having $\bigcup\limits_{n=1}^{\infty} f(K_n) = D$.

Before carrying out this proof, it is necessary for future developments to point out that the only property of a simply connected region which will be used is the fact that every non-vanishing analytic function has an analytic square root. (Actually it will be proved in Theorem VIII. 2.2 that this property is equivalent to simple connectedness.) So the Riemann Mapping Theorem will be completely proved by proving the following.

4.3 Lemma. *Let G be a region which is not the whole plane and such that every non-vanishing analytic function on G has an analytic square root. If $a \in G$ then there is an analytic function f on G such that*:

(a) $f(a) = 0$ and $f'(a) > 0$;
(b) f is one-one;
(c) $f(G) = D = \{z: |z| < 1\}$.

Proof. Define $\mathscr{F}$ by letting

$$\mathscr{F} = \{f \in H(G): f \text{ is one-one}, f(a) = 0, f'(a) > 0, f(G) \subset D\}$$

Since $f(G) \subset D$, $\sup \{|f(z)|: z \in G\} \le 1$ for f in $\mathscr{F}$; by Montel's Theorem $\mathscr{F}$ is normal if it is non-empty. So the first fact to be proved is

4.4 $\mathscr{F} \ne \square$.

It will be shown that

4.5 $\mathscr{F}^- = \mathscr{F} \cup \{0\}$.

Once these facts are known the proof can be completed. Indeed, suppose (4.4) and (4.5) hold and consider the function $f \to f'(a)$ of $H(G) \to \mathbb{C}$. This is a continuous function (Theorem 2.1) and, since $\mathscr{F}^-$ is compact, there is an f in $\mathscr{F}^-$ with $f'(a) \ge g'(a)$ for all g in $\mathscr{F}$. Because $\mathscr{F} \ne \square$, (4.5) implies that $f \in \mathscr{F}$. It remains to show that $f(G) = D$. Suppose $\omega \in D$ such that $\omega \notin f(G)$. Then the function

$$\frac{f(z) - \omega}{1 - \bar{\omega} f(z)}$$

is analytic in G and never vanishes. By hypothesis there is an analytic function $h: G \to \mathbb{C}$ such that

4.6 $[h(z)]^2 = \dfrac{f(z) - \omega}{1 - \bar{\omega} f(z)}.$

Since the Mobius transformation $T\zeta = \dfrac{\zeta - \omega}{1 - \bar{\omega}\zeta}$ maps D onto D, $h(G) \subset D$.

Define $g\colon G \to \mathbb{C}$ by

$$g(z) = \frac{|h'(a)|}{h'(a)} \frac{h(z) - h(a)}{1 - \overline{h(a)}h(z)}$$

Then $g(G) \subset D$, $g(a) = 0$, and g is one-one (why?). Also

$$g'(a) = \frac{|h'(a)|}{h'(a)} \cdot \frac{h'(a)\,[1 - |h(a)|^2]}{[1 - |h(a)|^2]^2}$$

$$= \frac{|h'(a)|}{1 - |h(a)|^2}$$

But $|h(a)|^2 = |-\omega| = |\omega|$ and differentiating (4.6) gives (since $f(a) = 0$) that

$$2h(a)h'(a) = f'(a)\,(1 - |\omega|^2).$$

Therefore

$$g'(a) = \frac{f'(a)\,(1 - |\omega|^2)}{2\sqrt{|\omega|}} \cdot \frac{1}{1 - |\omega|}$$

$$= f'(a)\left(\frac{1 + |\omega|}{2\sqrt{|\omega|}}\right)$$

$$> f'(a)$$

This gives that g is in $\mathscr{F}$ and contradicts the choice of f. Thus it must be that $f(G) = D$.

Now to establish (4.4) and (4.5). Since $G \neq \mathbb{C}$, let $b \in \mathbb{C} - G$ and let g be a function analytic on G such that $[g(z)]^2 = z - b$. If z_1 and z_2 are points in G and $g(z_1) = \pm g(z_2)$ then it follows that $z_1 = z_2$. In particular, g is one-one. By the Open Mapping Theorem there is a number $r > 0$ such that

4.7 $g(G) \supset B(g(a); r)$

So if there is a point z in G such that $g(z) \in B(-g(a); r)$ then $r > |g(z) + g(a)|$ $= |-g(z) - g(a)|$. According to (4.7) there is a w in G with $g(w) = -g(z)$; but the remarks preceding (4.7) show that $w = z$ which gives $g(z) = 0$. But then $z - b = [g(z)]^2 = 0$ implies b is in G, a contradiction. Hence

4.8 $g(G) \cap \{\zeta \colon |\zeta + g(a)| < r\} = \square.$

Let U be the disk $\{\zeta \colon |\zeta + g(a)| < r\} = B(-g(a); r)$. There is a Mobius transformation T such that $T(\mathbb{C}_\infty - U^-) = D$. Let $g_1 = T \circ g$; then g_1 is analytic and $g_1(G) \subset D$. If $\alpha = g_1(a)$ then let $g_2(z) = \varphi_\alpha \circ g_1(z)$; so we still have that $g_2(G) \subset D$ and g_2 is analytic, but we also have that $g_2(a) = 0$. Now it is a simple matter to find a complex number c, $|c| = 1$, such that $g_3(z) = cg_2(z)$ has positive derivative at $z = a$ and is, therefore, in $\mathscr{F}$. This establishes (4.4).

Suppose $\{f_n\}$ is a sequence in $\mathscr{F}$ and $f_n \to f$ in $H(G)$. Clearly $f(a) = 0$ and since $f_n'(a) \to f'(a)$ it follows that

4.9 $f'(a) \geq 0.$

Let z_1 be an arbitrary element of G and put $\zeta = f(z_1)$; let $\zeta_n = f_n(z_1)$. Let $z_2 \in G$, $z_2 \neq z_1$ and let K be a closed disk centered at z_2 such that $z_1 \notin K$. Then $f_n(z) - \zeta_n$ never vanishes on K since f_n is one-one. But $f_n(z) - \zeta_n \to f(z) - \zeta$ uniformly on K, so Hurwitz's Theorem gives that $f(z) - \zeta$ never vanishes on K or $f(z) \equiv \zeta$. If $f(z) \equiv \zeta$ on K then f is the constant function ζ throughout G; since $f(a) = 0$ we have that $f(z) \equiv 0$. Otherwise we get that $f(z_2) \neq f(z_1)$ for $z_2 \neq z_1$; that is, f is one-one. But if f is one-one then f' can never vanish; so (4.9) implies that $f'(a) > 0$ and f is in $\mathscr{F}$. This proves (4.5) and the proof of the lemma is complete. ∎

4.10 Corollary. *Among the simply connected regions there are only two equivalence classes; one consisting of $\mathbb{C}$ alone and the other containing all the proper simply connected regions.*

Exercises

1. Let G and Ω be open sets in the plane and let $f: G \to \Omega$ be a continuous function which is one-one, onto, and such that $f^{-1}: \Omega \to G$ is also continuous (a homeomorphism). Suppose $\{z_n\}$ is a sequence in G which converges to a point z in ∂G; also suppose that $w = \lim f(z_n)$ exists. Prove that $w \in \partial \Omega$.

2. (a) Let G be a region, let $a \in G$ and suppose that $f: (G - \{a\}) \to \mathbb{C}$ is an analytic function such that $f(G - \{a\}) = \Omega$ is bounded. Show that f has a removable singularity at $z = a$. If f is one-one, show that $f(a) \in \partial \Omega$.

   (b) Show that there is no one-one analytic function which maps $G = \{z: 0 < |z| < 1\}$ onto an annulus $\Omega = \{z: r < |z| < R\}$ where $r > 0$.

3. Let G be a simply connected region which is not the whole plane and suppose that $\bar{z} \in G$ whenever $z \in G$. Let $a \in G \cap \mathbb{R}$ and suppose that $f: G \to D = \{z: |z| < 1\}$ is a one-one analytic function with $f(a) = 0$, $f'(a) > 0$ and $f(G) = D$. Let $G_+ = \{z \in G: \text{Im } z > 0\}$. Show that $f(G_+)$ must lie entirely above or entirely below the real axis.

4. Find an analytic function f which maps $\{z: |z| < 1, \text{Re } z > 0\}$ onto $B(0; 1)$ in a one-one fashion.

5. Let f be analytic on $G = \{z: \text{Re } z > 0\}$, one-one, with $\text{Re } f(z) > 0$ for all z in G, and $f(a) = a$ for some real number a. Show that $|f'(a)| \leq 1$.

6. Let G_1 and G_2 be simply connected regions neither of which is the whole plane. Let f be a one-one analytic mapping of G_1 onto G_2. Let $a \in G_1$ and put $\alpha = f(a)$. Prove that for any one-one analytic map h of G_1 into G_2 with $h(a) = \alpha$ it follows that $|h'(a)| \leq |f'(a)|$. Suppose h is not assumed to be one-one; what can be said?

7. Let G be a simply connected region and suppose that G is not the whole plane. Let $\Delta = \{\xi: |\xi| < 1\}$ and suppose that f is an analytic, one-one map of G onto Δ with $f(a) = 0$ and $f'(a) > 0$ for some point a in G. Let g be any other analytic, one-one map of G onto Δ and express g in terms of f.

8. Let r_1, r_2, R_1, R_2, be positive numbers such that $R_1/r_1 = R_2/r_2$; show that ann $(0; r_1, R_1)$ and ann $(0; r_2, R_2)$ are conformally equivalent.

9. Show that there is an analytic function f defined on $g = \text{ann}(0; 0, 1)$ such that f' never vanishes and $f(G) = B(0; 1)$.

§5. The Weierstrass Factorization Theorem

The notion of convergence in $H(G)$ can be used to solve the following problem. Given a sequence $\{a_k\}$ in G which has no limit point in G and a sequence of integers $\{m_k\}$, is there a function f which is analytic on G and such that the only zeros of f are at the points a_k, with the multiplicity of the zero at a_k equal to m_k? The answer to the question is yes and the result is due to Weierstrass.

If there were only a finite number of points, $a_1, \ldots, a_n$ then $f(z) = (z - a_n)^{m_1} \ldots (z - a_n)^{m_n}$ would be the desired function. What happens if there are infinitely many points in this sequence? To answer this we must discuss the convergence of infinite products of numbers and functions.

Clearly one should define an infinite product of numbers z_n (denoted by $\prod_{n=1}^{\infty} z_n$) as the limit of the finite products. Observe, however, that if one of the numbers z_n is zero, then the limit is zero, regardless of the behavior of the remaining terms of the sequence. This does not present a difficulty, but it shows that when zeros appear, the existence of an infinite product is trivial.

5.1 Definition. If $\{z_n\}$ is a sequence of complex numbers and if $z = \lim_{k=1}^{n} z_k$ exists, then z is the *infinite product* of the numbers z_n and it is denoted by

$$z = \prod_{n=1}^{\infty} z_n.$$

Suppose that no one of the numbers z_n is zero, and that $z = \prod_{n=1}^{\infty} z_n$ exists and is also not zero. Let $p_n = \prod_{k=1}^{n} z_k$ for $n \geq 1$; then no p_n is zero and $\dfrac{p_n}{p_{n-1}} = z_n$. Since $z \neq 0$ and $p_n \to z$ we have that $\lim z_n = 1$. So that except for the cases where zero appears, a necessary condition for the convergence of an infinite product is that the n-th term must go to 1. On the other hand, note that for $z_n = a$ for all n and $|a| < 1$, $\prod z_n = 0$ although $\lim z_n = a \neq 0$.

Because of the fact that the exponential of a sum is the product of the exponentials of the individual terms, it is possible to discuss the convergence of an infinite product (when zero is not involved) by discussing the convergence of the series $\sum \log z_n$, where log is the principal branch of the logarithm. However, before this can be made meaningful the z_n must be restricted so that $\log z_n$ is meaningful. If the product is to be non-zero, then $z_n \to 1$. So it is no restriction to suppose that Re $z_n > 0$ for all n. Now suppose that the series $\sum \log z_n$ converges. If $s_n = \sum_{k=1}^{n} \log z_k$ and $s_n \to s$ then $\exp s_n \to \exp s$. But $\exp s_n = \prod_{k=1}^{n} z_k$ so that $\prod_{n=1}^{\infty} z_n$ is convergent to $z = e^s \neq 0$.

5.2 Proposition. *Let* Re $z_n > 0$ *for all* $n \geq 1$. *Then* $\prod_{n=1}^{\infty} z_n$ *converges to a non zero number iff the series* $\sum_{n=1}^{\infty} \log z_n$ *converges.*

Proof. Let $p_n = (z_1 \cdots z_n)$, $z = re^{i\theta}$, $-\pi < \theta \leq \pi$, and $\ell(p_n) = \log |p_n| + i\theta_n$ where $\theta - \pi < \theta_n \leq \theta + \pi$. If $s_n = \log z_1 + \cdots + \log z_n$ then $\exp(s_n) = p_n$ so that $s_n = \ell(p_n) + 2\pi i k_n$ for some integer k_n. Now suppose that $p_n \to z$. Then $s_n - s_{n-1} = \log z_n \to 0$; also $\ell(p_n) - \ell(p_{n-1}) \to 0$, Hence, $(k_n - k_{n-1}) \to 0$ as $n \to \infty$. Since each k_n is an integer this gives that there is an n_0 and a k such that $k_m = k_n = k$ for m, $n \geq n_0$. So $s_n \to \ell(z) + 2\pi i k$; that is, the series $\sum \log z_n$ converges. Since the converse was proved above, this completes the proof. ∎

Consider the power series expansion of $\log (1+z)$ about $z = 0$:

$$\log (1+z) = \sum_{n=1}^{\infty} (-1)^{n-1} \frac{z^n}{n} = z - \frac{z^2}{2} + \ldots ,$$

which has radius of convergence 1. If $|z| < 1$ then

$$\left| 1 - \frac{\log (1+z)}{z} \right| = |\tfrac{1}{2}z - \tfrac{1}{3}z^2 + \ldots|$$

$$\leq \tfrac{1}{2}(|z| + |z|^2 + \ldots)$$

$$= \tfrac{1}{2} \frac{|z|}{1-|z|}.$$

If we further require $|z| < \tfrac{1}{2}$ then

$$\left| 1 - \frac{\log (1+z)}{z} \right| \leq \tfrac{1}{2}.$$

This gives that for $|z| < \tfrac{1}{2}$

5.3 $\tfrac{1}{2}|z| \leq |\log (1+z)| \leq \tfrac{3}{2}|z|.$

This will be used to prove the following result.

5.4 Proposition. *Let* Re $z_n > -1$; *then the series* $\sum \log (1+z_n)$ *converges absolutely iff the series* $\sum z_n$ *converges absolutely.*

Proof. If $\sum |z_n|$ converges then $z_n \to 0$; so eventually $|z_n| < \tfrac{1}{2}$. By (5.3) $\sum |\log (1+z_n)|$ is dominated by a convergent series, and it must converge also. If, conversely, $\sum |\log (1+z_n)|$ converges, then it follows that $|z_n| < \tfrac{1}{2}$ for sufficiently large n (why?). Again (5.3) allows us to conclude that $\sum |z_n|$ converges. ∎

We wish to define the absolute convergence of an infinite product. The first temptation should be avoided. That is, we do not want to say that $\prod |z_n|$ converges. Why? If $\prod |z_n|$ converges it does not follow that $\prod z_n$ converges. In fact, let $z_n = -1$ for all n; then $|z_n| = 1$ for all n so that $\prod |z_n|$ converges to 1. However $\prod\limits_{k=1}^{n} z_k$ is ± 1 depending on whether n is even or odd, so that $\prod z_n$ does not converge. Thus, if absolute convergence is to imply convergence, we must seek a different definition.

On the basis of Proposition 5.2 the following definition is justified.

5.5 Definition. If Re $z_n > 0$ for all n then the infinite product $\prod z_n$ is said to *converge absolutely* if the series $\sum \log z_n$ converges absolutely.

According to Proposition 5.2 and the fact that absolute convergence of a series implies convergence, we have that absolute convergence of a product implies the convergence of the product. Similarly, if a product converges absolutely then any rearrangement of the terms of the product results in a product which is still absolutely convergent. If we combine Propositions 5.2 and 5.4 with the definition, the following fundamental criterion for convergence of a infinite product is obtained.

5.6 Corollary. *If* Re $z_n > 0$ *then the product* $\prod z_n$ *converges absolutely iff the series* $\sum (z_n - 1)$ *converges absolutely.*

Although the preceding corollary gives a necessary and sufficient condition for the absolute convergence of an infinite product phrased in terms with which we are familiar, it does not give a method for evaluating infinite products in terms of the corresponding infinite series. To evaluate a particular product one must often resort to trickery.

We now apply these results to the convergence of products of functions. A fundamental question to be answered is the following. Suppose $\{f_n\}$ is a sequence of functions on a set X and $f_n(x) \to f(x)$ uniformly for x in X; when will exp $(f_n(x)) \to$ exp $(f(x))$ uniformly for x in X? Below is a partial answer which is sufficient to meet our needs.

5.7 Lemma. *Let X be a set and let $f, f_1, f_2, \ldots$ be functions from X into $\mathbb{C}$ such that $f_n(x) \to f(x)$ uniformly for x in X. If there is a constant a such that* Re $f(x) \le a$ *for all x in X then* exp $f_n(x) \to$ exp $f(x)$ *uniformly for x in X.*

Proof. If $\epsilon > 0$ is given then choose $\delta > 0$ such that $|e^z - 1| < \epsilon e^{-a}$ whenever $|z| < \delta$. Now choose n_0 such that $|f_n(x) - f(x)| < \delta$ for all x in X whenever $n \ge n_0$. Thus

$$\epsilon e^{-a} > |\exp [f_n(x) - f(x)] - 1|$$

$$= \left| \frac{\exp f_n(x)}{\exp f(x)} - 1 \right|$$

It follows that for any x in X and for $n \ge n_0$,

$$|\exp f_n(x) - \exp f(x)| < \epsilon e^{-a} |\exp f(x)| \le \epsilon. \blacksquare$$

5.8 Lemma. *Let (X, d) be a compact metric space and let $\{g_n\}$ be a sequence* *of continuous functions from X into* $\mathbb{C}$ *such that* $\sum g_n(x)$ *converges absolutely and uniformly for x in X. Then the product*

$$f(x) = \prod_{n=1}^{\infty} (1 + g_n(x))$$

converges absolutely and uniformly for x in X. Also there is an integer n_0 such that $f(x) = 0$ iff $g_n(x) = -1$ for some n, $1 \le n \le n_0$.

Proof. Since $\sum g_n(x)$ converges uniformly for x in X there is an integer n_0 such that $|g_n(x)| < \frac{1}{2}$ for all x in X and $n > n_0$. This implies that Re $[1 + g_n(x)]$ > 0 and also, according to inequality (5.3), $|\log (1 + g_n(x))| \le \frac{3}{2} |g_n(x)|$ for all $n > n_0$ and x in X. Thus

$$h(x) = \sum_{n=n_0+1}^{\infty} \log (1 + g_n(x))$$

converges uniformly for x in X. Since h is continuous and X is compact it follows that h must be bounded; in particular, there is a constant a such that Re $h(x) < a$ for all x in X. Thus, Lemma 5.7 applies and gives that

$$\exp h(x) = \prod_{n=n_0+1}^{\infty} (1 + g_n(x))$$

converges uniformly for x in X.

Finally,

$$f(x) = [1 + g_1(x)] \cdots [1 + g_{n_0}(x)] \exp h(x)$$

and $\exp h(x) \ne 0$ for any x in X. So if $f(x) = 0$ it must be that $g_n(x) = -1$ for some n with $1 \le n \le n_0$. ∎

We now leave this general situation to discuss analytic functions.

5.9 Theorem. *Let G be a region in* $\mathbb{C}$ *and let* $\{f_n\}$ *be a sequence in H(G) such that no f_n is identically zero. If $\sum [f_n(z) - 1]$ converges absolutely and uniformly on compact subsets of G then $\prod_{n=1}^{\infty} f_n(z)$ converges in H(G) to an analytic function $f(z)$. If a is a zero of f then a is a zero of only a finite number of the functions f_n, and the multiplicity of the zero of f at a is the sum of the multiplicities of the zeros of the functions f_n at a.*

Proof. Since $\sum [f_n(z) - 1]$ converges uniformly and absolutely on compact subsets of G, it follows from the preceding theorem that $f(z) = \prod f_n(z)$ converges uniformly and absolutely on compact subsets of G. That is, the infinite product converges in $H(G)$.

Suppose $f(a) = 0$ and let $r > 0$ be chosen such that $\bar{B}(a; r) \subset G$. By hypothesis, $\sum [f_n(z) - 1]$ converges uniformly on $\bar{B}(a; r)$. According to Lemma 5.8 there is an integer n such that $f(z) = f_1(z) \ldots f_n(z)g(z)$ where g does not vanish in $\bar{B}(a; r)$. The proof of the remainder of the theorem now follows. ∎

Let us now return to a discussion of the original problem. If $\{a_n\}$ is a sequence in a region G with no limit point in G (but possibly some point may be repeated in the sequence a finite number of times), consider the functions $(z - a_n)$. According to Theorem 5.9 if we can find functions $g_n(z)$ which are analytic on G, have no zeros in G, and are such that $\Sigma |(z - a_n) g_n(z) - 1|$ converges uniformly on compact subsets of G; then $f(z) = \Pi(z - a_n) g_n(z)$ is analytic and has its zeros only at the points $z = a_n$. The safest way to guarantee that $g_n(z)$ never vanishes is to express it as $g_n(z) = \exp h_n(z)$ for some analytic function $h_n(z)$. In fact, if G is simply connected it follows that $g_n(z)$ must be of this form. The functions we are looking for were introduced by Weierstrass.

5.10 Definition. An *elementary factor* is one of the following functions $E_p(z)$ for $p = 0, 1, \ldots$:

$$E_0(z) = 1 - z,$$

$$E_p(z) = (1 - z) \exp\left(z + \frac{z^2}{2} + \ldots + \frac{z^p}{p} \right), p \geq 1.$$

The function $E_p(z/a)$ has a simple zero at $z = a$ and no other zero. Also if b is a point in $\mathbb{C} - G$ then $E_p\left(\dfrac{a-b}{z-b} \right)$ has a simple zero at $z = a$ and is analytic in G. These functions will be used to manufacture analytic functions with prescribed zeros of prescribed multiplicity, but first an inequality must be proved which will enable us to apply Theorem 5.9 and obtain a convergent infinite product.

5.11 Lemma. *If* $|z| \leq 1$ *and* $p \geq 0$ *then* $|1 - E_p(z)| \leq |z|^{p+1}$.

Proof. We may restrict our attention to the case where $p \geq 1$. For a fixed p let

$$E_p(z) = 1 + \sum_{k=1}^{\infty} a_k z^k$$

be its power series expansion about $z = 0$. By differentiating the power series as well as the original expression for $E_p(z)$ we obtain

$$E_p'(z) = \sum_{k=1}^{\infty} k a_k z^{k-1}$$

$$= -z^p \exp\left(z + \ldots + \frac{z^p}{p} \right)$$

Comparing the two expressions gives two pieces of information about the coefficients a_k. First, $a_1 = a_2 = \ldots = a_p = 0$; second, since the coefficients of the expansion of $\exp\left(z + \ldots + \dfrac{z^p}{p} \right)$ are all positive, $a_k \leq 0$ for $k \geq p+1$. Thus, $|a_k| = -a_k$ for $k \geq p+1$; this gives

$$0 = E_p(1) = 1 + \sum_{k=p+1}^{\infty} a_k,$$

or

$$\sum_{k=p+1}^{\infty} |a_k| = -\sum_{k=p+1}^{\infty} a_k = 1.$$

Hence, for $|z| \leq 1$,

$$
\begin{aligned}
|E_p(z) - 1| &= |\sum_{k=p+1} a_k z^k| \\
&= |z|^{p+1} |\sum_{k=p+1}^{\infty} a_k z^{k-p-1}| \\
&\leq |z|^{p+1} \sum_{k=p+1}^{\infty} |a_k| \\
&= |z|^{p+1}
\end{aligned}
$$

which is the desired inequality. ■

Before solving the general problem of finding a function with prescribed zeros, the problem for the case where $G = \mathbb{C}$ will be solved. This is done for several reasons. In a later chapter on entire functions the specific information obtained when G is the whole plane is needed. Moreover, the proof of the general case, although similar to the proof for $\mathbb{C}$, tends to obscure the rather simple idea behind the proof.

5.12 Theorem. *Let $\{a_n\}$ be a sequence in $\mathbb{C}$ such that $\lim |a_n| = \infty$ and $a_n \neq 0$ for all $n \geq 1$. (This is not a sequence of distinct points; but, by hypothesis, no point is repeated an infinite number of times.) If $\{p_n\}$ is any sequence of integers such that*

5.13
$$
\sum_{n=1}^{\infty} \left(\frac{r}{|a_n|}\right)^{p_n+1} < \infty
$$

for all $r > 0$ then

$$
f(z) = \prod_{n=1}^{\infty} E_{p_n}(z/a_n)
$$

converges in $H(\mathbb{C})$. The function f is an entire function with zeros only at the points a_n. If z_0 occurs in the sequence $\{a_n\}$ exactly m times then f has a zero at $z = z_0$ of multiplicity m. Furthermore, if $p_n = n-1$ then (5.13) will be satisfied.

Proof. Suppose there are integers p_n such that (5.13) is satisfied. Then, according to Lemma 5.11,

$$
|1 - E_{p_n}(z/a_n)| \leq \left|\frac{z}{a_n}\right|^{p_n+1} \leq \left(\frac{r}{|a_n|}\right)^{p_n+1}
$$

whenever $|z| \leq r$ and $r \leq |a_n|$. For a fixed $r > 0$ there is an integer N such that $|a_n| \geq r$ for all $n \geq N$ (because $\lim |a_n| = \infty$). Thus for each $r > 0$ the series $\sum |1 - E_p(z/a_n)|$ is dominated by the convergent series (5.13) on the disk $\bar{B}(0; r)$. This gives that $\sum [1 - E_{p_n}(z/a_n)]$ converges absolutely in $H(\mathbb{C})$. By Theorem 5.9, the infinite product $\prod_{n=1}^{\infty} E_{p_n}(z/a_n)$ converges in $H(\mathbb{C})$.

To show that $\{p_n\}$ can be found so that (5.13) holds for all r is a trivial matter. For any r there is an integer N such that $|a_n| > 2r$ for all $n \geq N$. This gives that $\left(\dfrac{r}{|a_n|}\right) < \frac{1}{2}$ for all $n \geq N$; so if $p_n = n-1$ for all n, the tail end of the series (5.13) is dominated by $\sum (\frac{1}{2})^n$. Thus, (5.13) converges. ∎

There is, of course, a great latitude in picking the integers p_n. If p_n were bigger than $n-1$ we would have the same conclusion. However, there is an advantage in choosing the p_n as small as possible. After all, the smaller the integer p_n the more elementary the elementary factor $E_{p_n}(z/a_n)$. As is evident in considering the series (5.13), the size of the integers p_n depends on the rate at which $\{|a_n|\}$ converges to infinity. This will be explored later in Chapter XI.

5.14 The Weierstrass Factorization Theorem. *Let f be an entire function and let $\{a_n\}$ be the non-zero zeros of f repeated according to multiplicity; suppose f has a zero at $z=0$ of order $m \geqslant 0$ (a zero of order $m=0$ at $z=0$ means $f(0) \neq 0$). Then there is an entire function g and a sequence of integers $\{p_n\}$ such that*

$$f(z) = z^m e^{g(z)} \prod_{n=1}^{\infty} E_{p_n}\left(\frac{z}{a_n}\right).$$

Proof. According to the preceding theorem integers $\{p_n\}$ can be chosen such that

$$h(z) = z^m \prod_{n=1}^{\infty} E_{p_n}\left(\frac{z}{a_n}\right)$$

has the same zeros as f with the same multiplicities. It follows that $f(z)/h(z)$ has removable singularities at $z = 0, a_1, a_2, \ldots$. Thus f/h is an entire function and, furthermore, has no zeros. Since $\mathbb{C}$ is simply connected there is an entire function g such that

$$\frac{f(z)}{h(z)} = e^{g(z)}$$

The result now follows. ∎

5.15 Theorem. *Let G be a region and let $\{a_j\}$ be a sequence of distinct points in G with no limit point in G; and let $\{m_j\}$ be a sequence of integers. Then there is an analytic function f defined on G whose only zeros are at the points a_j; furthermore, a_j is a zero of f of multiplicity m_j.*

Proof. We begin by showing that it suffices to prove this theorem for the special case where there is a number $R > 0$ such that

5.16 $\qquad \{z: |z| > R\} \subset G$ and $|a_j| \leq R$ for all $j \geq 1$.

It must be shown that with this hypothesis there is a function f in $H(G)$ with the a_j's as its only zeros and $m_j =$ the multiplicity of the zero at $z = a_j$;

and with the further property that

5.17 $$\lim_{z \to \infty} f(z) = 1.$$

In fact, if such an f can always be found for a set satisfying (5.16), let G_1 be an arbitrary open set in $\mathbb{C}$ with $\{\alpha_j\}$ a sequence of distinct points in G_1 with no limit point, and let $\{m_j\}$ be a sequence of integers. Now if $\bar{B}(a; r)$ is a disk in G_1 such that $\alpha_j \notin B(a; r)$ for all $j \geq 1$, consider the Mobius transformation $Tz = (z-a)^{-1}$. Put $G = T(G_1)$; it is easy to see that G satisfies condition (5.16) where $a_j = T\alpha_j = (\alpha_j - a)^{-1}$. If there is a function f in $H(G)$ with a zero at each a_j of multiplicity m_j, with no other zeros, and such that f satisfies (5.17); then $g(z) = f(Tz)$ is analytic in $G_1 - \{a\}$ with a removable singularity at $z = a$. Furthermore, g has the prescribed zero at each α_j of multiplicity m_j.

So assume that G satisfies (5.16). Define a second sequence $\{z_n\}$ consisting of the points in $\{a_j\}$, but such that each a_j is repeated according to its multiplicity m_j. Now, for each $n \geq 1$ there is a point w_n in $\mathbb{C} - G$ such that

$$|w_n - z_n| = d(z_n, \mathbb{C} - G).$$

Notice that the hypothesis (5.16) excludes the possibility that $G = \mathbb{C}$ unless the sequence $\{a_j\}$ were finite. In fact, if $\{a_j\}$ were finite the theorem could be easily proved so it suffices to assume that $\{a_j\}$ is infinite. Since $|a_j| \leq R$ for all j and $\{a_j\}$ has no limit point in G it follows that $\mathbb{C} - G$ is non-empty as well as compact. Also,

$$\lim |z_n - w_n| = 0.$$

Consider the functions

$$E_n\left(\frac{z_n - w_n}{z - w_n}\right);$$

each has a simple zero at $z = z_n$. It must be shown that the infinite product of these functions converges in $H(G)$.

To do this let K be a compact subset of G so that $d(\mathbb{C} - G, K) > 0$. For any point z in K

$$\left|\frac{z_n - w_n}{z - w_n}\right| \leq |z_n - w_n| \, [d(w_n, K)]^{-1}$$

$$\leq |z_n - w_n| \, [d(\mathbb{C} - G, K)]^{-1}$$

It follows that for any δ, $0 < \delta < 1$, there is an integer N such that

$$\left|\frac{z_n - w_n}{z - w_n}\right| < \delta$$

for all z in K and $n \geq N$. But then Lemma 5.11 gives that

5.18 $$\left|E_n\left(\frac{z_n - w_n}{z - w_n}\right) - 1\right| \leq \delta^{n+1}$$

for all z in K and $n \geq N$. But this gives that the series

$$\sum_{n=1}^{\infty} \left[E_n \left(\frac{z_n - w_n}{z - w_n} \right) - 1 \right]$$

converges uniformly and absolutely on K. According to Theorem 5.9

$$f(z) = \prod_{n=1}^{\infty} E_n \left(\frac{z_n - w_n}{z - w_n} \right)$$

converges in $H(G)$, so that f is an analytic function on G. Also, Theorem 5.9 implies that the points $\{a_j\}$ are the only zeros of f and m_j is the order of the zero at $z = a_j$ (because a_j occurs m_j times in the sequence $\{z_n\}$). To show that $\lim_{z \to \infty} f(z) = 1$, let $\epsilon > 0$ be an arbitrary number and let $R_1 > R$ (R_1 will be further specified shortly). If $|z| \geq R_1$ then, because $|z_n| \leq R$ and $w_n \in \mathbb{C} - G \subset B(0; R)$,

$$\left| \frac{z_n - w_n}{z - w_n} \right| \leq \frac{2R}{R_1 - R}.$$

So, if we choose $R_1 > R$ so that $2R < \delta(R_1 - R)$ for some δ, $0 < \delta < 1$, (5.18) holds for $|z| \geq R_1$ and for *all* $n \geq 1$. In particular, $\text{Re } E_n \left(\frac{z_n - w_n}{z - w_n} \right) > 0$ for all n and $|z| \geq R_1$; so that

5.19 $$|f(z) - 1| = \left| \exp \left(\sum_{n=1}^{\infty} \log E_n \left(\frac{z_n - w_n}{z - w_n} \right) \right) - 1 \right|$$

is a meaningful equation. On the other hand (5.3) and (5.18) give that

$$\left| \sum_{n=1}^{\infty} \log E_n \left(\frac{z_n - w_n}{z - w_n} \right) \right| \leq \sum_{n=1}^{\infty} \left| \log E_n \left(\frac{z_n - w_n}{z - w_n} \right) \right|$$

$$\leq \sum_{n=1}^{\infty} \frac{3}{2} \left| E_n \left(\frac{z_n - w_n}{z - w_n} \right) - 1 \right|$$

$$\leq \sum_{n=1}^{\infty} \frac{3}{2} \delta^{n+1}$$

$$= \frac{3}{2} \frac{\delta^2}{1 - \delta}$$

for $|z| \geq R_1$. If we further restrict δ so that $|e^w - 1| < \epsilon$ whenever

$$|w| < \frac{3}{2} \left(\frac{\delta^2}{1 - \delta} \right),$$

then equation (5.19) gives that $|f(z)-1| < \epsilon$ whenever $|z| \geq R_1$. That is, $\lim_{z \to \infty} f(z) = 1.$ ∎

One of the more interesting results that follows from the above theorem says (in algebraic terms) that $M(G)$ is the quotient field of the integral domain $H(G)$. Avoiding this language the result is as follows.

5.20 Corollary. *If f is a meromorphic function on an open set G then there are analytic functions g and h on G such that $f = g/h$.*

Proof. Let $\{a_j\}$ be the poles of f and let m_j be the order of the pole at a_j. According to the preceding theorem there is an analytic function h with a zero of multiplicity m_j at each $z = a_j$ and with no other zeros. Thus hf has removable singularities at each point a_j. It follows that $g = hf$ is analytic in $G.$ ∎

Exercises

1. Show that $\prod (1+z_n)$ converges absolutely iff $\prod (1+|z_n|)$ converges.

2. Prove that $\lim_{z \to 0} \dfrac{\log(1+z)}{z} = 1.$

3. Let f and g be analytic functions on a region G and show that there are analytic functions f_1, g_1, and h on G such that $f(z) = h(z)f_1(z)$ and $g(z) = h(z)g_1(z)$ for all z in G; and f_1 and g_1 have no common zeros.

4. (a) Let $0 < |a| < 1$ and $|z| \leq r < 1$; show that

$$\left| \frac{a+|a| z}{(1-\bar{a}z)a} \right| \leq \frac{1+r}{1-r}$$

(b) Let $\{a_n\}$ be a sequence of complex numbers with $0 < |a_n| < 1$ and $\sum (1-|a_n|) < \infty$. Show that the infinite product

$$B(z) = \prod_{n=1}^{\infty} \frac{|a_n|}{a_n} \left(\frac{a_n - z}{1 - \bar{a}_n z} \right)$$

converges in $H(B(0; 1))$ and that $|B(z)| \leq 1$. What are the zeros of B? ($B(z)$ is called a *Blaschke Product*.)

(c) Find a sequence $\{a_n\}$ in $B(0; 1)$ such that $\sum (1-|a_n|) < \infty$ and every number $e^{i\theta}$ is a limit point of $\{a_n\}$.

5. Discuss the convergence of the infinite product $\prod_{n=1}^{\infty} \dfrac{1}{n^p}$ for $p > 0$.

6. Discuss the convergence of the infinite products $\prod \left[1 + \dfrac{i}{n} \right]$ and $\prod \left| 1 + \dfrac{i}{n} \right|$.

7. Show that $\prod_{n=2}^{\infty} \left(1 - \dfrac{1}{n^2} \right) = \dfrac{1}{2}$.

8. For which values of z do the products $\prod (1-z^n)$ and $\prod (1+z^{2n})$ converge? Is there an open set G such that the product converges uniformly on each compact subset of G? If so, give the largest such open set.

9. Use Theorem 5.15 to show there is an analytic function f on $D = \{z: |z| < 1\}$ which is not analytic on any open set G which properly contains D.

10. Suppose G is an open set and $\{f_n\}$ is a sequence in $H(G)$ such that $f(z) = \prod f_n(z)$ converges in $H(G)$. (a) Show that

$$\sum_{k=1}^{\infty} \left[f_k'(z) \prod_{n \neq k} f_n(z) \right]$$

converges in $H(G)$ and equals $f'(z)$. (b) Assume that f is not the identically zero function and let K be a compact subset of G such that $f(z) \neq 0$ for all z in K. Show that

$$\frac{f'(z)}{f(z)} = \sum_{n=1}^{\infty} \frac{f_n'(z)}{f_n(z)}$$

and the convergence is uniform over K.

11. A subset $\mathscr{I}$ of $H(G)$, G a region, is an *ideal* iff: (i) f and g in $\mathscr{I}$ implies $af + bg$ is in $\mathscr{I}$ for all complex numbers a and b; (ii) f in $\mathscr{I}$ and g any function in $H(G)$ implies fg is in $\mathscr{I}$. $\mathscr{I}$ is called a *proper ideal* if $\mathscr{I} \neq (0)$ and $\mathscr{I} \neq H(G)$; $\mathscr{I}$ is a *maximal ideal* if $\mathscr{I}$ is a proper ideal and whenever $\mathscr{J}$ is an ideal with $\mathscr{I} \subset \mathscr{J}$ then either $\mathscr{J} = \mathscr{I}$ or $\mathscr{J} = H(G)$; $\mathscr{I}$ is a *prime ideal* if whenever f and $g \in H(G)$ and $fg \in \mathscr{I}$ then either $f \in \mathscr{I}$ or $g \in \mathscr{I}$. If $f \in H(G)$ let $\mathscr{Z}(f)$ be the set of zeros of f counted according to their multiplicity. So $\mathscr{Z}((z-a)^3) = \{a, a, a\}$. If $\mathscr{S} \subset H(G)$ then $\mathscr{Z}(\mathscr{S}) = \cap \{\mathscr{Z}(f): f \in \mathscr{S}\}$, where the zeros are again counted according to their multiplicity. So if $\mathscr{S} = \{(z-a)^3 (z-b), (z-a)^2\}$ then $\mathscr{Z}(\mathscr{S}) = \{a, a\}$.

   (a) If f and $g \in H(G)$ then f *divides* g (in symbols, $f|g$) if there is an h in $H(G)$ such that $g = fh$. Show that $f|g$ iff $\mathscr{Z}(f) \subset \mathscr{Z}(g)$.

   (b) If $\mathscr{S} \subset H(G)$ and $\mathscr{S} \neq \square$ then f is a *greatest common divisor* of $\mathscr{S}$ if: (i) $f|g$ for each g in $\mathscr{S}$ and (ii) whenever $h|g$ for each g in $\mathscr{S}$, $h|f$. In symbols, $f = \text{g.c.d.} \mathscr{S}$. Prove that $f = \text{g.c.d.} \mathscr{S}$. iff $\mathscr{Z}(f) = \mathscr{Z}(\mathscr{S})$ and show that each non-empty subset of $H(G)$ has a g.c.d.

   (c) If $A \subset G$ let $\mathscr{I}(A) = \{f \in H(G): \mathscr{Z}(f) \supset A\}$. Show that $\mathscr{I}(A)$ is a closed ideal in $H(G)$ and $\mathscr{I}(A) = (0)$ iff A has a limit point in G.

   (d) Let $a \in G$ and $\mathscr{I} = \mathscr{I}(\{a\})$. Show that $\mathscr{I}$ is a maximal ideal.

   (e) Show that every maximal ideal in $H(G)$ is a prime ideal.

   (f) Give an example of an ideal which is not a prime ideal.

12. Find an entire function f such that $f(n + in) = 0$ for every integer n (positive, negative or zero). Give the most elementary example possible (i.e., choose the p_n to be as small as possible).

13. Find an entire function f such that $f(m + in) = 0$ for all possible integers m, n. Find the most elementary solution possible.

§6. *Factorization of the sine function*

In this section an application of the Weierstrass Factorization Theorem to $\sin \pi z$ is given. If an infinite sum or product is followed by a prime (apostrophe) (i.e., $\sum'$ or $\prod'$), then the sum or product is to be taken over all the indicated indices n except $n = 0$. For example,

$$\sum_{n=-\infty}^{\infty}{}' \, a_n = \sum_{n=1}^{\infty} a_{-n} + \sum_{n=1}^{\infty} a_n.$$

The zeros of $\sin \pi z = \dfrac{1}{2i}(e^{i\pi z} - e^{-i\pi z})$ are precisely the integers; moreover, each zero is simple. Since

$$\sum_{n=-\infty}^{\infty}{}' \left(\frac{r}{n}\right)^2 < \infty$$

for all $r > 0$, one can (5.13) choose $p_n = 1$ for all n in the Weierstrass Factorization Theorem. Thus

$$\sin \pi z = [\exp g(z)] \, z \prod_{n=-\infty}^{\infty}{}' \left(1 - \frac{z}{n}\right) e^{z/n};$$

or, because the terms of the infinite product can be rearranged,

6.1
$$\sin \pi z = [\exp g(z)] \, z \prod_{n=1}^{\infty} \left(1 - \frac{z^2}{n^2}\right)$$

for some entire function $g(z)$. If $f(z) = \sin \pi z$ then, according to Theorem 2.1,

$$\pi \cot \pi z = \frac{f'(z)}{f(z)}$$

$$= g'(z) + \frac{1}{z} + \sum_{n=1}^{\infty} \frac{2z}{z^2 - n^2}$$

and the convergence is uniform over compact subsets of the plane that contain no integers (actually, a small additional argument is necessary to justify this—see Exercise 5.10). But according to Exercise V. 2.8,

$$\pi \cot \pi z = \frac{1}{z} + \sum_{n=1}^{\infty} \frac{2z}{z^2 - n^2}$$

for z not an integer. So it must be that g is a constant, say $g(z) = a$ for all z. It follows from (6.1) that for $0 < |z| < 1$

$$\frac{\sin \pi z}{\pi z} = \frac{e^a}{\pi} \prod_{n=1}^{\infty} \left(1 - \frac{z^2}{n^2}\right).$$

Letting z approach zero gives that $e^a = \pi$. This gives the following:

6.2
$$\sin \pi z = \pi z \prod_{n=1}^{\infty} \left(1 - \frac{z^2}{n^2}\right)$$

and the convergence is uniform over compact subsets of $\mathbb{C}$.

Exercises

1. Show that $\cos \pi z = \displaystyle\prod_{n=1}^{\infty}\left[1 - \frac{4z^2}{(2n-1)^2}\right]$.

2. Find a factorization for $\sinh z$ and $\cosh z$.

3. Show that $\cos\left(\dfrac{\pi z}{4}\right) - \sin\left(\dfrac{\pi z}{4}\right) = \displaystyle\prod_{n=1}^{\infty}\left(\frac{1+(-1)^n z}{2n-1}\right)$.

4. Prove Wallis's formula: $\dfrac{\pi}{2} = \displaystyle\prod_{n=1}^{\infty}\frac{(2n)^2}{(2n-1)(2n+1)}$.

§7. *The gamma function*

Let G be an open set in the plane and let $\{f_n\}$ be a sequence of analytic functions on G. If $\{f_n\}$ converges in $H(G)$ to f and f is not identically zero, then it easily follows that $\{f_n\}$ converges to f in $M(G)$. Since $d(z_1, z_2) = d\left(\dfrac{1}{z_1}, \dfrac{1}{z_2}\right)$, where d is the spherical metric on $\mathbb{C}_\infty$ (see (3.1)), it follows that $\left\{\dfrac{1}{f_n}\right\}$ converges to $\dfrac{1}{f}$ in $M(G)$. It is an easy exercise to show that $\left\{\dfrac{1}{f_n}\right\}$ converges uniformly to $\dfrac{1}{f}$ on any compact set K on which no f_n vanishes. (What does Hurwitz's Theorem have to say about this situation). Since, according to Theorem 5.12, the infinite product

$$\prod_{n=1}^{\infty}\left(1 + \frac{z}{n}\right)e^{-z/n}$$

converges in $H(\mathbb{C})$ to an entire function which only has simple zeros at $z = -1, -2, \ldots$, the above discussion yields that

7.1
$$\prod_{n=1}^{\infty}\left(1 + \frac{z}{n}\right)^{-1}e^{z/n}$$

converges on compact subsets of $\mathbb{C} - \{-1, -2, \ldots\}$ to a function with simple poles at $z = -1, -2, \ldots$.

7.2 Definition. The *gamma function*, $\Gamma(z)$, is the meromorphic function on $\mathbb{C}$ with simple poles at $z = 0, -1, \ldots$ defined by

7.3
$$\Gamma(z) = \frac{e^{-\gamma z}}{z}\prod_{n=1}^{\infty}\left(1 + \frac{z}{n}\right)^{-1}e^{z/n},$$

where γ is a constant chosen so that $\Gamma(1) = 1$.

The first thing that must be done is to show that the constant γ exists; this is an easy matter. Substituting $z = 1$ in (7.1) yields a finite number

$$c = \prod_{n=1}^{\infty} \left(1 + \frac{1}{n}\right)^{-1} e^{1/n}$$

which is clearly positive. Let $\gamma = \log c$; it follows that with this choice of γ, equation (7.3) for $z = 1$ gives $\Gamma(1) = 1$. This constant γ is called *Euler's constant* and it satisfies

7.4
$$e^{\gamma} = \prod_{n=1}^{\infty} \left(1 + \frac{1}{n}\right)^{-1} e^{1/n}$$

Since both sides of (7.4) involve only real positive numbers and the real logarithm is continuous, we may apply the logarithm function to both sides of (7.4) and obtain

$$\gamma = \sum_{k=1}^{\infty} \log\left[\left(1 + \frac{1}{k}\right)^{-1} e^{1/k}\right]$$

$$= \sum_{k=1}^{\infty} \left[\frac{1}{k} - \log(k+1) + \log k\right]$$

$$= \lim_{n \to \infty} \sum_{k=1}^{n} \left[\frac{1}{k} - \log(k+1) + \log k\right]$$

$$= \lim_{n \to \infty} \left[\left(1 + \frac{1}{2} + \ldots + \frac{1}{n}\right) - \log(n+1)\right].$$

Adding and subtracting $\log n$ to each term of this sequence and using the fact that $\lim \log\left(\frac{n+1}{n}\right) = 0$ yields

7.5
$$\gamma = \lim_{n \to \infty} \left[\left(1 + \frac{1}{2} + \ldots + \frac{1}{n}\right) - \log n\right].$$

This last formula can be used to approximate γ. Equation (7.5) is also used to derive another expression for $\Gamma(z)$. From the definition of $\Gamma(z)$ it follows that

$$\Gamma(z) = \frac{e^{-\gamma z}}{z} \lim_{n \to \infty} \prod_{k=1}^{n} \left(1 + \frac{z}{k}\right)^{-1} e^{z/k}$$

$$= \frac{e^{-\gamma z}}{z} \lim_{n \to \infty} \prod_{k=1}^{n} \frac{k e^{z/k}}{z+k}$$

$$= \lim_{n \to \infty} \frac{e^{-\gamma z} n!}{z(z+1)\ldots(z+n)} \exp\left(z\left(1 + \tfrac{1}{2} + \ldots + \frac{1}{n}\right)\right)$$

However

$$e^{-\gamma z} \exp\left[\left(1 + \tfrac{1}{2} + \ldots + \frac{1}{n}\right)z\right] = n^z \exp\left[z\left(-\gamma + 1 + \tfrac{1}{2} + \ldots + \frac{1}{n} - \log n\right)\right].$$

So that the following is obtained

7.6 Gauss's Formula. For $z \neq 0, -1, \ldots$

$$\Gamma(z) = \lim_{n \to \infty} \frac{n! \, n^z}{z(z+1) \ldots (z+n)}$$

The formula of Gauss yields a simple derivation of the functional equation satisfied by the gamma function.

7.7 Functional Equation. For $z \neq 0, -1, \ldots$

$$\Gamma(z+1) = z\Gamma(z)$$

To obtain this important equation substitute $z+1$ for z in (7.6); this gives

$$\Gamma(z+1) = \lim_{n \to \infty} \frac{n! \, n^{z+1}}{(z+1) \ldots (z+n+1)}$$

$$= z \lim_{n \to \infty} \left[\frac{n! \, n^z}{z(z+1) \ldots (z+n)}\right]\left[\frac{n}{z+n+1}\right]$$

$$= z\Gamma(z)$$

since $\lim\left(\dfrac{n}{z+n+1}\right) = 1$.

Now consider $\Gamma(z+2)$; we have $\Gamma(z+2) = \Gamma((z+1)+1) = (z+1)\,\Gamma(z+1)$ by the functional equation. A second application of (7.7) gives $\Gamma(z+2) = z(z+1)\Gamma(z)$. In fact, by reiterating this procedure

7.8 $\Gamma(z+n) = z(z+1) \ldots (z+n-1)\Gamma(z)$

for n a non negative integer and $z \neq 0, -1, \ldots$. In particular setting $z = 1$ gives that

7.9 $\Gamma(n+1) = n!$

That is, the Γ function is analytic in the right half plane and agrees with the factorial function at the integers. We may therefore consider the gamma function as an extension of the factorial to the complex plane; alternately, if $z \neq -1, -2, \ldots$ then letting $z! = \Gamma(z+1)$ is a justifiable definition of $z!$.

As has been pointed out, Γ has simple poles at $z = 0, -1, \ldots$; we wish to find the residue of Γ at each of its poles. To do this recall from Proposition V. 2.4 that

$$\mathrm{Res}\,(\Gamma; -n) = \lim_{z \to -n} (z+n)\Gamma(z)$$

for each non-negative integer n. But from (7.8)

$$(z+n)\Gamma(z) = \frac{\Gamma(z+n+1)}{z(z+1)\dots(z+n-1)}.$$

So letting z approach $-n$ gives that

7.10 $$\text{Res}\,(\Gamma;\, -n) = \frac{(-1)^n}{n!},\ n \geq 0.$$

According to Exercise 5.10 we can calculate Γ'/Γ by

7.11 $$\frac{\Gamma'(z)}{\Gamma(z)} = -\gamma - \frac{1}{z} + \sum_{n=1}^{\infty} \frac{z}{n(n+z)}$$

for $z \neq 0, -1, \dots$ and convergence is uniform on every compact subset of $\mathbb{C} - \{0, -1, \dots\}$. It follows from Theorem 2.1 that to calculate the derivative of Γ'/Γ we may differentiate the series (7.11) term by term. Thus when z is not a negative integer

7.12 $$\left(\frac{\Gamma'(z)}{\Gamma(z)}\right)' = \frac{1}{z^2} + \sum_{n=1}^{\infty} \frac{1}{(n+z)^2}.$$

At this time the reader may well be asking when this process will stop. Will we calculate the second derivative of Γ'/Γ? The answer to this question is no. The answer to the implied question of why anyone would want to derive formulas (7.11) and (7.12) is that they allow us to characterize the gamma function in a particularly beautiful way.

Notice that the definition of $\Gamma(z)$ gives that $\Gamma(x) > 0$ if $x > 0$. Thus, $\log \Gamma(x)$ is well defined for $x > 0$ and, according to formula (7.12), the second derivative of $\log \Gamma(x)$ is always positive. According to Proposition VI. 3.4 this implies that the gamma function is logarithmically convex on $(0, \infty)$; that is, $\log \Gamma(x)$ is convex there. It turns out that this property together with the functional equation and the fact that $\Gamma(1) = 1$ completely characterize the gamma function.

7.13 Bohr-Mollerup Theorem. *Let f be a function defined on $(0, \infty)$ such that $f(x) > 0$ for all $x > 0$. Suppose that f has the following properties*:

(a) $\log f(x)$ *is a convex function*;
(b) $f(x+1) = x f(x)$ *for all x*;
(c) $f(1) = 1$.

Then $f(x) = \Gamma(x)$ for all x.

Proof. Begin by noting that since f has properties (b) and (c), the function also satisfies

7.14 $$f(x+n) = x(x+1)\dots(x+n-1)f(x).$$

for every non-negative integer n. So if $f(x) = \Gamma(x)$ for $0 < x \leq 1$, this equation will give that f and Γ are everywhere identical. Let $0 < x \leq 1$ and let n be an integer larger than 2. From Exercise VI. 3.3

$$\frac{\log f(n-1) - \log f(n)}{(n-1) - n} \leq \frac{\log f(x+n) - \log f(n)}{(x+n) - n} \leq \frac{\log f(n+1) - \log f(n)}{(n+1) - n}$$

Since (7.14) holds we have that $f(m) = (m-1)!$ for every integer $m \geq 1$. Thus the above inequalities become

$$-\log (n-2)! + \log (n-1)! \leq \frac{\log f(x+n) - \log (n-1)!}{x} \leq \log n! - \log (n-1)!;$$

or

$$x \log (n-1) \leq \log f(x+n) - \log (n-1)! \leq x \log n.$$

Adding $\log (n-1)!$ to each side of this inequality and applying the exponential (exp is a monotone increasing function and therefore preserves inequalities) gives

$$(n-1)^x (n-1)! \leq f(x+n) \leq n^x (n-1)!$$

Applying (7.14) to calculate $f(x+n)$ yields

$$\frac{(n-1)^x (n-1)!}{x(x+1) \ldots (x+n-1)} \leq f(x) \leq \frac{n^x (n-1)!}{x(x+1) \ldots (x+n-1)}$$

$$= \frac{n^x n!}{x(x+1) \ldots (x+n)} \left[\frac{x+n}{n} \right].$$

Since the term in the middle of this sandwich, $f(x)$, does not involve the integer n and since the inequality holds for all integers $n \geq 2$, we may vary the integers on the left and right hand side independently of one another and preserve the inequality. In particular, $n+1$ may be substituted for n on the left while allowing the right hand side to remain unchanged. This gives

$$\frac{n^x n!}{x(x+1) \ldots (x+n)} \leq f(x) \leq \frac{n^x n!}{x(x+1) \ldots (x+n)} \left[\frac{x+n}{n} \right]$$

for all $n \geq 2$ and x in $[0, 1)$. Now take the limit as $n \to \infty$. Since $\lim \left(\frac{x+n}{n} \right)$ $= 1$, Gauss's formula implies that $\Gamma(x) = f(x)$ for $0 < x \leq 1$. The result now follows by applying (7.14) and the Functional Equation. $\blacksquare$

7.15 Theorem. *If* Re $z > 0$ *then*

$$\Gamma(z) = \int_0^\infty e^{-t} t^{z-1} \, dt$$

The integrand in 7.15 behaves badly at $t = 0$ and $t = \infty$, so that the meaning of the above equation must be explicitly stated. Rather than give a formal definition of the convergence of an improper integral, the properties of this particular integral are derived in Lemma 7.16 below (see also Exercise 2.2).

7.16 Lemma. *Let* $S = \{z: a \leq \text{Re } z \leq A\}$ *where* $0 < a < A < \infty$.

(a) *For every* $\epsilon > 0$ *there is a* $\delta > 0$ *such that for all* z *in* S

$$\left| \int_\alpha^\beta e^{-t} t^{z-1} \, dt \right| < \epsilon$$

whenever $0 < \alpha < \beta < \delta$.

(b) *For every* $\epsilon > 0$ *there is a number* κ *such that for all* z *in* S

$$\left| \int_\alpha^\beta e^{-t} t^{z-1} \, dt \right| < \epsilon$$

whenever $\beta > \alpha > \kappa$.

Proof. To prove (a) note that if $0 < t \leq 1$ and z is in S then $(\text{Re } z - 1) \log t \leq (a-1) \log t$; since $e^{-t} \leq 1$,

$$|e^{-t} t^{z-1}| \leq t^{\text{Re } z - 1} \leq t^{a-1}.$$

So if $0 < \alpha < \beta < 1$ then

$$\left| \int_\alpha^\beta e^{-t} t^{z-1} \, dt \right| \leq \int_\alpha^\beta t^{a-1} \, dt$$

$$= \frac{1}{a} (\beta^a - \alpha^a)$$

for all z in S. If $\epsilon > 0$ then we can choose δ, $0 < \delta < 1$, such that $a^{-1}(\beta^a - \alpha^a) < \epsilon$ for $|\alpha - \beta| < \delta$. This proves part (a).

To prove part (b) note that for z in S and $t \geq 1$, $|t^{z-1}| \leq t^{A-1}$. Since $t^{A-1} \exp(-\frac{1}{2}t)$ is continuous on $[1, \infty)$ and converges to zero as $t \to \infty$, there is a constant c such that $t^{A-1} \exp(-\frac{1}{2}t) \leq c$ for all $t \geq 1$. This gives that

$$|e^{-t} t^{z-1}| \leq c e^{-\frac{1}{2}t}$$

for all z in S and $t \geq 1$. If $\beta > \alpha > 1$ then

$$\left| \int_\alpha^\beta e^{-t} t^{z-1} \, dt \right| \leq c \int_\alpha^\beta e^{-\frac{1}{2}t}$$

$$= 2c(e^{-\frac{1}{2}\alpha} - e^{-\frac{1}{2}\beta}).$$

Again, for any $\epsilon > 0$ there is a number $\kappa > 1$ such that $|2c(e^{-\frac{1}{2}\alpha} - e^{-\frac{1}{2}\beta})| < \epsilon$ whenever $\alpha, \beta > \kappa$, giving part (b). ∎

The results of the preceding lemma embody exactly the concept of a uniformly convergent integral. In fact, if we consider the integrals

$$\int_\alpha^1 e^{-t} t^{z-1} \, dt$$

for $0 < \alpha < 1$, then part (a) of Lemma 7.16 says that these integrals satisfy a Cauchy criterion as $\alpha \to 0$. That is, the difference between any two will be arbitrarily small if α and β are taken sufficiently close to zero. A similar interpretation is available for the integrals

$$\int_1^\alpha e^{-t}t^{z-1}\, dt$$

for $\alpha > 1$. The next proposition formalizes this discussion.

7.17 Proposition. *If* $G = \{z: \text{Re } z > 0\}$ *and*

$$f_n(z) = \int_{1/n}^n e^{-t}t^{z-1}\, dt$$

for $n \geq 1$ *and* z *in* G, *then each* f_n *is analytic on* G *and the sequence is convergent in* $H(G)$.

Proof. Think of $f_n(z)$ as the integral of $\varphi(t, z) = e^{-t}t^{z-1}$ along the straight line segment $\left[\dfrac{1}{n}, n\right]$ and apply Exercise IV. 2.2 to conclude that f_n is analytic. Now if K is a compact subset of G there are positive real numbers a and A such that $K \subset \{z: a \leq \text{Re } z \leq A\}$. Since

$$f_m(z)-f_n(z) = \int_{1/m}^{1/n} e^{-t}t^{z-1}\, dt + \int_n^m e^{-t}t^{z-1}\, dt$$

for $m > n$, Lemma 7.16 and Lemma 1.7 imply that $\{f_n\}$ is a Cauchy sequence in $H(G)$. But $H(G)$ is complete (Corollary 2.3) so that $\{f_n\}$ must converge. ∎

If f is the limit of the functions $\{f_n\}$ from the above proposition then define the integral to be this function. That is,

7.18 $$f(z) = \int_0^\infty e^{-t}t^{z-1}\, dt, \text{ Re } z > 0.$$

To show that this function $f(z)$ is indeed the gamma function for $\text{Re } z > 0$ we only have to show that $f(x) = \Gamma(x)$ for $x \geq 1$. Since $[1, \infty)$ has limit points in the right half plane and both f and Γ are analytic then it follows that f must be Γ (Corollary IV. 3.8). Now observe that successive performing of integration by parts on $(1-t/n)^n t^{x-1}$ yields

$$\int_0^n \left(1 - \frac{t}{n}\right)^n t^{x-1}\, dt = \frac{n!\, n^x}{x(x+1)\ldots(x+n)}$$

which converges to $\Gamma(x)$ as $n \to \infty$ by Gauss's formula. If we can show that the integral in this equation converges to $\int_0^\infty e^{-t}t^{x-1}\, dt = f(x)$ as $n \to \infty$ then

Theorem 7.15 is proved. This is indeed the case and it follows from the following lemma.

7.19 Lemma. (a) $\left\{\left(1 + \dfrac{z}{n}\right)^n\right\}$ *converges to* e^z *in* $H(\mathbb{C})$.

(b) *If* $t \geq 0$ *then* $\left(1 - \dfrac{t}{n}\right)^n \leq e^{-t}$ *for all* $n \geq t$.

Proof. (a) Let K be a compact subset of the plane. Then $|z| < n$ for all z in K and n sufficiently large. It suffices to show that

$$\lim_{n \to \infty} n \log\left(1 + \frac{z}{n}\right) = z$$

uniformly for z in K by Lemma 5.7. Recall that

$$\log(1+w) = \sum_{k=1}^{\infty} (-1)^{k-1} \frac{w^k}{k}$$

for $|w| < 1$. Let $n > |z|$ for all z in K; if z is any point in K then

$$n \log\left(1 + \frac{z}{n}\right) = z - \frac{1}{2}\frac{z^2}{n} + \frac{1}{3}\frac{z^3}{n^2} - \cdots.$$

So

7.20 $n \log\left(1 + \dfrac{z}{n}\right) - z = z\left[-\dfrac{1}{2}\left(\dfrac{z}{n}\right) + \dfrac{1}{3}\left(\dfrac{z}{n}\right)^2 - \cdots\right];$

taking absolute values gives that

$$\left| n \log\left(1 + \frac{z}{n}\right) - z \right| \leq |z| \sum_{k=2}^{\infty} \frac{1}{k}\left|\frac{z}{n}\right|^{k-1}$$

$$\leq |z| \sum_{k=1}^{\infty} \left|\frac{z}{n}\right|^k$$

$$= \frac{|z|^2}{n} \frac{1}{1 - |z/n|}$$

$$\leq \frac{R^2}{n - R}$$

where $R \geq |z|$ for all z in K. If $n \to \infty$ then this difference goes to zero uniformly for z in K.

(b) Now let $t \geq 0$ and substitute $-t$ for z in (7.20) where $t \leq n$. This gives

$$n \log\left(1 - \frac{t}{n}\right) + t = -t \sum_{k=2}^{\infty} \frac{1}{k}\left(\frac{t}{n}\right)^{k-1} \leq 0$$

Thus

$$n \log\left(1 - \frac{t}{n}\right) \le -t;$$

and since exp is a monotone function part (b) is proved. ∎

Proof of Theorem 7.15. Fix $x > 1$ and let $\epsilon > 0$. According to Lemma 7.16 (b) we can choose $\kappa > 0$ such that

7.21
$$\int_\kappa^r e^{-t}t^{x-1}\, dt < \frac{\epsilon}{4}$$

whenever $r > \kappa$. Let n be any integer larger than κ and let f_n be the function defined in Proposition 7.17. Then

$$f_n(x) - \int_0^n \left(1 - \frac{t}{n}\right)^n t^{x-1}\, dt = -\int_0^{1/n} \left(1 - \frac{t}{n}\right)^n t^{x-1}\, dt +$$

$$\int_{1/n}^n \left[e^{-t} - \left(1 + \frac{t}{n}\right)^n\right] t^{x-1}\, dt$$

Now by Lemma 7.19 (b) and Lemma 7.16 (a)

7.22
$$\int_0^{1/n} \left(1 - \frac{t}{n}\right)^n t^{x-1}\, dt \le \int_0^{1/n} e^{-t}t^{x-1}\, dt < \frac{\epsilon}{4}$$

for sufficiently large n. Also, if n is sufficiently large, part (a) of the preceding lemma gives

$$\left|\left(1 - \frac{t}{n}\right)^n - e^{-t}\right| \le \frac{\epsilon}{4M\kappa}$$

for t in $[0, \kappa]$ where $M = \int_0^\kappa t^{x-1}\, dt$. Thus

7.23
$$\left|\int_{1/n}^\kappa \left[e^{-t} - \left(1 - \frac{t}{n}\right)^n\right] t^{x-1}\, dt\right| \le \frac{\epsilon}{4}$$

Using Lemma 7.19 (b) and (7.21)

$$\left|\int_\kappa^n \left[e^{-t} - \left(1 - \frac{t}{n}\right)^n\right] t^{x-1}\, dt\right| \le 2\int_\kappa^n e^{-t}t^{x-1}\, dt \le \frac{\epsilon}{2}$$

for $n > \kappa$. If we combine this inequality with (7.22) and (7.23), we get

$$\left|f_n(x) - \int_0^n \left(1 - \frac{t}{n}\right)^n t^{x-1}\, dt\right| < \epsilon$$

for n sufficiently large. That is

$$0 = \lim \left[f_n(x) - \int\limits_0^n \left(1 - \frac{t}{n}\right)^n t^{x-1} \, dt \right]$$

$$= \lim \left[f_n(x) - \frac{n! \, n^x}{x(x+1) \dots (x+n)} \right]$$

$$= f(x) - \Gamma(x).$$

This completes the proof of Theorem 7.15. ∎

As an application of Theorem 7.15 and the fact that $\Gamma(\tfrac{1}{2}) = \sqrt{\pi}$ (Exercise 2) notice that

$$\sqrt{\pi} = \int\limits_0^\infty e^{-t} t^{-\frac{1}{2}} \, dt.$$

Performing a change of variables by putting $t = s^2$ gives

$$\sqrt{\pi} = \int\limits_0^\infty e^{-s^2} s^{-1} \, (2s) \, ds$$

$$= 2 \int\limits_0^\infty e^{-s^2} \, ds$$

That is,

$$\int\limits_0^\infty e^{-s^2} \, ds = \frac{\sqrt{\pi}}{2}$$

This integral is often used in probability theory.

Exercises

1. Show that $0 < \gamma < 1$. (An approximation to γ is .57722. It is unknown whether γ is rational or irrational.)
2. Show that $\Gamma(z) \, \Gamma(1-z) = \pi \csc \pi z$ for z not an integer. Deduce from this that $\Gamma(\tfrac{1}{2}) = \sqrt{\pi}$.
3. Show: $\sqrt{\pi} \, \Gamma(2z) = 2^{2z-1} \Gamma(z) \, \Gamma(z+\tfrac{1}{2})$. (Hint: Consider the function $\Gamma(z)$ $\Gamma(z+\tfrac{1}{2}) \, \Gamma(2z)^{-1}$.)
4. Show that $\log \Gamma(z)$ is defined for z in $\mathbb{C} - (-\infty, 0]$ and that

$$\log \Gamma(z) = -\log z - \gamma z - \sum_{n=1}^\infty \left[\log\left(1 + \frac{z}{n}\right) - \frac{z}{n} \right].$$

5. Let f be analytic on the right half plane Re $z > 0$ and satisfy: $f(1) = 1$, $f(z+1) = zf(z)$, and $\lim\limits_{n \to \infty} \dfrac{f(z+n)}{n^z f(n)} = 1$ for all z. Show that $f = \Gamma$.

6. Show that

$$\Gamma(z) = \sum_{n=0}^{\infty} \frac{(-1)^n}{n!(z+n)} + \int_1^{\infty} e^{-t} t^{z-1} \, dt$$

for $z \neq 0, -1, -2, \ldots$ (not for $\operatorname{Re} z > 0$ alone).

7. Show that

$$\int_0^{\infty} \sin(t^2) \, dt = \int_0^{\infty} \cos(t^2) \, dt = \tfrac{1}{2}\sqrt{\tfrac{1}{2}\pi}.$$

8. Let $u > 0$ and $v > 0$ and express $\Gamma(u)\,\Gamma(v)$ as a double integral over the first quadrant of the plane. By changing to polar coordinates show that

$$\Gamma(u)\,\Gamma(v) = 2\Gamma(u+v) \int_0^{\pi/2} (\cos\theta)^{2u-1} (\sin\theta)2^{v-1} \, d\theta.$$

The function

$$B(u, v) = \frac{\Gamma(u)\,\Gamma(v)}{\Gamma(u+v)}$$

is called the *beta function*. By changes of variables show that

$$B(u, v) = \int_0^1 t^{u-1} (1-t)^{v-1} \, dt$$

$$= \int_0^{\infty} \frac{t^{u-1}}{(1+t)^{u+v}} \, dt$$

Can this be generalized to the case when u and v are complex numbers with positive real part?

9. Let α_n be the volume of the ball of radius one in $\mathbb{R}^n$ ($n \geq 1$). Prove by induction and iterated integrals that

$$\alpha_n = 2\alpha_{n-1} \int_0^1 (1-t^2)^{(n-1)/2} \, dt$$

10. Show that

$$\alpha_n = \frac{\pi^{n/2}}{(n/2)\Gamma(n/2)}$$

where α_n is defined in problem 9. Show that if $n = 2k$, $k \geq 1$, then $\alpha_n = \pi^k/k!$

11. The Gaussian psi function is defined by

$$\Psi(z) = \frac{\Gamma'(z)}{\Gamma(z)}$$

(a) Show that Ψ is meromorphic in $\mathbb{C}$ with simple poles at $z = 0, -1, \ldots$ and $\operatorname{Res}(\Psi; -n) = -1$ for $n \geq 0$.

(b) Show that $\Psi'(1) = -\gamma$.

(c) Show that $\Psi(z+1) - \Psi(z) = \dfrac{1}{z}$.

(d) Show that $\Psi(z) - \Psi(1-z) = -\pi \cot \pi z$.

(e) State and prove a characterization of Ψ analogous to the Bohr-Mollerup Theorem.

§8. The Riemann zeta function

Let z be a complex number and n a positive integer. Then $|n^z| = |\exp (z \log n)| = \exp (\operatorname{Re} z \log n)$. Thus

$$\sum_{k=1}^{n} |k^{-z}| = \sum_{k=1}^{n} \exp (-\operatorname{Re} z \log k)$$

$$= \sum_{k=1}^{n} k^{-\operatorname{Re} z}$$

So if $\operatorname{Re} z \geq 1 + \epsilon$ then

$$\sum_{k=1}^{n} |k^{-z}| \leq \sum_{k=1}^{n} k^{-(1+\epsilon)};$$

that is, the series

$$\sum_{n=1}^{\infty} n^{-z}$$

converges uniformly and absolutely on $\{z: \operatorname{Re} z \geq 1 + \epsilon\}$. In particular, this series converges in $H(\{z: \operatorname{Re} z > 1\})$ to an analytic function $\zeta(z)$.

8.1 Definition. The *Riemann zeta function* is defined for $\operatorname{Re} z > 1$ by the equation

$$\zeta(z) = \sum_{n=1}^{\infty} n^{-z}.$$

The zeta function, as well as the gamma function, has been the subject of an enormous amount of mathematical research since their introduction. The analysis of the zeta function has had a profound effect on number theory and this has, in turn, inspired more work on ζ. In fact, one of the most famous unsolved problems in Mathematics is the location of the zeros of the zeta function.

We wish to demonstrate a relationship between the zeta function and the gamma function. To do this we appeal to Theorem 7.15 and write

$$\Gamma(z) = \int_{0}^{\infty} e^{-t} t^{z-1}\, dt.$$

for Re $z > 0$. Performing a change of variable in this integral by letting $t = nu$ gives

$$\Gamma(z) = n^z \int_0^\infty e^{-nt} t^{z-1} \, dt;$$

that is

$$n^{-z}\Gamma(z) = \int_0^\infty e^{-nt} t^{z-1} \, dt.$$

If Re $z > 1$ and we sum this equation over all positive n, then

8.2
$$\zeta(z)\Gamma(z) = \sum_{n=1}^\infty n^{-z}\Gamma(z)$$

$$= \sum_{n=1}^\infty \int_0^\infty e^{-nt} t^{z-1} \, dt.$$

We wish to show that this infinite sum can be taken inside the integral sign. But first, an analogue of Lemma 7.16.

8.3 Lemma. (a) *Let* $S = \{z: \text{Re } z \geq a\}$ *where* $a > 1$. *If* $\epsilon > 0$ *then there is a number* $\delta, 0 < \delta < 1$, *such that for all z in S*

$$\left| \int_\alpha^\beta (e^t - 1)^{-1} t^{z-1} \, dt \right| < \epsilon$$

whenever $\delta > \beta > \alpha$.

(b) *Let* $S = \{z: \text{Re } z \leq A\}$ *where* $-\infty < A < \infty$. *If* $\epsilon > 0$ *then there is a number* $\kappa > 1$ *such that for all z in S*

$$\left| \int_\alpha^\beta (e^t - 1)^{-1} t^{z-1} \, dt \right| < \epsilon$$

whenever $\beta > \alpha > \kappa$.

Proof. (a) Since $e^t - 1 \geq t$ for all $t \geq 0$ we have that for $0 < t \leq 1$ and z in S

$$|(e^t - 1)^{-1} t^{z-1}| \leq t^{a-2}.$$

Since $a > 1$ the integral $\int_0^1 t^{a-2} \, dt$ is finite so that δ can be found to satisfy (a).

(b) If $t \geq 1$ and z is any point in S then, as in the proof of Lemma 7.16 (b), there is a constant c such that

$$|(e^t - 1)^{-1} t^{z-1}| \leq (e^t - 1)^{-1} t^{A-1} \leq c \, e^{\frac{1}{2}t} (e^t - 1)^{-1}.$$

Since $e^{\frac{1}{2}t}(e^t - 1)^{-1}$ is integrable on $[1, \infty)$ the required number κ can be found. ∎

8.4 Corollary. (a) *If* $S = \{z: a \leq \mathrm{Re}\ z \leq A\}$ *where* $1 < a < A < \infty$ *then the integral*

$$\int_0^\infty (e^t - 1)^{-1} t^{z-1}\, dt$$

converges uniformly on S.
  (b) *If* $S = \{z: \mathrm{Re}\ z \leq A\}$ *where* $-\infty < A < \infty$ *then the integral*

$$\int_1^\infty (e^t - 1)^{-1} t^{z-1}\, dt$$

converges uniformly on S.

8.5 Proposition. *For* $\mathrm{Re}\ z > 1$

$$\zeta(z)\Gamma(z) = \int_0^\infty (e^t - 1)^{-1} t^{z-1}\, dt$$

Proof. According to the above corollary this integral is an analytic function in the region $\{z: \mathrm{Re}\ z > 1\}$. Thus, it suffices to show that $\zeta(z)\Gamma(z)$ equals this integral for $z = x > 1$.
  From Lemma 8.3 there are numbers α and β, $0 < \alpha < \beta < \infty$, such that:

$$\int_0^\alpha (e^t - 1)^{-1} t^{x-1}\, dt < \frac{\epsilon}{4},$$

$$\int_\beta^\infty (e^t - 1)^{-1} t^{x-1}\, dt < \frac{\epsilon}{4}.$$

Since

$$\sum_{k=1}^n e^{-kt} \leq \sum_{k=1}^\infty e^{-kt} = (e^t - 1)^{-1}$$

for all $n \geq 1$,

$$\sum_{n=1}^\infty \int_0^\alpha e^{-nt} t^{x-1}\, dt < \frac{\epsilon}{4},$$

$$\sum_{n=1}^\infty \int_\beta^\infty e^{-nt} t^{x-1}\, dt < \frac{\epsilon}{4}.$$

Using equation (8.2) yields

$$\left| \zeta(x)\Gamma(x) - \int_0^\infty (e^t - 1)^{-1} t^{x-1}\, dt \right| \leq \epsilon + \left| \sum_{n=1}^\infty \int_\alpha^\beta e^{-nt} t^{x-1}\, dt \right.$$

$$\left. - \int_\alpha^\beta (e^t - 1)^{-1} t^{x-1}\, dt \right|$$

But $\sum e^{-nt}$ converges to $(e^t - 1)^{-1}$ uniformly on $[\alpha, \beta]$, so that the right hand side is exactly ϵ. ∎

We wish to use Proposition 8.5 to extend the domain of definition of ζ to $\{z: \text{Re } z > -1\}$ (and eventually to all $\mathbb{C}$). To do this, consider the Laurent expansion of $(e^z - 1)^{-1}$; this is

8.6
$$\frac{1}{e^z - 1} = \frac{1}{z} - \frac{1}{2} + \sum_{n=1}^{\infty} a_n z^n$$

for some constants $a_1, a_2, \ldots$. Thus $[(e^t - 1)^{-1} - t^{-1}]$ remains bounded in a neighborhood of $t = 0$. But this implies that the integral

$$\int_0^1 \left(\frac{1}{e^t - 1} - \frac{1}{t} \right) t^{z-1} \, dt$$

converges uniformly on compact subsets of the right half plane $\{z: \text{Re } z > 0\}$ and therefore represents an analytic function there. Hence

8.7
$$\zeta(z)\Gamma(z) = \int_0^1 \left(\frac{1}{e^t - 1} - \frac{1}{t} \right) t^{z-1} \, dt + (z-1)^{-1} + \int_1^{\infty} \frac{t^{z-1}}{e^t - 1} \, dt,$$

and (using Corollary 8.4(b)) each of these summands, except $(z-1)^{-1}$, is analytic in the right half plane. Thus one may define $\zeta(z)$ for $\text{Re } z > 0$ by setting it equal to $[\Gamma(z)]^{-1}$ times the right hand side of (8.7). In this manner ζ is meromorphic in the right half plane with a simple pole at $z = 1$ ($\sum n^{-1}$ diverges) whose residue is 1.

Now suppose $0 < \text{Re } z < 1$; then

$$(z-1)^{-1} = - \int_1^{\infty} t^{z-2} \, dt.$$

Applying this to equation (8.7) gives

8.8
$$\zeta(z)\Gamma(z) = \int_0^{\infty} \left(\frac{1}{e^t - 1} - \frac{1}{t} \right) t^{z-1} \, dt, \ 0 < \text{Re } z < 1.$$

Again considering the Laurent expansion of $(e^z - 1)^{-1}$ (8.6) we see that $[(e^t - 1)^{-1} - t^{-1} + \frac{1}{2}] \leq ct$ for some constant c and all t in the unit interval $[0, 1]$. Thus the integral

$$\int_0^1 \left(\frac{1}{e^t - 1} - \frac{1}{t} + \frac{1}{2} \right) t^{z-1} \, dt$$

is uniformly convergent on compact subsets of $\{z: \text{Re } z > -1\}$. Also, since

$$\lim_{t \to \infty} t \left(\frac{1}{t} - \frac{1}{e^t - 1} \right) = 1$$

there is a constant c' such that

$$\left(\frac{1}{t} - \frac{1}{e^t - 1}\right) \le \frac{c'}{t}, \quad t \ge 1.$$

This gives that the integral

$$\int_1^\infty \left(\frac{1}{e^t-1} - \frac{1}{t}\right) t^{z-1} \, dt$$

converges uniformly on compact subsets of $\{z: \operatorname{Re} z < 1\}$. Using these last two integrals with equation (8.8) gives

8.9 $\zeta(z)\Gamma(z) = \displaystyle\int_0^1 \left(\frac{1}{e^t-1} - \frac{1}{t} + \frac{1}{2}\right) t^{z-1} \, dt - \frac{1}{2z} + \int_1^\infty \left(\frac{1}{e^t-1} - \frac{1}{t}\right) t^{z-1} \, dt$

for $0 < \operatorname{Re} z < 1$. But since both integrals converge in the strip $-1 < \operatorname{Re} z < 1$ (8.9) can be used to define $\zeta(z)$ in $\{z: -1 < \operatorname{Re} z < 1\}$. What happens at $z = 0$? Since the term $(2z)^{-1}$ appears on the right hand side of (8.9) will ζ have a pole at $z = 0$? The answer is no. To define $\zeta(z)$ we must divide (8.9) by $\Gamma(z)$. When this happens the term in question becomes $[2z\Gamma(z)]^{-1} = [2\Gamma(z+1)]^{-1}$ which is analytic at $z = 0$. Thus, if ζ is so defined in the strip $\{z: -1 < \operatorname{Re} z < 1\}$ it is analytic there. If this is combined with (8.7), $\zeta(z)$ is defined for $\operatorname{Re} z > -1$ with a simple pole at $z = 1$.

Now if $-1 < \operatorname{Re} z < 0$ then

$$\int_1^\infty t^{z-1} \, dt = -\frac{1}{z};$$

inserting this in (8.9) gives

8.10 $\zeta(z)\Gamma(z) = \displaystyle\int_0^\infty \left(\frac{1}{e^t-1} - \frac{1}{t} + \frac{1}{2}\right) t^{z-1} \, dt, \quad -1 < \operatorname{Re} z < 0.$

But

$$\frac{1}{e^t-1} + \frac{1}{2} = \frac{1}{2}\left(\frac{e^t+1}{e^t-1}\right) = \frac{i}{2} \cot\left(\tfrac{1}{2} it\right).$$

A straightforward computation with Exercise V. 2.8 gives

$$\cot\left(\tfrac{1}{2} it\right) = \frac{2}{it} - 4it \sum_{n=1}^\infty \frac{1}{t^2 + 4n^2\pi^2}$$

for $t \ne 0$. Thus

$$\left(\frac{1}{e^t-1} - \frac{1}{t} + \frac{1}{2}\right)\frac{1}{t} = 2 \sum_{n=1}^\infty \frac{1}{t^2 + 4n^2\pi^2}$$

Applying this to (8.10) gives

8.11
$$\zeta(z)\Gamma(z) = 2\int_0^\infty \left(\sum_{n=1}^\infty \frac{1}{t^2+4n^2\pi^2}\right) t^z \, dt$$

$$= 2\sum_{n=1}^\infty \int_0^\infty \frac{t^z}{t^2+4n^2\pi^2}\, dt$$

$$= 2\sum_{n=1}^\infty (2\pi n)^{z-1} \int_0^\infty \frac{t^z}{t^2+1}\, dt$$

$$= 2(2\pi)^{z-1}\zeta(1-z)\int_0^\infty \frac{t^z}{t^2+1}\, dt,$$

for $-1 < \operatorname{Re} z < 0$. (It is left to the reader to justify the interchanging of the sum and the integral.) Now for x a real number with $-1 < x < 0$, the change of variable $s = t^2$ gives (by Example V. 2.12)

8.12
$$\int_0^\infty \frac{t^x}{t^2+1}\, dt = \frac{1}{2}\int_0^\infty \frac{s^{\frac{1}{2}(x-1)}}{s+1}\, ds$$

$$= \frac{1}{2}\pi \operatorname{cosec} [\tfrac{1}{2}\pi(1-x)]$$

$$= \frac{1}{2}\pi \sec (\tfrac{1}{2}\pi x).$$

But Exercise 7.2 gives

$$\frac{1}{\Gamma(x)} = \frac{\Gamma(1-x)}{\pi}\sin \pi x = \frac{\Gamma(1-x)}{\pi}[2\sin (\tfrac{1}{2}\pi x)\cos (\tfrac{1}{2}\pi x)]$$

Combining this with (8.11) and (8.12) yields the following.

8.13 Riemann's Functional Equation.

$$\zeta(z) = 2(2\pi)^{z-1}\Gamma(1-z)\zeta(1-z)\sin (\tfrac{1}{2}\pi z)$$

for $-1 < \operatorname{Re} z < 0$.

Actually this was shown for x real and in $(-1, 0)$; but since both sides of (8.13) are analytic in the strip $-1 < \operatorname{Re} z < 0$, (8.13) follows. The same type of reasoning gives that (8.13) holds for $-1 < \operatorname{Re} z < 1$ (what happens at $z = 0$?). But we wish to do more than this. We notice that the right hand side of (8.13) is analytic in the left hand plane $\operatorname{Re} z < 0$. Thus, use (8.13) to extend the definition of $\zeta(z)$ to $\operatorname{Re} z < 0$. We summarize what was done as follows.

8.14 Theorem. *The zeta function can be defined to be meromorphic in the plane with only a simple pole at $z = 1$ and* Res $(\zeta; 1) = 1$. *For $z \neq 1$ ζ satisfies Riemann's functional equation.*

Since $\Gamma(1-z)$ has a pole at $z = 1, 2, \ldots$ and since ζ is analytic at $z = 2, 3, \ldots$ we know, from Riemann's functional equation, that

8.15
$$\zeta(1-z) \sin (\tfrac{1}{2}\pi z) = 0$$

for $z = 2, 3, \ldots$. Furthermore, since the pole of $\Gamma(1-z)$ at $z = 2, 3, \ldots$ is simple, each of the zeros of (8.15) must be simple. Since $\sin (\tfrac{1}{2}\pi z) = 0$ whenever z is an even integer, $\zeta(1-z) = 0$ for $z = 3, 5, \ldots$. That is $\zeta(z) = 0$ for $z = -2, -4, -6, \ldots$. Similar reasoning gives that ζ has no other zeros outside the closed strip $\{z: 0 \leq \text{Re } z \leq 1\}$.

8.16 Definition. The points $z = -2, -4, \ldots$ are called the *trivial zeros of ζ* and the strip $\{z: 0 \leq \text{Re } z \leq 1\}$ is called the *critical strip*.

We now are in a position to state one of the most celebrated open questions in all of Mathematics. Is the following true?

The Riemann Hypothesis. *If z is a zero of the zeta function in the critical strip then* Re $z = \tfrac{1}{2}$.

It is known that there are no zeros of ζ on the line Re $z = 1$ (and hence none on Re $z = 0$ by the functional equation) and there are an infinite number of zeros on the line Re $z = \tfrac{1}{2}$. But no one has been able to show that ζ has any zeros off the line Re $z = \tfrac{1}{2}$ and no one has been able to show that all zeros must lie on the line.

A positive resolution of the Riemann Hypothesis will have numerous beneficial effects on number theory. Perhaps the best way to realize the connection between the zeta function and number theory is to prove the following theorem.

8.17 Euler's Theorem. *If* Re $z > 1$ *then*

$$\zeta(z) = \prod_{n=1}^{\infty} \left(\frac{1}{1-p_n^{-z}} \right)$$

where $\{p_n\}$ is the sequence of prime numbers.

Proof. First use the geometric series to find

8.18
$$\frac{1}{1-p_n^{-z}} = \sum_{m=0}^{\infty} p_n^{-mz}$$

for all $n \geq 1$. Now if $n \geq 1$ and we take the product of the terms $(1-p_k^{-z})^{-1}$ for $1 \leq k \leq n$, then by the distributive law of multiplication and by (8.18),

8.19
$$\prod_{k=1}^{n} \left(\frac{1}{1-p_k^{-z}} \right) = \sum_{j=1}^{\infty} n_j^{-z}$$

where the integers $n_1, n_2, \ldots$ are all the integers which can be factored as a product of powers of the prime numbers $p_1, \ldots, p_n$ alone. (The reason that no number n_j^{-z} has a coefficient in this expansion other than 1 is that the factorization of n_j into the product of primes is unique.) By letting $n \to \infty$ the result is achieved. ∎

Exercises

1. Let $\xi(z) = z(z-1)\pi^{-\frac{1}{2}z}\zeta(z)\Gamma(\frac{1}{2}z)$ and show that ξ is an entire function which satisfies the functional equation $\xi(z) = \xi(1-z)$.

2. Use Theorem 8.17 to prove that $\sum p_n^{-1} = \infty$. Notice that this implies that there are an infinite number of primes.

3. Prove that $\zeta^2(z) = \displaystyle\sum_{n=1}^{\infty} \frac{d(n)}{n^z}$ for Re $z > 1$, where $d(n)$ is the number of divisors of n.

4. Prove that $\zeta(z)\zeta(z-1) = \displaystyle\sum_{n=1}^{\infty} \frac{\sigma(n)}{n^z}$, for Re $z > 1$, where $\sigma(n)$ is the sum of the divisors of n.

5. Prove that $\dfrac{\zeta(z-1)}{\zeta(z)} = \displaystyle\sum_{n=1}^{\infty} \frac{\varphi(n)}{n^z}$ for Re $z > 1$, where $\varphi(n)$ is the number of integers less than n and which are relatively prime to n.

6. Prove that $\dfrac{1}{\zeta(z)} = \displaystyle\sum_{n=1}^{\infty} \frac{\mu(n)}{n^z}$ for Re $z > 1$, where $\mu(n)$ is defined as follows. Let $n = p_1^{k_1} p_2^{k_2} \ldots p_m^{k_m}$ be the factorization of n into a product of primes $p_1, \ldots, p_m$ and suppose that these primes are distinct. Let $\mu(1) = 1$; if $k_1 = \ldots = k_m = 1$ then let $\mu(n) = (-1)^m$; otherwise let $\mu(n) = 0$.

7. Prove that $\dfrac{\zeta'(z)}{\zeta(z)} = -\displaystyle\sum_{n=1}^{\infty} \frac{\Lambda(n)}{n^z}$ for Re $z > 1$, where $\Lambda(n) = \log p$ if $n = p^m$ for some prime p and $m \geq 1$; and $\Lambda(n) = 0$ otherwise.

8. (a) Let $\eta(z) = \zeta'(z)/\zeta(z)$ for Re $z > 1$ and show that $\lim_{z \to z_0} (z-z_0)\eta(z)$ is always an integer for Re $z_0 \geq 1$. Characterize the point z_0 (in its relation to ζ) in terms of the sign of this integer.

(b) Show that for $\epsilon > 0$

$$\mathrm{Re}\,\eta\,(1+\epsilon+it) = -\sum_{n=1}^{\infty} \Lambda(n)n^{-(1+\epsilon)} \cos{(t \log n)}$$

where $\Lambda(n)$ is defined in Exercise 7.

(c) Show that for all $\epsilon > 0$,

$$3\mathrm{Re}\,\eta\,(1+\epsilon)+4\mathrm{Re}\,\eta\,(1+\epsilon+it)+\mathrm{Re}\,\eta\,(1+\epsilon+2it) \leq 0.$$

(d) Show that $\zeta(z) \neq 0$ if Re $z = 1$ (or 0).

Chapter VIII

Runge's Theorem

In this chapter we will prove Runge's Theorem and investigate simple connectedness. Also proved is a Theorem of Mittag-Leffler on the existence of meromorphic functions with prescribed poles and singular parts.

§1. Runge's Theorem

In Chapter IV we saw that an analytic function in an open disk is given by a power series. Furthermore, on proper subdisks the power series converges uniformly to the function. As a corollary to this result, an analytic function on a disk D is the limit in $H(D)$ of a sequence of polynomials. We ask the question: Can this be generalized to arbitrary regions G? The answer is no. As one might expect the counter-example is furnished by $G = \{z: 0 < |z| < 2\}$. If $\{p_n(z)\}$ is a sequence of polynomials which converges to an analytic function f on G, and γ is the circle $|z| = 1$ then $\int_\gamma f = \lim \int_\gamma p_n = 0$. But z^{-1} is in $H(G)$ and $\int_\gamma z^{-1} \neq 0$.

The fact that functions analytic on a disk are limits of polynomials is due to the fact that disks are simply connected. If G is a punctured disk then the Laurent series development shows that each analytic function on G is the uniform limit of rational functions whose poles lie outside G (in fact at the center of G). That is, each f in $H(G)$ is the limit of a sequence of rational functions which also belong to $H(G)$. This is what can be generalized to arbitrary regions, and it is part of the content of Runge's Theorem.

We begin by proving a version of the Cauchy Integral Formula. Unlike the former version, however, the next proposition says that there exist curves such that the formula holds; not that the formula holds for every curve.

1.1 Proposition. *Let K be a compact subset of the region G; then there are straight line segments $\gamma_1, \ldots, \gamma_n$ in $G - K$ such that for every function f in $H(G)$,*

$$f(z) = \sum_{k=1}^{n} \frac{1}{2\pi i} \int_{\gamma_k} \frac{f(w)}{w-z} \, dw$$

for all z in K. The line segments form a finite number of closed polygons.

Proof. Observe that by enlarging K a little we may assume that $K = (\operatorname{int} K)^-$. Let $0 < \delta < \frac{1}{2} d(K, \mathbb{C} - G)$ and place a "grid" of horizontal and vertical lines in the plane such that consecutive lines are less than a distance δ apart. Let $R_1, \ldots, R_m$ be the resulting rectangles that intersect K

(there are only a finite number of them because K is compact). Also let ∂R_j be the boundary of R_j, $1 \le j \le m$, considered as a polygon with the counter-clockwise direction.

If $z \in R_j$, $1 \le j \le m$, then $d(z, K) < \sqrt{2}\delta$ so that $R_j \subset G$ by the choice of δ. Also, many of the sides of the rectangles $R_1, \ldots, R_m$ will intersect. Suppose R_j and R_i have a common side and let σ_j and σ_i be the line segments in ∂R_j, and ∂R_i respectively, such that $R_i \cap R_j = \{\sigma_j\} = \{\sigma_i\}$. From the direction given ∂R_j and ∂R_i, σ_j and σ_i are directed in the opposite sense. So if φ is any continuous function on $\{\sigma_j\}$,

$$\int_{\sigma_j} \varphi + \int_{\sigma_i} \varphi = 0.$$

Let $\gamma_1, \ldots, \gamma_n$ be those directed line segments that constitute a side of exactly one of the R_j, $1 \le j \le m$. Thus

1.2
$$\sum_{k=1}^{n} \int_{\gamma_k} \varphi = \sum_{j=1}^{m} \int_{\partial R_j} \varphi$$

for every continuous function φ on $\bigcup_{j=1}^{m} \partial R_j$.

We claim that each γ_k is in $G - K$. In fact, if one of the γ_k intersects K, it is easy to see that there are two rectangles in the grid with γ_k as a side and so both meet K. That is, γ_k is the common side of two of the rectangles $R_1, \ldots, R_m$ and this contradicts the choice of γ_k.

If z belongs to K and is not on the boundary of any R_j then

$$\varphi(w) = \frac{1}{2\pi i} \left(\frac{f(w)}{w - z} \right)$$

is continuous on $\bigcup_{j=1}^{m} \partial R_j$ for f in $H(G)$. It follows from (1.2) that

1.3
$$\sum_{j=1}^{m} \frac{1}{2\pi i} \int_{\partial R_j} \frac{f(w)}{w - z} dw = \sum_{k=1}^{n} \frac{1}{2\pi i} \int_{\gamma_k} \frac{f(w)}{w - z} dw.$$

But z belongs to the interior of exactly one R_j. If $z \notin R_j$,

$$\frac{1}{2\pi i} \int_{\partial R_j} \frac{f(w)}{w - z} dw = 0;$$

and if z is in R_j, this integral equals $f(z)$ by Cauchy's Formula. Thus (1.3) becomes

1.4
$$f(z) = \sum_{k=1}^{n} \frac{1}{2\pi i} \int_{\gamma_k} \frac{f(w)}{w - z} dw$$

whenever $z \in K - \bigcup_{j=1}^{m} \partial R_j$. But both sides of (1.4) are continuous functions on K (because each γ_k misses K) and they agree on a dense subset of K. Thus, (1.4) holds for all z in K. The remainder of the proof follows. ∎

This next lemma provides the first step in obtaining approximation by rational functions.

1.5 Lemma. *Let γ be a rectifiable curve and let K be a compact set such that $K \cap \{\gamma\} = \square$. If f is a continuous function on $\{\gamma\}$ and $\epsilon > 0$ then there is a rational function $R(z)$ having all its poles on $\{\gamma\}$ and such that*

$$\left| \int_\gamma \frac{f(w)}{w-z} \, dw - R(z) \right| < \epsilon$$

for all z in K.

Proof. Since K and $\{\gamma\}$ are disjoint there is a number r with $0 < r < d(K, \{\gamma\})$. If γ is defined on $[0, 1]$ then for $0 \leq s, t \leq 1$ and z in K

$$\left| \frac{f(\gamma(t))}{\gamma(t)-z} - \frac{f(\gamma(s))}{\gamma(s)-z} \right| \leq \frac{1}{r^2} \left| f(\gamma(t))\gamma(s) - f(\gamma(s))\gamma(t) - z[f(\gamma(t)) - f(\gamma(s))] \right|$$

$$\leq \frac{1}{r^2} \left| f(\gamma(t)) \right| \left| \gamma(s) - \gamma(t) \right| + \frac{1}{r^2} \left| \gamma(t) \right| \left| f(\gamma(s)) \right.$$

$$\left. - f(\gamma(t)) \right| + \frac{|z|}{r^2} \left| f(\gamma(s)) - f(\gamma(t)) \right|$$

There is a constant $c > 0$ such that $|z| \leq c$ for all z in K, $|\gamma(t)| \leq c$ and $|f(\gamma(t))| \leq c$ for all t in $[0, 1]$. This gives that for all s and t in $[0, 1]$ and z in K,

$$\left| \frac{f(\gamma(t))}{\gamma(t)-z} - \frac{f(\gamma(s))}{\gamma(s)-z} \right| \leq \frac{c}{r^2} |\gamma(s) - \gamma(t)| + \frac{2c}{r^2} |f(\gamma(s)) - f(\gamma(t))|$$

Since both γ and $f \circ \gamma$ are uniformly continuous on $[0, 1]$, there is a partition $\{0 = t_0 < t_1 < \ldots < t_n = 1\}$ such that

1.6 $$\left| \frac{f(\gamma(t))}{\gamma(t)-z} - \frac{f(\gamma(t_j))}{\gamma(t_j)-z} \right| < \frac{\epsilon}{V(\gamma)}$$

for $t_{j-1} \leq t \leq t_j$, $1 \leq j \leq n$, and z in K. Define $R(z)$ to be the rational function

$$R(z) = \sum_{j=1}^n f(\gamma(t_{j-1})) [\gamma(t_j) - \gamma(t_{j-1})] [\gamma(t_{j-1}) - z]^{-1}$$

The poles of $R(z)$ are $\gamma(0), \gamma(t_1), \ldots, \gamma(t_{n-1})$. Using (1.6) yields that

$$\left| \int_\gamma \frac{f(w)}{w-z} \, dw - R(z) \right| = \left| \sum_{j=1}^n \int_{t_{j-1}}^{t_j} \left[\frac{f(\gamma(t))}{\gamma(t)-z} - \frac{f(\gamma(t_{j-1}))}{\gamma(t_{j-1})-z} \right] d\gamma(t) \right|$$

$$\leq \frac{\epsilon}{V(\gamma)} \sum_{j=1}^n \int_{t_{j-1}}^{t_j} d|\gamma|(t)$$

$$= \epsilon$$

for all z in K. ■

Before stating Runge's Theorem let us agree to say that a polynomial is a rational function with a pole at ∞. It is easy to see that a rational function whose only pole is at ∞ is a polynomial.

1.7 Runge's Theorem. *Let K be a compact subset of $\mathbb{C}$ and let E be a subset of $\mathbb{C}_\infty - K$ that meets each component of $\mathbb{C}_\infty - K$. If f is analytic in an open set containing K and $\epsilon > 0$ then there is a rational function $R(z)$ whose only poles lie in E and such that*

$$|f(z) - R(z)| < \epsilon$$

for all z in K.

The proof that will be given here was obtained by S. Grabiner (*Amer. Math. Monthly*, **83** (1976), 807–808). For this proof we place the result in a different setting. On the space $C(K, \mathbb{C})$ we define a distance function ρ by

$$\rho(f, g) = \sup\{|f(z) - g(z)| : z \in K\}$$

for f and g in $C(K, \mathbb{C})$. It is easy to see that $\rho(f_n, f) \to 0$ iff $f = u - \lim f_n$ on K. Hence $C(K, \mathbb{C})$ is a complete metric space.

So Runge's Theorem says that if f is analytic on a neighborhood of K and $\epsilon > 0$ then there is a rational function $R(z)$ with poles in E such that $\rho(f, R) < \epsilon$. By taking $\epsilon = 1/n$ it is seen that we want to find a sequence of rational functions $\{R_n(z)\}$ with poles in E such that $\rho(f, R_n) \to 0$; that is, such that $f = u - \lim R_n$ on K.

Let $B(E) = $ all functions f in $C(K, \mathbb{C})$ such that there is a sequence $\{R_n\}$ of rational functions with poles in E such that $f = u - \lim R_n$ on K. Runge's Theorem states that if f is analytic in a neighborhood of K then $f|K$, the restriction of f to K, is in $B(E)$.

1.8 Lemma. *$B(E)$ is a closed subalgebra of $C(K, \mathbb{C})$ that contains every rational function with a pole in E.*

To say that $B(E)$ is an algebra is to say that if f and g are in $B(E)$ and $\alpha \in \mathbb{C}$ then $\alpha f, f + g$, and fg are in $B(E)$. The proof of Lemma 1.8 is left to the reader.

1.9 Lemma. *Let V and U be open subsets of $\mathbb{C}$ with $V \subset U$ and $\partial V \cap U = \square$. If H is a component of U and $H \cap V \neq \square$ then $H \subset V$.*

Proof. Let $a \in H \cap V$ and let G be the component of V such that $a \in G$. Then $H \cup G$ is connected (II.2.6) and contained in U. Since H is a component of U, $G \subset H$. But $\partial G \subset \partial V$ and so the hypothesis of the lemma says that $\partial G \cap H = \square$. This implies that

$$H - G = H \cap [(\mathbb{C} - G^-) \cup \partial G]$$

$$= H \cap (\mathbb{C} - G^-)$$

so that $H - G$ is open in H. But G is open implies that $H - G = H \cap (\mathbb{C} - G)$ is closed in H. Since H is connected and $G \neq \square$, $H - G = \square$. That is, $H = G \subset V$. ∎

1.10 Lemma. *If $a \in \mathbb{C} - K$ then $(z-a)^{-1} \in B(E)$.*

Proof. Case 1. $\infty \notin E$.

Let $U = \mathbb{C} - K$ and let $V = \{a \in \mathbb{C} : (z-a)^{-1} \in B(E)\}$; so $E \subset V \subset U$.

1.11   *If $a \in V$ and $|b-a| < d(a,K)$ then $b \in V$.*

The condition on b gives the existence of a number r, $0 < r < 1$, such that $|b-a| < r|z-a|$ for all z in K. But

1.12
$$(z-b)^{-1} = (z-a)^{-1} \left[1 - \frac{b-a}{z-a} \right]^{-1}$$

Hence $|b-a| |z-a|^{-1} < r < 1$ for all z in K gives that

1.13
$$\left[1 - \frac{b-a}{z-a} \right]^{-1} = \sum_{n=0}^{\infty} \left(\frac{b-a}{z-a} \right)^n$$

converges uniformly on K by the Weierstrass M-test.

If $Q_n(z) = \sum_{k=0}^{n} \left(\frac{b-a}{z-a} \right)^k$ then $(z-a)^{-1} Q_n(z) \in B(E)$ since $a \in V$ and $B(E)$ is an algebra. Since $B(E)$ is closed (1.12) and the uniform convergence of (1.13) imply that $(z-b)^{-1} \in B(E)$. That is, $b \in V$.

Note that (1.11) implies that V is open.

If $b \in \partial V$ then let $\{a_n\}$ be a sequence in V with $b = \lim a_n$. Since $b \notin V$ it follows from (1.11) that $|b - a_n| \geq d(a_n, K)$. Letting $n \to \infty$ gives (by II.5.7) that $0 = d(b, K)$, or $b \in K$. Thus $\partial V \cap U = \square$.

If H is a component of $U = \mathbb{C} - K$ then $H \cap E \neq \square$, so $H \cap V \neq \square$. By Lemma 1.9, $H \subset V$. But H was arbitrary so $U \subset V$, or $V = U$.

Case 2. $\infty \in E$.

Let $d = $ the metric on $\mathbb{C}_\infty$. Choose a_0 in the unbounded component of $\mathbb{C} - K$ such that $d(a_0, \infty) \leq \frac{1}{2} d(\infty, K)$ and $|a_0| > 2 \max\{|z| : z \in K\}$. Let $E_0 = (E - \{\infty\}) \cup \{a_0\}$; so E_0 meets each component of $\mathbb{C}_\infty - K$. If $a \in \mathbb{C} - K$ then Case 1 gives that $(z-a)^{-1} \in B(E_0)$. If we can show that $(z-a_0)^{-1} \in B(E)$ then it follows that $B(E_0) \subset B(E)$ and so $(z-a)^{-1} \in B(E)$ for each a in $\mathbb{C} - K$.

Now $|z/a_0| \leq \frac{1}{2}$ for all z in K so

$$\frac{1}{z-a_0} = -\frac{1}{a_0(1-z/a_0)} = -\frac{1}{a_0} \sum_{n=0}^{\infty} (z/a_0)^n$$

converges uniformly on K. So $Q_n(z) = -a_0^{-1} \sum_{k=0}^{n} (z/a_0)^k$ is a polynomial and $(z-a_0)^{-1} = u - \lim Q_n$ on K. Since Q_n has its only pole at ∞, $Q_n \in B(E)$. Thus $(z-a_0)^{-1} \in B(E)$. ∎

The Proof of Runge's Theorem. If f is analytic on an open set G and $K \subset G$ then for each $\epsilon > 0$ Proposition 1.1 and Lemma 1.5 imply the existence of a rational function $R(z)$ with poles in $\mathbb{C} - K$ such that $|f(z) - R(z)| < \epsilon$ for all z in K. But Lemma 1.10 and the fact that $B(E)$ is an algebra gives that $R \in B(E)$. ∎

1.14 Corollary. *Let G be an open subset of the plane and let E be a subset of $\mathbb{C}_\infty - G$ such that E meets every component of $\mathbb{C}_\infty - G$. Let $R(G,E)$ be the set of rational functions with poles in E and consider $R(G,E)$ as a subspace of $H(G)$. If $f \in H(G)$ then there is a sequence $\{R_n\}$ in $R(G,E)$ such that $f = \lim R_n$. That is, $R(G,E)$ is dense in $H(G)$.*

Proof. Let K be a compact subset of G and $\epsilon > 0$; it must be show that there is an R in $R(G, E)$ such that $|f(z) - R(z)| < \epsilon$ for all z in K. According to Proposition VII.1.2 there is a compact set K_1 such that $K \subset K_1 \subset G$ and each component of $\mathbb{C}_\infty - K_1$ contains a component of $\mathbb{C}_\infty - G$. Hence, E meets each component of $\mathbb{C}_\infty - K_1$. The corollary now follows from Runge's Theorem. ∎

The next corollary follows by letting $E = \{\infty\}$ and using the fact that a rational function whose only pole is at ∞ is a polynomial.

1.15 Corollary. *If G is an open subset of $\mathbb{C}$ such that $\mathbb{C}_\infty - G$ is connected then for each analytic function f on G there is a sequence of polynomials $\{p_n\}$ such that $f = \lim p_n$ in $H(G)$.*

Corollary 1.14 can be strengthened a little by requiring only that E^- meets each component of $\mathbb{C}^\infty - G$. The reader is asked to do this (Exercise 1).

The condition that E^- meet every component of $\mathbb{C}_\infty - G$ cannot be relaxed. This can be seen by considering the punctured plane $\mathbb{C} - \{0\} = G$. So $\mathbb{C}_\infty - G = \{0, \infty\}$. Suppose that for this case we could weaken Runge's Theorem by assuming that E consisted of ∞ alone. Then for each integer $n \geq 1$ we could find a polynomial $p_n(z)$ such that

1.16
$$\left| \frac{1}{z} - p_n(z) \right| < \frac{1}{n}$$

for $\frac{1}{n} \leq |z| \leq n$. Then $|1 - zp_n(z)| \leq \frac{|z|}{n} \leq 1$ for $\frac{1}{n} \leq |z| \leq n$. But if $|z| = n$ then

$$|p_n(z)| = \frac{1}{n} |zp_n(z)| \leq \frac{1}{n} |zp_n(z) - 1| + \frac{1}{n} \leq \frac{2}{n}.$$

By the Maximum Modulus Theorem, $|p_n(z)| \leq \frac{2}{n}$ for $|z| \leq n$. In particular, $p_n(z) \to 0$ uniformly on $|z| \leq 1$. This clearly contradicts (1.16) and shows that E must be the set $\{0, \infty\}$.

Of course, the point in the above paragraph could have been made by appealing to what was said about this same example at the beginning of this section. However, this further exposition gives an introduction to a concept whose connection with Runge's Theorem is quite intimate.

1.21 Definition. Let K be a compact subset of the plane; the *polynomially* *convex hull* of K, denoted by $\hat{K}$, is defined to be the set of all points w such that for every polynomial p

$$|p(w)| \leq \max \{|p(z)|: z \in K\}.$$

That is, if the right hand side of this inequality is denoted by $\|p\|_K$, then

$$\hat{K} = \{w: |p(w)| \leq \|p\|_K \text{ for all polynomials } p\}.$$

If K is an annulus then $\hat{K}$ is the disk obtained by filling in the interior hole. In fact, if K is any compact set the Maximum Modulus Theorem gives that $\hat{K}$ is obtained by filling in any "holes" that may exist in K.

Exercises

1. Prove Corollary 1.14 if it is only assumed that E^- meets each component of $\mathbb{C}_\infty - G$.
2. Let G be the open unit disk $B(0; 1)$ and let $K = \{z: \frac{1}{4} \leq |z| \leq \frac{3}{4}\}$. Show that there is a function f analytic on some open subset G_1 containing K which cannot be approximated on K by functions in $H(G)$.

Remarks. The next two problems are concerned with the following question. Given a compact set K contained in an open set $G_1 \subset G$, can functions in $H(G_1)$ be approximated on K by functions in $H(G)$? Exercise 2 says that for an arbitrary choice of K, G, and G_1 this is not true. Exercise 4 below gives criteria for a fixed K and G such that this can be done for any G_1. Exercise 3 is a lemma which is useful in proving Exercise 4.

3. Let K be a compact subset of the open set G and suppose that any bounded component D of $G - K$ has $D^- \cap \partial G \neq \square$. Then every component of $\mathbb{C}_\infty - K$ contains a component of $\mathbb{C}_\infty - G$.
4. Let K be a compact subset of the open set G; then the following are equivalent:
    (a) If f is analytic in a neighborhood of K and $\epsilon > 0$ then there is a g in $H(G)$ with $|f(z) - g(z)| < \epsilon$ for all z in K;
    (b) If D is a bounded component of $G - K$ then $D^- \cap \partial G \neq \square$;
    (c) If z is any point in $G - K$ then there is a function f in $H(G)$ with

$$|f(z)| > \sup \{|f(w)|: w \text{ in } K\}.$$

5. Can you interpret part (c) of Exercise 4 in terms of $\hat{K}$?
6. Let K be a compact subset of the region G and define $\hat{K}_G = \{z \in G: |f(z)| \leq \|f\|_K \text{ for all } f \text{ in } H(G)\}$.
    (a) Show that if $\mathbb{C}_\infty - G$ is connected then $\hat{K}_G = \hat{K}$.
    (b) Show that $d(K, \mathbb{C} - G) = d(\hat{K}_G, \mathbb{C} - G)$.
    (c) Show that $\hat{K}_G \subset$ the convex hull of $K \equiv$ the intersection of all convex subsets of $\mathbb{C}$ which contain K.

(d) If $\hat{K}_G \subset G_1 \subset G$ and G_1 is open then for every g in $H(G_1)$ and $\epsilon > 0$ there is a function f in $H(G)$ such that $|f(z) - g(z)| < \epsilon$ for all z in $\hat{K}_G$. (Hint: see Exercise 4.)

(e) $\hat{K}_G$ = the union of K and all bounded components of $G - K$ whose closure does not intersect ∂G.

§2. Simple connectedness

Recall that an open connected set G is simply connected if and only if every closed rectifiable curve in G is homotopic to zero. The purpose of this section is to prove some equivalent formulations of simple connectedness.

2.1 Definition. Let X and Ω be metric spaces; a *homeomorphism* between X and Ω is a continuous map $f: X \to \Omega$ which is one-one, onto, and such that $f^{-1}: \Omega \to X$ is also continuous.

Notice that if $f: X \to \Omega$ is one-one, onto, and continuous then f is a homeomorphism if and only if f is open (or, equivalently, f is closed).

If there is a homeomorphism between X and Ω then the metric spaces X and Ω are *homeomorphic*.

We claim that $\mathbb{C}$ and $D = \{z: |z| < 1\}$ are homeomorphic. In fact $f(z) = z(1 + |z|)^{-1}$ maps $\mathbb{C}$ onto D in a one-one fashion and its inverse, $f^{-1}(\omega) = \omega(1 - |\omega|)^{-1}$, is clearly continuous. Also, if f is a one-one analytic function on an open set G and $\Omega = f(G)$ then G and Ω are homeomorphic. Finally, all annuli are homeomorphic to the punctured plane.

2.2 Theorem. *Let G be an open connected subset of $\mathbb{C}$. Then the following are equivalent*:

(a) *G is simply connected*;

(b) *$n(\gamma; a) = 0$ for every closed rectifiable curve γ in G and every point a in $\mathbb{C} - G$*;

(c) *$\mathbb{C}_\infty - G$ is connected*;

(d) *For any f in $H(G)$ there is a sequence of polynomials that converges to f in $H(G)$*;

(e) *For any f in $H(G)$ and any closed rectifiable curve γ in G, $\int_\gamma f = 0$*;

(f) *Every function f in $H(G)$ has a primitive*;

(g) *For any f in $H(G)$ such that $f(z) \neq 0$ for all z in G there is a function g in $H(G)$ such that $f(z) = \exp g(z)$*;

(h) *For any f in $H(G)$ such that $f(z) \neq 0$ for all z in G there is a function g in $H(G)$ such that $f(z) = [g(z)]^2$*;

(i) *G is homeomorphic to the unit disk*;

(j) *If $u: G \to R$ is harmonic then there is a harmonic function $v: G \to R$ such that $f = u + iv$ is analytic on G*.

Proof. The plan is to show that (a) $\Rightarrow$ (b) $\Rightarrow \ldots \Rightarrow$ (i) $\Rightarrow$ (a) and (h) $\Rightarrow$ (j) $\Rightarrow$ (g). Many of these implications have already been done.

(a) $\Rightarrow$ (b) If γ is a closed rectifiable curve in G and a is a point in the complement of G then $(z-a)^{-1}$ is analytic in G, and part (b) follows by Cauchy's Theorem.

(b) $\Rightarrow$ (c) Suppose $\mathbb{C}_\infty - G$ is not connected; then $\mathbb{C}_\infty - G = A \cup B$ where A and B are disjoint, non-empty, closed subsets of $\mathbb{C}_\infty$. Since ∞ must be either in A or in B, suppose that ∞ is in B; thus, A must be a compact subset of $\mathbb{C}$ (A is compact in $\mathbb{C}_\infty$ and doesn't contain ∞). But then $G_1 = G \cup A = \mathbb{C}_\infty - B$ is an open set in $\mathbb{C}$ and contains A. According to Proposition 1.1 there are a finite number of polygons $\gamma_1, \ldots, \gamma_m$ in $G_1 - A = G$ such that for every analytic function f on G_1

$$f(z) = \sum_{k=1}^m \frac{1}{2\pi i} \int_{\gamma_k} \frac{f(w)}{w-z}\, dw$$

for all z in A. In particular, if $f(z) \equiv 1$ then

$$1 = \sum_{k=1}^m n(\gamma_k; z)$$

for all z in A. Thus for any z in A there is at least one polygon γ_k in G such that $n(\gamma_k; z) \neq 0$. This contradicts (b).

(c) $\Rightarrow$ (d) See Corollary 1.15.

(d) $\Rightarrow$ (e) Let γ be a closed rectifiable curve in G, let f be an analytic function on G, and let $\{p_n\}$ be a sequence of polynomials such that $f = \lim p_n$ in $H(G)$. Since each polynomial is analytic in $\mathbb{C}$ and $\gamma \sim 0$ in $\mathbb{C}$, $\int_\gamma p_n = 0$ for every n. But $\{p_n\}$ converges to f uniformly on $\{\gamma\}$ so that $\int_\gamma f = \lim \int_\gamma p_n = 0$.

(e) $\Rightarrow$ (f) Fix a in G. From condition (e) it follows that there is a function $F: G \to \mathbb{C}$ defined by letting $F(z) = \int_\gamma f$ where γ is any rectifiable curve in G from a to z. It follows that $F' = f$ (see the proof of Corollary IV. 6.16).

(f) $\Rightarrow$ (g) If $f(z) \neq 0$ for all z in G then f'/f is analytic on G. Part (f) implies there is a function F such that $F' = f'/f$. It follows (see the proof of Corollary IV. 6.17) that there is an appropriate constant c such that $g = F+c$ satisfies $f(z) = \exp g(z)$ for all z in G.

(g) $\Rightarrow$ (h) This is trivial.

(h) $\Rightarrow$ (i) If $G = \mathbb{C}$ then the function $z(1+|z|)^{-1}$ was shown to be a homeomorphism immediately prior to this theorem. If $G \neq \mathbb{C}$ then Lemma VII. 4.3 implies that there is an analytic mapping f of G onto D which is one-one. Such a map is a homeomorphism.

(i) $\Rightarrow$ (a) Let $h: G \to D = \{z: |z| < 1\}$ be a homeomorphism and let γ be a closed curve in G (note that γ is not assumed to be rectifiable). Then $\sigma(s) = h(\gamma(s))$ is a closed curve in D. Thus, there is a continuous function $\Lambda: I^2 \to D$ such that $\Lambda(s, 0) = \sigma(s)$ for $0 \le s \le 1$, $\Lambda(s, 1) = 0$ for $0 \le s \le 1$ and $\Lambda(0, t) = \Lambda(1, t)$ for $0 \le t \le 1$. It follows that $\Gamma = h^{-1} \circ \Lambda$ is a continuous map of I^2 into G and demonstrates that γ is homotopic to the curve which is constantly equal to $h^{-1}(0)$. The details are left to the reader.

(h) $\Rightarrow$ (j) Suppose that $G \neq \mathbb{C}$; then the Riemann Mapping Theorem implies there is an analytic function h on G such that h is one-one and $h(G) = D$. If $u: G \to R$ is harmonic then $u_1 = u \circ h^{-1}$ is a harmonic function on D. By Theorem III. 2.30 there is a harmonic function $v_1: D \to R$ such that $f_1 = u_1 + iv_1$ is analytic on D. Let $f = f_1 \circ h$. Then f is analytic on G and u is the real part of f. Thus $v = \operatorname{Im} f = v_1 \circ h$ is the sought after harmonic conjugate. Since Theorem III. 2.30 also applies to $\mathbb{C}$, (j) follows from (h).

(j) $\Rightarrow$ (g) Suppose $f: G \to \mathbb{C}$ is analytic and never vanishes, and let $u = \operatorname{Re} f$, $v = \operatorname{Im} f$. If $U: G \to R$ is defined by $U(x, y) = \log |f(x+iy)| = \log [u(x, y)^2 + v(x, y)^2]^{\frac{1}{2}}$ then a computation shows that U is harmonic. Let V be a harmonic function on G such that $g = U+iV$ is analytic on G and let $h(z) = \exp g(z)$. Then h is analytic, never vanishes, and $\left|\dfrac{f(z)}{h(z)}\right| = 1$ for all z in G. That is, f/h is an analytic function whose range is not open. It follows that there is a constant c such that $f(z) = c\, h(z) = c \exp g(z) = \exp [g(z) + c_1]$. Thus, $g(z) + c_1$ is a branch of $\log f(z)$.

This completes the proof of the theorem. ∎

This theorem constitutes an aesthetic peak in Mathematics. Notice that it says that a topological condition (simple connectedness) is equivalent to analytical conditions (e.g., the existence of harmonic conjugates and Cauchy's Theorem) as well as an algebraic condition (the existence of a square root) and other topological conditions. This certainly was not expected when simple connectedness was first defined. Nevertheless, the value of the theorem is somewhat limited to the fact that simple connectedness implies these nine properties. Although it is satisfying to have the converse of these implications, it is only the fact that the connectedness of $\mathbb{C}_\infty - G$ implies that G is simply connected which finds wide application. No one ever verifies one of the other properties in order to prove that G is simply connected.

For an example consider the set $G = \mathbb{C} - \{z = re^{ir}: 0 \le r < \infty\}$; that is, G is the complement of the infinite spiral $r = \theta$, $0 \le \theta < \infty$. Then $\mathbb{C}_\infty - G$ is the spiral together with the point at infinity. Since this is connected, G is simply connected.

Exercise

1. The set $G = \{re^{it}: -\infty < t < 0$ and $1+e^t < r < 1+2e^t\}$ is called a *cornucopia*. Show that G is simply connected. Let $K = G^-$; is $\operatorname{int} \hat{K}$ connected?

§3 *Mittag-Leffler's Theorem*

Consider the following problem: Let G be an open subset of $\mathbb{C}$ and let $\{a_k\}$ be a sequence of distinct points in G such that $\{a_k\}$ has no limit point in G. For each integer $k \ge 1$ consider the rational function

$$3.1 \qquad\qquad S_k(z) = \sum_{j=1}^{m_k} \frac{A_{jk}}{(z-a_k)^j},$$

where m_k is some positive integer and $A_{1k}, \ldots, A_{m_k k}$ are arbitrary complex coefficients. Is there a meromorphic function f on G whose poles are exactly the points $\{a_k\}$ and such that the singular part of f at $z = a_k$ is $S_k(z)$? The answer is yes and this is the content of Mittag-Leffler's Theorem.

3.2 Mittag-Leffler's Theorem. *Let G be an open set, $\{a_k\}$ a sequence of distinct points in G without a limit point in G, and let $\{S_k(z)\}$ be the sequence of rational functions given by equation (3.1). Then there is a meromorphic function f on G whose poles are exactly the points $\{a_k\}$ and such that the singular part of f at a_k is $S_k(z)$.*

Proof. Although the details of this proof are somewhat cumbersome, the idea is simple. We use Runge's Theorem to find rational functions $\{R_k(z)\}$ with poles in $\mathbb{C}_\infty - G$ such that $\left\{\sum_{k=1}^{n} S_k(z) - R_k(z)\right\}$ is a Cauchy sequence in $M(G)$. The resulting limit is the sought after meromorphic function. (Actually we must do a little more than this.)

Use Proposition VII. 1.2 to find compact subsets of G such that

$$G = \bigcup_{n=1}^{\infty} K_n, \ K_n \subset \text{int } K_{n+1},$$

and each component of $\mathbb{C}_\infty - K_n$ contains a component of $\mathbb{C}_\infty - G$. Since each K_n is compact and $\{a_k\}$ has no limit point in G, there are only a finite number of points a_k in each K_n. Define the sets of integers I_n as follows:

$$I_1 = \{k : a_k \in K_1\},$$

$$I_n = \{k : a_k \in K_n - K_{n-1}\}$$

for $n \geq 2$. Define functions f_n by

$$f_n(z) = \sum_{k \in I_n} S_k(z)$$

for $n \geq 1$. Then f_n is rational and its poles are the points $\{a_k : k \in I_n\} \subset K_n - K_{n-1}$. (If I_n is empty let $f_n = 0$.) Since f_n has no poles in K_{n-1} (for $n \geq 2$) it is analytic in a neighborhood of K_{n-1}. According to Runge's Theorem there is a rational function $R_n(z)$ with its poles in $\mathbb{C}_\infty - G$ and that satisfies

$$|f_n(z) - R_n(z)| < (\tfrac{1}{2})^n$$

for all z in K_{n-1}. We claim that

3.3 $$f(z) = f_1(z) + \sum_{n=2}^{\infty} [f_n(z) - R_n(z)]$$

is the desired meromorphic function. It must be shown that f is a meromorphic function and that it has the desired properties. Start by showing that the series in (3.3) converges uniformly on every compact subset of $G - \{a_k : k \geq 1\}$. This will give that f is analytic on $G - \{a_k : k \leq 1\}$ and it will only remain to show that each a_k is a pole with singular part $S_k(z)$. So let K be a compact subset of $G - \{a_k : k \geq 1\}$; then K is a compact subset of G and, therefore, there is an integer N such that $K \subset K_N$. If $n \geq N$ then $|f_n(z) - R_n(z)| < (\frac{1}{2})^n$ for all z in K. That is, the series (3.3) is dominated on K by a convergent series of numbers; by the Weierstrass M-test (II. 6.2) the series (3.3) converges uniformly on K. Thus f is analytic on $G - \{a_k : k \geq 1\}$.

Now consider a fixed integer $k \geq 1$; there is a number $R > 0$ such that $|a_j - a_k| > R$ for $j \neq k$. Thus $f(z) = S_k(z) + g(z)$ for $0 < |z - a_k| < R$, where g is analytic in $B(a_k; R)$. Hence, $z = a_k$ is a pole of f and $S_k(z)$ is its singular part. This completes the proof of the theorem. ∎

Just as there is merit in choosing the integers p_n in Weierstrass's Theorem (VII. 5.2) as small as possible, there is merit in choosing the rational functions $R_n(z)$ in (3.3) to be as simple as possible. As an example let us calculate the simplest meromorphic function in the plane with a pole at every integer n.

The simplest singular part is $(z - n)^{-1}$ but $\sum_{-\infty}^{\infty} (z - n)^{-1}$ does not converge in $M(\mathbb{C})$. However $(z - n)^{-1} + (z + n)^{-1} = 2z(z^2 - n^2)^{-1}$ and

$$\frac{1}{z} + \sum_{n=1}^{\infty} \frac{2z}{z^2 - n^2}$$

does converge in $M(\mathbb{C})$. The singular part of this function at $z = n$ is $(z - n)^{-1}$. In fact, from Exercise V. 2.8 we have that this function is $\pi \cot \pi z$.

Exercises

1. Let G be a region and let $\{a_n\}$ and $\{b_m\}$ be two sequences of distinct points in G such that $a_n \neq b_m$ for all n, m. Let $S_n(z)$ be a singular part at a_n and let p_m be a positive integer. Show that there is a meromorphic function f on G whose only poles and zeros are $\{a_n\}$ and $\{b_m\}$ respectively, the singular part at $z = a_n$ is $S_n(z)$, and $z = b_m$ is a zero of multiplicity p_m.

2. Let $\{a_n\}$ be a sequence of points in the plane such that $|a_n| \to \infty$, and let $\{b_n\}$ be an arbitrary sequence of complex numbers.

(a) Show that if integers $\{k_n\}$ can be chosen such that

3.4
$$\sum_{n=1}^{\infty} \left(\frac{r}{a_n}\right)^{k_n} \frac{b_n}{a_n}$$

converges absolutely for all $r > 0$ then

3.5
$$\sum_{n=1}^{\infty} \left(\frac{z}{a_n}\right)^{k_n} \frac{b_n}{z - a_n}$$

converges in $M(\mathbb{C})$ to a function f with poles at each point $z = a_n$.

(b) Show that if lim sup $|b_n| < \infty$ then (3.4) converges absolutely if $k_n = n$ for all n.

(c) Show that if there is an integer k such that the series

$$\text{3.6} \qquad \sum_{n=1}^{\infty} \frac{b_n}{a_n^{k+1}}$$

converges absolutely, then (3.4) converges absolutely if $k_n = k$ for all n.

(d) Suppose there is an $r > 0$ such that $|a_n - a_m| \geq r$ for all $n \neq m$. Show that $\sum |a_n|^{-3} < \infty$. In particular, if the sequence $\{b_n\}$ is bounded then the series (3.6) with $k = 2$ converges absolutely. (This is somewhat involved and the reader may prefer to prove part (f) directly since this is the only application.)

(e) Show that if the series (3.5) converges in $M(\mathbb{C})$ to a meromorphic function f then

$$f(z) = \sum_{n=1}^{\infty} \left[\frac{b_n}{z - a_n} + \frac{b_n}{a_n} \left\{ 1 + \left(\frac{z}{a_n} \right) + \ldots + \left(\frac{z}{a_n} \right)^{k_n - 1} \right\} \right]$$

(f) Let ω and ω' be two complex numbers such that Im $(\omega'/\omega) \neq 0$. Using the previous parts of this exercise show that the series

$$\zeta(z) = \frac{1}{z} + \sum' \left(\frac{1}{z - w} + \frac{1}{w} + \frac{z}{w^2} \right),$$

where the sum is over all $w = 2n\omega + 2n'\omega$ for $n, n' = 0, \pm 1, \pm 2, \ldots$ but not $w = 0$, is convergent in $M(\mathbb{C})$ to a meromorphic function ζ with simple poles at the points $2n\omega + 2n'\omega'$. This function is called the *Weierstrass zeta function*.

(g) Let $\wp(z) = -\zeta'(z)$; $\wp$ is called the *Weierstrass pe function*. Show that

$$\wp(z) = \frac{1}{z^2} + \sum' \left(\frac{1}{(z-w)^2} - \frac{1}{w^2} \right)$$

where the sum is over the same w as in part (f). Also show that

$$\wp(z) = \wp(z + 2n\omega + 2n'\omega')$$

for all integers n and n'. That is, $\wp$ is doubly periodic with periods 2ω and $2\omega'$.

3. This exercise shows how to deduce Weierstrass's Theorem for the plane (Theorem VII. 5.12) from Mittag-Leffler's Theorem.

(a) Deduce from Exercises 2(a) and 2(b) that for any sequence $\{a_n\}$ in $\mathbb{C}$ with lim $a_n = \infty$ and $a_n \neq 0$ there is a sequence of integers $\{k_n\}$ such that

$$h(z) = \sum_{n=1}^{\infty} \left[\frac{1}{z - a_n} + \frac{1}{a_n} + \frac{1}{a_n} \left(\frac{z}{a_n} \right) + \ldots + \frac{1}{a_n} \left(\frac{z}{a_n} \right)^{k_n - 1} \right]$$

is a meromorphic function on $\mathbb{C}$ with simple poles at $a_1, a_2, \ldots$.

The remainder of the proof consists of showing that there is a function f such that $h = f'/f$. This function f will then have the appropriate zeros.

(b) Let z be an arbitrary but fixed point in $\mathbb{C} - \{a_1, a_2, \ldots\}$. Show that if γ_1 and γ_2 are any rectifiable curves in $\mathbb{C} - \{a_1, a_2, \ldots\}$ from 0 to z and h is the function obtained in part (a), then there is an integer m such that

$$\int_{\gamma_1} h - \int_{\gamma_2} h = 2\pi i m.$$

(c) Again let h be the meromorphic function from part (a). Prove that for $z \neq a_1, a_2, \ldots$ and γ any rectifiable curve in $\mathbb{C} - \{a_1, a_2, \ldots\}$,

$$f(z) = \exp\left(\int_\gamma h\right)$$

defines an analytic function on $\mathbb{C} - \{a_1, a_2, \ldots\}$ with $f'/f = h$. (That is, the value of $f(z)$ is independent of the curve γ and the resulting function f is analytic.

(d) Suppose that $z \in \{a_1, a_2, \ldots\}$; show that z is a removable singularity of the function f defined in part (c). Furthermore, show that $f(z) = 0$ and that the multiplicity of this zero equals the number of times that z appears in the sequence $\{a_1, a_2, \ldots\}$.

(e) Show that

3.7 $\qquad f(z) = \prod_{n=1}^{\infty}\left(1 - \frac{z}{a_n}\right) \exp\left[\frac{z}{a_n} + \frac{1}{2}\left(\frac{z}{a_n}\right)^2 + \ldots + \frac{1}{k_n}\left(\frac{z}{a_n}\right)^{k_n}\right]$

Remark. We could have skipped parts (b), (c), and (d) and gone directly from (a) to (e). However this would have meant that we must show that (3.7) converges in $H(\mathbb{C})$ and it could hardly be classified as a new proof. The steps outlined in parts (a) through (d) give a proof of Weierstrass's Theorem without introducing infinite products.

4. This exercise assumes a knowledge of the terminology and results of Exercise VII. 5.11.

(a) Define two functions f and g in $H(G)$ to be *relatively prime* (in symbols, $(f, g) = 1$) if the only common divisors of f and g are non-vanishing functions in $H(G)$. Show that $(f, g) = 1$ iff $\mathscr{Z}(f) \cap \mathscr{Z}(g) = \square$.

(b) If $(f, g) = 1$, show that there are functions f_1, g_1 in $H(G)$ such that $ff_1 + gg_1 = 1$. (Hint: Show that there is a meromorphic function φ on G such that $f_1 = \varphi g \in H(G)$ and $g|(1 - ff_1)$.)

(c) Let $f_1, \ldots, f_n \in H(G)$ and $g = $ g.c.d $\{f_1, \ldots, f_n\}$. Show that there are functions $\varphi_1, \ldots, \varphi_n$ in $H(G)$ such that $g = \varphi_1 f_1 + \ldots + \varphi_n f_n$. (Hint: Use (b) and induction.)

(d) If $\{\mathscr{I}_\alpha\}$ is a collection of ideals in $H(G)$, show that $\mathscr{I} = \bigcap_\alpha \mathscr{I}_\alpha$ is also an ideal. If $\mathscr{S} \subset H(G)$ then let $\mathscr{I} = \cap \{\mathscr{J}: \mathscr{J}$ is an ideal of $H(G)$ and $\mathscr{S} \subset \mathscr{J}\}$. Prove that $\mathscr{I}$ is the smallest ideal in $H(G)$ that contains $\mathscr{S}$ and $\mathscr{I} = \{\varphi_1 f_1 + \ldots + \varphi_n f_n: \varphi_k \in H(G), f_k \in \mathscr{S}$ for $1 \le k \le n\}$. $\mathscr{I}$ is called the *ideal generated by* $\mathscr{S}$ and is denoted by $\mathscr{I} = (\mathscr{S})$. If $\mathscr{S}$ is finite then $(\mathscr{S})$ is called a *finitely generated ideal*. If $\mathscr{S} = \{f\}$ for a single function f then (f) is called a *principal ideal*.

(e) Show that every finitely generated ideal in $H(G)$ is a principal ideal.

(f) An ideal $\mathscr{I}$ is called a *fixed ideal* if $\mathscr{Z}(\mathscr{I}) \neq \Box$; otherwise it is called a *free ideal*. Prove that if $\mathscr{I} = (\mathscr{S})$ then $\mathscr{Z}(\mathscr{I}) = \mathscr{Z}(\mathscr{S})$ and that a principal ideal is fixed.

(g) Let $f_n(z) = \sin(2^{-n}z)$ for all $n \geq 0$ and let $\mathscr{I} = (\{f_1, f_2, \ldots\})$. Show that $\mathscr{I}$ is a fixed ideal in $H(\mathbb{C})$ which is not a principal ideal.

(h) Let $\mathscr{I}$ be a fixed ideal and prove that there is an f in $H(G)$ with $\mathscr{Z}(f) = \mathscr{Z}(\mathscr{I})$ and $\mathscr{I} \subset (f)$. Also show that $\mathscr{I} = (f)$ if $\mathscr{I}$ is finitely generated.

(i) Let $\mathscr{M}$ be a maximal ideal that is fixed. Show that there is a point a in G such that $\mathscr{M} = ((z-a))$.

(j) Let $\{a_n\}$ be a sequence of distinct points in G with no limit point in G. Let $\mathscr{I} = \{f \in H(G): f(a_n) = 0 \text{ for all but a finite number of the } a_n\}$. Show that $\mathscr{I}$ is a proper free ideal in $H(G)$.

(k) If $\mathscr{I}$ is a free ideal show that for any finite subset $\mathscr{S}$ of $\mathscr{I}$, $\mathscr{Z}(\mathscr{S}) \neq \Box$. Use this to show that $\mathscr{I}$ can contain no polynomials.

(l) Let $\mathscr{I}$ be a free ideal; then $\mathscr{I}$ is a maximal ideal iff whenever $g \in H(G)$ and $\mathscr{Z}(g) \cap \mathscr{Z}(f) \neq \Box$ for all f in $\mathscr{I}$ then $g \in \mathscr{I}$.

5. Let G be a region and let $\{a_n\}$ be a sequence of distinct points in G with no limit point in G. For each integer $n \geq 1$ choose integers $k_n \geq 0$ and constants $A_n^{(k)}$, $0 \leq k \leq k_n$. Show that there is an analytic function f on G such that $f^{(k)}(a_n) = k! A_n^{(k)}$. (Hint: Let g be an analytic function on G with a zero at a_n of multiplicity k_n. Let h be a meromorphic function on G with poles at each a_n of order k_n and with singular part $S_n(z)$. Choose the S_n so that $f = gh$ has the desired property.)

6. Find a meromorphic function with poles of order 2 at $1, \sqrt{2}, \sqrt{3}, \ldots$ such that the residue at each pole is 0 and $\lim_{z \to \sqrt{n}} (z - \sqrt{n})^2 f(z) = 1$ for all n.

Chapter IX

Analytic Continuation and Riemann Surfaces

Consider the following problem. Let f be an analytic function on a region G; when can f be extended to an analytic function f_1 on an open set G_1 which properly contains G? If G_1 is obtained by adjoining to G a disjoint open set so that G becomes a component of G_1, f can be extended to G_1 by defining it in any way we wish on $G_1 - G$ so long as the result is analytic. So to eliminate such trivial cases it is required that G_1 also be a region.

Actually, this process has already been encountered. Recall that in the discussion of the Riemann zeta function (Section VII. 8) $\zeta(z)$ was initially defined for Re $z > 1$. Using various identities, principal among which was Riemann's functional equation, ζ was extended so that it was defined and analytic in $\mathbb{C} - \{1\}$ with a simple pole at $z = 1$. That is, ζ was analytically continued from a smaller region to a larger one.

Another example was in the discussion that followed the proof of the Argument Principle (V.3.4). There a meromorphic function f and a closed rectifiable curve γ not passing through any zero or pole of f was given. If $z = a$ is the initial point of γ (and the final point), we put a disk D_1 about a on which it was possible to define a branch ℓ_1 of $\log f$. Continuing, we covered γ by a finite number of disks $D_1, D_2, \ldots, D_n$, where consecutive disks intersect and such that there is a branch ℓ_j of $\log f$ on D_j. Furthermore, the functions ℓ_j were chosen so that $\ell_j(z) = \ell_{j-1}(z)$ for z in $D_{j-1} \cap D_j$, $2 \le j \le n$. The process analytically continues ℓ_1 to $D_1 \cup D_2$, then $D_1 \cup D_2 \cup D_3$, and so on. However, an unfortunate thing (for this continuation) happened when the last disk D_n was reached. According to the Argument Principle it is distinctly possible that $\ell_n(z) \ne \ell_1(z)$ for z in $D_n \cap D_1$. In fact, $\ell_n(z) - \ell_1(z) = 2\pi i K$ for some (possibly zero) integer K.

This last example is a particularly fruitful one. This process of continuing a function along a path will be examined and a criterion will be derived which ensures that continuation around a closed curve results in the same function that begins the continuation. Also, the fact that continuation around a closed path can lead to a different function than the one started with, will introduce us to the concept of a Riemann Surface.

This chapter begins with the Schwarz Reflection Principle which is more like the process used to continue the zeta function than the process of continuing along an arc.

§1. Schwarz Reflection Principle

If G is a region and $G^* = \{z : \bar{z} \in G\}$ and if f is an analytic function on G, then $f^* : G^* \to \mathbb{C}$ defined by $f^*(z) = \overline{f(\bar{z})}$ is also analytic. Now suppose that

$G = G^*$; that is, G is symmetric with respect to the real axis. Then $g(z) = f(z) - \overline{f(\bar{z})}$ is analytic on G. Since G is connected it must be that G contains an open interval of the real line. Suppose $f(x)$ is real for all x in $G \cap \mathbb{R}$; then $g(x) \equiv 0$ for x in $G \cap \mathbb{R}$. But $G \cap \mathbb{R}$ has a limit point in G so that $f(z) = \overline{f(\bar{z})}$ for all z in G.

The fact that f must satisfy this equation is used to extend a function defined on $G \cap \{z: \operatorname{Im} z \geq 0\}$ to all of G.

If G is a symmetric region (i.e., $G = G^*$) then let $G_+ = \{z \in G: \operatorname{Im} z > 0\}$, $G_- = \{z \in G: \operatorname{Im} z < 0\}$, and $G_0 = \{z \in G: \operatorname{Im} z = 0\}$.

1.1 Schwarz Reflection Principle. *Let G be a region such that $G = G^*$. If $f: G_+ \cup G_0 \to \mathbb{C}$ is a continuous function which is analytic on G_+ and if $f(x)$ is real for x in G_0, then there is an analytic function $g: G \to \mathbb{C}$ such that $g(z) = f(z)$ for z in $G_+ \cup G_0$.*

Proof. For z in G_- define $g(z) = \overline{f(\bar{z})}$ and for z in $G_+ \cup G_0$ let $g(z) = f(z)$. It is easy to see that $g: G \to \mathbb{C}$ is continuous; it must be shown that g is analytic. It is trivial that g is analytic on $G_+ \cup G_-$ so fix a point x_0 in G_0 and let $R > 0$ with $B(x_0; R) \subset G$. It suffices to show that g is analytic on $B(x_0; R)$; to do this apply Morera's Theorem. Let $T = [a, b, c, a]$ be a triangle in $B(x_0; R)$. To show that $\int_T f = 0$ it is sufficient to show that $\int_P f = 0$

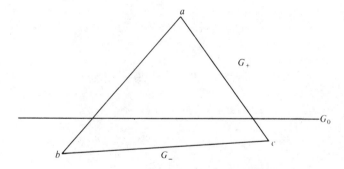

whenever P is a triangle or a quadrilateral lying entirely in $G_+ \cup G_0$ or $G_- \cup G_0$. In fact, this is easily seen by considering various pictures such as the one above. Therefore assume that $T \subset G_+ \cup G_0$ and $[a, b] \subset G_0$. The proof of the other cases is similar and will be left to the reader. (See Exercise 1 for a general proposition which proves all these cases at once.)

Let Δ designate T together with its inside; then $g(z) = f(z)$ for all z in Δ. By hypothesis f is continuous on $G_+ \cup G_0$ and so f is uniformly continuous on Δ. So if $\epsilon > 0$ there is a $\delta > 0$ such that when z and $z' \in \Delta$ and $|z - z'| < \delta$ then $|f(z) - f(z')| < \epsilon$. Now choose α and β on the line segments $[c, a]$ and $[b, c]$ respectively, so that $|\alpha - a| < \delta$ and $|\beta - b| < \delta$. Let $T_1 = [\alpha, \beta, c, \alpha]$ and $Q = [a, b, \beta, \alpha, a]$. Then $\int_T f = \int_{T_1} f + \int_Q f$, but T_1 and its inside are contained in G_+ and f is analytic there; hence

1.2
$$\int_T f = \int_Q f.$$

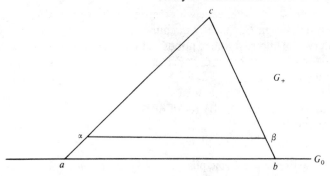

But if $0 \le t \le 1$ then

$$|[t\beta+(1-t)\alpha]-[tb+(1-t)a]| < \delta$$

so that

$$|f(t\beta+(1-t)\alpha)-f(tb+(1-t)a)| < \epsilon.$$

If $M = \max \{|f(z)|: z \in \Delta\}$ and $\ell =$ the perimeter of T then

$$\textbf{1.3} \quad \left| \int_{[a,b]} f + \int_{[\beta,\alpha]} f \right| = \left| (b-a) \int_0^1 f(tb+(1-t)a)dt - (\beta-\alpha) \int_0^1 f(t\beta+(1-t)\alpha)dt \right|$$

$$\le |b-a| \left| \int_0^1 [f(tb+(1-t)a)-f(t\beta+(1-t)\alpha)dt] \right|$$

$$+ |(b-a)-(\beta-\alpha)| \left| \int_0^1 f(t\beta+(1-t)\alpha)dt \right|$$

$$\le \epsilon|b-a| + M|(b-\beta)+(\alpha-a)|$$

$$\le \epsilon\ell + 2M\delta$$

Also

$$\left| \int_{[\alpha,a]} f \right| \le M|a-\alpha| \le M\delta$$

and

$$\left| \int_{[b,\beta]} f \right| \le M\delta.$$

Combining these last two inequalities with (1.2) and (1.3) gives that

$$\left| \int_T f \right| \le \epsilon\ell + 4M\delta.$$

Since it is possible to choose $\delta < \epsilon$ and since ϵ is arbitrary, it follows that $\int_T f = 0$; thus f must be analytic. ∎

Can the reflection principle be generalized? For example, instead of requiring that G be a region which is symmetric with respect to the real axis, suppose that G is symmetric with respect to a circle. (Definition III. 3.17).

Can the reflection principle be formulated and proved in this setting? The answer is provided in the exercises below.

Exercises

1. Let γ be a simple closed rectifiable curve with the property that there is a point a such that for all z on γ the line segment $[a, z]$ intersects $\{\gamma\}$ only at z; i.e. $[a, z] \cap \{\gamma\} = \{z\}$. Define a point w to be inside γ if $[a, w] \cap \{\gamma\} = \square$ and let G be the collection of all points that are inside γ.

(a) Show that G is a region and $G^- = G \cup \{\gamma\}$.

(b) Let $f: G^- \to \mathbb{C}$ be a continuous function such that f is analytic on G. Show that $\int_\gamma f = 0$.

(c) Show that $n(\gamma; z) = \pm 1$ if z is inside γ and $n(\gamma; z) = 0$ if $z \notin G^-$.

Remarks. It is not necessary to assume that γ has such a point a as above; each part of this exercise remains true if γ is only assumed to be a simple closed rectifiable curve. Of course, we must define what is meant by the inside of γ. This is difficult to obtain. The fact that a simple closed curve divides the plane into two pieces (an inside and an outside) is the content of the Jordan Curve Theorem. This is a very deep result of topology.

2. Let G be a region in the plane that does not contain zero and let G^* be the set of all points z such that there is a point w in G where z and w are symmetric with respect to the circle $|\xi| = 1$. (See III. 3.17.)

(a) Show that $G^* = \{z: (1/\bar{z}) \in G\}$.

(b) If $f: G \to \mathbb{C}$ is analytic, define $f^*: G^* \to \mathbb{C}$ by $f^*(z) = \overline{f(1/\bar{z})}$. Show that f^* is analytic.

(c) Suppose that $G = G^*$ and f is an analytic function defined on G such that $f(z)$ is real for z in G with $|z| = 1$. Show that $f = f^*$.

(d) Formulate and prove a version of the Schwarz Reflection Principle where the circle $|\xi| = 1$ replaces $\mathbb{R}$. Do the same thing for an arbitrary circle.

3. Let G, G_+, G_-, G_0 be as in the statement of the Schwarz Reflection Principle and let $f: G_+ \cup G_0 \to \mathbb{C}_\infty$ be a continuous function such that f is meromorphic on G_+. Also suppose that for x in G_0 $f(x) \in \mathbb{R}$. Show that there is a meromorphic function $g: G \to \mathbb{C}_\infty$ such that $g(z) = f(z)$ for z in $G + \cup G_0$. Is it possible to allow f to assume the value ∞ on G_0?

§2. *Analytic Continuation Along a Path*

Let us begin this section by recalling the definition of a function. We use the somewhat imprecise statement that a function is a triple (f, G, Ω) where G and Ω are sets and f is a "rule" which assigns to each element of G a unique element of Ω. Thus, for two functions to be the same not only must the rule be the same but the domains and the ranges must coincide. If we enlarge the range Ω to a set Ω_1 then (f, G, Ω_1) is a different function. However, this point should not be emphasized here; we do wish to emphasize that a change in the domain results in a new function. Indeed, the purpose of analytic continuation is to enlarge the domain. Thus, let $G = \{z: \text{Re } z > -1\}$ and

$f(z) = \log(1+z)$ for z in G, where log is the principal branch of the logarithm. Let $D = B(0; 1)$ and let

$$g(z) = \sum_{n=1}^{\infty} (-1)^{n-1} \frac{z^n}{n}$$

for z in D. Then $(f, G, \mathbb{C}) \neq (g, D, \mathbb{C})$ even though $f(z) = g(z)$ for all z in D.

Nevertheless, it is desirable to recognize the relationship between f and g. This leads, therefore, to the concept of a germ of analytic functions.

2.1 Definition. A *function element* is a pair (f, G) where G is a region and f is an analytic function on G. For a given function element (f, G) define the *germ of f at a* to be the collection of all function elements (g, D) such that $a \in D$ and $f(z) = g(z)$ for all z in a neighborhood of a. Denote the germ by $[f]_a$.

Notice that $[f]_a$ is a collection of function elements and it is not a function element itself. Also $(g, D) \in [f]_a$ if and only if $(f, G) \in [g]_a$. (Verify!). It should also be emphasized that it makes no sense to talk of the equality of two germs $[f]_a$ and $[g]_b$ unless the points a and b are the same. For example, if (f, G) is a function element then it makes no sense to say that $[f]_a = [f]_b$ for two distinct points a and b in G.

2.2 Definition. Let $\gamma: [0, 1] \to \mathbb{C}$ be a path and suppose that for each t in $[0, 1]$ there is a function element (f_t, D_t) such that:

(a) $\gamma(t) \in D_t$;
(b) for each t in $[0, 1]$ there is a $\delta > 0$ such that $|s-t| < \delta$ implies $\gamma(s) \in D_t$ and

2.3 $$[f_s]_{\gamma(s)} = [f_t]_{\gamma(s)}.$$

Then (f_1, D_1) is the *analytic continuation of (f_0, D_0) along the path γ*; or, (f_1, D_1) *is obtained from (f_0, D_0) by analytic continuation along γ*.

Before proceeding, examine part (b) of this definition. Since γ is a continuous function and $\gamma(t)$ is in the open set D_t, it follows that there is a $\delta > 0$ such that $\gamma(s) \in D_t$ for $|s-t| < \delta$. The important content of part (b) is that (2.3) is satisfied whenever $|s-t| < \delta$. That is,

$$f_s(z) = f_t(z), \quad z \in D_s \cap D_t$$

whenever $|s-t| < \delta$.

Whether for a given curve and a given function element there is an analytic continuation along the curve can be a difficult question. Since no degree of generality can be achieved which justifies the effort, no existence theorems for analytic continuations will be proved. Each individual case will be considered by itself. Instead uniqueness theorems for continuations are proved. One such theorem is the Monodromy Theorem of the next section. This theorem gives a criterion by which one can tell when a continuation along two different curves connecting the same points results in the same function element.

The next proposition fixes a curve and shows that two different continuations along this curve of the same function element result in the same function element. Alternately, this result can be considered as an affirmative answer to the following question: Is it possible to define the concept of "the continuation of a germ along a curve?"

2.4 Proposition. *Let* $\gamma: [0, 1] \to \mathbb{C}$ *be a path from a to b and let* $\{(f_t, D_t): 0 \le t \le 1\}$ *and* $\{(g_t, B_t): 0 \le t \le 1\}$ *be analytic continuations along γ such that* $[f_0]_a = [g_0]_a$. *Then* $[f_1]_b = [g_1]_b$.

Proof. This proposition will be proved by showing that the set

$$T = \{t \in [0, 1]: [f_t]_{\gamma(t)} = [g_t]_{\gamma(t)}\}$$

is both open and closed in $[0, 1]$; since T is non-empty $(0 \in T)$ it will follow that $T = [0, 1]$ so that, in particular, $1 \in T$.

The easiest part of the proof is to show that T is open. So fix t in T and assume $t \ne 0$ or 1. (If $t = 1$ the proof is complete; if $t = 0$ then the argument about to be given will also show that $[a, a+\delta) \subset T$ for some $\delta > 0$.) By the definition of analytic continuation there is a $\delta > 0$ such that for $|s-t| < \delta$, $\gamma(s) \in D_t \cap B_t$ and

2.5 $$\begin{cases} [f_s]_{\gamma(s)} = [f_t]_{\gamma(s)}. \\ [g_s]_{\gamma(s)} = [g_t]_{\gamma(s)}. \end{cases}$$

But since $t \in T$, $f_t(z) = g_t(z)$ for all z in $D_t \cap B_t$. Hence $[f_t]_{\gamma(s)} = [g_t]_{\gamma(s)}$ for all $\gamma(s)$ in $D_t \cap B_t$. So it follows from (2.5) that $[f_s]_{\gamma(s)} = [g_s]_{\gamma(s)}$ whenever $|s-t| < \delta$. That is, $(t-\delta, t+\delta) \subset T$ and so T is open.

To show that T is closed let t be a limit point of T, and again choose $\delta > 0$ so that $\gamma(s) \in D_t \cap B_t$ and (2.5) is satisfied whenever $|s-t| < \delta$. Since t is a limit point of T there is a point s in T with $|s-t| < \delta$; so $G = D_t \cap B_t \cap D_s \cap B_s$ contains $\gamma(s)$ and, therefore, is a non-empty open set. Thus, $f_s(z) = g_s(z)$ for all z in G by the definition of T. But, according to (2.5), $f_s(z) = f_t(z)$ and $g_s(z) = g_t(z)$ for all z in G. So $f_t(z) = g_t(z)$ for all z in G and, because G has a limit point in $D_t \cap B_t$, this gives that $[f_t]_{\gamma(t)} = [g_t]_{\gamma(t)}$. That is, $t \in T$ and so T is closed. ∎

2.6 Definition. If $\gamma: [0, 1] \to \mathbb{C}$ is a path from a to b and $\{(f_t, D_t): 0 \le t \le 1\}$ is an analytic continuation along γ then *the germ* $[f_1]_b$ *is the analytic continuation of* $[f_0]_a$ *along* γ.

The preceding proposition implies that Definition 2.6 is unambiguous. As stated this definition seems to depend on the choice of the continuation $\{(f_t, D_t)\}$. However, Proposition 2.4 says that if $\{(g_t, B_t)\}$ is another continuation along γ with $[f_0]_a = [g_0]_a$ then $[f_1]_b = [g_1]_b$. So in fact the definition does not depend on the choice of continuation.

2.7 Definition. If (f, G) is a function element then *the complete analytic function obtained from* (f, G) is the collection $\mathcal{F}$ of all germs $[g]_b$ for which there is a point a in G and a path γ from a to b such that $[g]_b$ is the analytic continuation of $[f]_a$ along γ.

A collection of germs $\mathcal{F}$ is called *a complete analytic function* if there is a function element (f, G) such that $\mathcal{F}$ is the complete analytic function obtained from (f, G).

Notice that the point a in the definition is immaterial; any point in G can be chosen since G is an open connected subset of $\mathbb{C}$ (see II. 2.3). Also, if $\mathcal{F}$ is the complete analytic function associated with (f, G) then $[f]_z \in \mathcal{F}$ for all z in G.

Although there is no ambiguity in the definition of a complete analytic function there is an incompleteness about it. Is it a function? We should refrain from calling an object a function unless it is indeed a function. To make $\mathcal{F}$ a function one must manufacture a domain (the range will be $\mathbb{C}$) and show that $\mathcal{F}$ gives a "rule". This is easy. In a sense we let $\mathcal{F}$ be its own domain; more precisely, let

$$\mathcal{R} = \{(z, [f]_z) : [f]_z \in \mathcal{F}\}.$$

Define $\mathcal{F} : \mathcal{R} \to \mathbb{C}$ by $\mathcal{F}(z, [f]_z) = f(z)$. In this way $\mathcal{F}$ becomes an "honest" function. Nevertheless there is still a lingering dissatisfaction. To have a satisfying solution a structure will be imposed on $\mathcal{R}$ which will make it possible to discuss the concept of analyticity for functions defined on $\mathcal{R}$. In this setting, the function $\mathcal{F}$ defined above becomes analytic; moreover, it reflects the behavior of each function element belonging to a germ that is in $\mathcal{F}$. The introduction of this structure is postponed until Section 5.

Exercises

1. The collection $\{D_0, D_1, \ldots, D_n\}$ of open disks is called a *chain of disks* if $D_{j-1} \cap D_j \neq \square$ for $1 \leq j \leq n$. If $\{(f_j, D_j) : 0 \leq j \leq n\}$ is a collection of function elements such that $\{D_0, D_1, \ldots, D_n\}$ is a chain of disks and $f_{j-1}(z) = f_j(z)$ for z in $D_{j-1} \cap D_j$, $1 \leq j \leq n$; then $\{(f_j, D_j) : 0 \leq j \leq n\}$ is called an *analytic continuation along a chain of disks*. We say that (f_n, D_n) is obtained by an analytic continuation of (f_0, D_0) along a chain of disks.

(a) Let $\{(f_j, D_j) : 0 \leq j \leq n\}$ be an analytic continuation along a chain of disks and let a and b be the centers of the disks D_0 and D_n respectively. Show that there is a path γ from a to b and an analytic continuation $\{(g_t, B_t)\}$ along γ such that $\{\gamma\} \subset \bigcup_{j=0}^{n} D_j$, $[f_0]_a = [g_0]_a$, and $[f_n]_b = [g_1]_b$.

(b) Conversely, let $\{(f_t, D_t) : 0 \leq t \leq 1\}$ be an analytic continuation along a path $\gamma : [0, 1] \to \mathbb{C}$ and let $a = \gamma(0)$, $b = \gamma(1)$. Show that there is an analytic continuation along a chain of disks $\{(g_j, B_j) : 0 \leq j \leq n\}$ such that $\{\gamma\} \subset \bigcup_{j=0}^{n} B_j$, $[f_0]_a = [g_0]_a$, and $[f_1]_b = [g_n]_b$.

2. Let $D_0 = B(1; 1)$ and let f_0 be the restriction of the principal branch of $\sqrt{z}$ to D_0. Let $\gamma(t) = \exp(2\pi i t)$ and $\sigma(t) = \exp(4\pi i t)$ for $0 \leq t \leq 1$.

(a) Find an analytic continuation $\{(f_t, D_t) : 0 \leq t \leq 1\}$ of (f_0, D_0) along γ and show that $[f_1]_1 = [-f_0]_1$.

(b) Find an analytic continuation $\{(g_t, B_t): 0 \le t \le 1\}$ of (f_0, D_0) along σ and show that $[g_1]_1 = [g_0]_1$.

3. Let f be an entire function, $D_0 = B(0; 1)$, and let γ be a path from 0 to b. Show that if $\{(f_t, D_t): 0 \le t \le 1\}$ is a continuation of (f, D_0) along γ then $f_1(z) = f(z)$ for all z in D_1. (This exercise is rather easy; it is actually an exercise in the use of the terminology.)

4. Let $\gamma: [0, 1] \to \mathbb{C}$ be a path and let $\{(f_t, D_t): 0 \le t \le 1\}$ be an analytic continuation along γ. Show that $\{(f_t', D_t): 0 \le t \le 1\}$ is also a continuation along γ.

5. Suppose $\gamma: [0, 1] \to \mathbb{C}$ is a closed path with $\gamma(0) = \gamma(1) = a$ and let $\{(f_t, D_t): 0 \le t \le 1\}$ be an analytic continuation along γ such that $[f_1]_a = [f_0]_a$ and $f_0 \ne 0$. What can be said about (f_0, D_0)?

6. Let $D_0 = B(1; 1)$ and let f_0 be the restriction to D_0 of the principal branch of the logarithm. For an integer n let $\gamma(t) = \exp(2\pi i n t)$, $0 \le t \le 1$. Find a continuation $\{(f_t, D_t): 0 \le t \le 1\}$ along γ of (f_0, D_0) and show that $[f_1]_1 = [f_0 + 2\pi i n]_1$.

7. Let $\gamma: [0, 1] \to \mathbb{C}$ be a path and let $\{(f_t, D_t): 0 \le t \le 1\}$ be an analytic continuation along γ. Suppose G is a region such that $f_t(D_t) \subset G$ for all t, and suppose there is an analytic function $h: G \to \mathbb{C}$ such that $h(f_0(z)) = z$ for all z in D_0. Show that $h(f_t(z)) = z$ for all z in D_t and for all t.

Hint: Show that $T = \{t: h(f_t(z)) = z \text{ for all } z \text{ in } D_t\}$ is both open and closed in $[0, 1]$.

8. Let $\gamma: [0, 1] \to \mathbb{C}$ be a path with $\gamma(0) = 1$ and $\gamma(t) \ne 0$ for any t. Suppose that $\{(f_t, D_t): 0 \le t \le 1\}$ is an analytic continuation of $f_0(z) = \log z$. Show that each f_t is a branch of the logarithm.

§3. Monodromy Theorem

Let a and b be two complex numbers and suppose γ and σ are two paths from a to b. Suppose $\{(f_t, D_t)\}$ and $\{(g_t, B_t)\}$ are analytic continuations along γ and σ respectively, and also suppose that $[f_0]_a = [g_0]_a$. Does it follow that $[f_1]_b = [g_1]_b$? If γ and σ are the same path then Proposition 2.4 gives an affirmative answer. However, if γ and σ are distinct then the answer can be no. In fact, Exercises 2.2 and 2.6 furnish examples that illustrate the possibility that $[f_1]_b \ne [g_1]_b$. Since both of these examples involve curves that wind around the origin, the reader might believe that a sufficient condition for $[f_1]_b$ and $[g_1]_b$ to be equal can be couched in the language of homotopy. However, since all curves in the plane are homotopic the result would have to be phrased in terms of homotopy in a proper subregion of $\mathbb{C}$. For the examples in Exercises 2.2 and 2.6, this sought after criterion must involve the homotopy of the curves in the punctured plane. This is indeed the case. The origin is discarded in the above examples because there is no germ $[h]_0$ centered at zero that belongs to the complete analytic function obtained from (f_0, D_0).

If (f, D) is a function element and $a \in D$ then f has a power series expansion at $z = a$. The first step in proving the Monodromy Theorem is to investigate the behavior of the radius of convergence for an analytic continuation along a curve.

3.1 Lemma. *Let $\gamma\colon [0, 1] \to \mathbb{C}$ be a path and let $\{(f_t, D_t)\colon 0 \le t \le 1\}$ be an analytic continuation along γ. For $0 \le t \le 1$ let $R(t)$ be the radius of convergence of the power series expansion of f_t about $z = \gamma(t)$. Then either $R(t) \equiv \infty$ or $R\colon [0, 1] \to (0, \infty)$ is continuous.*

Proof. If $R(t) = \infty$ for some value of t then it is possible to extend f_t to an entire function. It follows that $f_s(z) = f_t(z)$ for all z in D_s so that $R(s) = \infty$ for each s in $[0, 1]$; that is $R(s) \equiv \infty$. So suppose that $R(t) < \infty$ for all t. Fix t in $[0, 1]$ and let $\tau = \gamma(t)$; let

$$f_t(z) = \sum_{n=0}^{\infty} \tau_n (z-\tau)^n$$

be the power series expansion of f_t about τ. Now let $\delta_1 > 0$ be such that $|s-t| < \delta_1$ implies that $\gamma(s) \in D_t \cap B(\tau; R(t))$ and $[f_s]_{\gamma(s)} = [f_t]_{\gamma(s)}$. Fix s with $|s-t| < \delta_1$ and let $\sigma = \gamma(s)$. Now f_t can be extended to an analytic function on $B(\tau; R(t))$. Since f_s agrees with f_t on a neighborhood of σ, f_s can be extended so that it is also analytic on $B(\tau; R(t)) \cup D_s$. If f_s has power series expansion

$$f_s(z) = \sum_{n=0}^{\infty} \sigma_n (z-\sigma)^n$$

about $z = \sigma$, then the radius of convergence $R(s)$ must be at least as big as the distance from σ to the circle $|z-\tau| = R(t)$: that is, $R(s) \ge d(\sigma, \{z\colon |z-\tau| = R(t)\}) \ge R(t) - |\tau - \sigma|$. But this gives that $R(t) - R(s) \le |\gamma(t) - \gamma(s)|$. A similar argument gives that $R(s) - R(t) \le |\gamma(t) - \gamma(s)|$; hence

$$|R(s) - R(t)| \le |\gamma(t) - \gamma(s)|$$

for $|s-t| < \delta_1$. Since $\gamma\colon [0, 1] \to \mathbb{C}$ is continuous it follows that R must be continuous at t. ∎

3.2 Lemma. *Let $\gamma\colon [0, 1] \to \mathbb{C}$ be a path from a to b and let $\{(f_t, D_t)\colon 0 \le t \le 1\}$ be an analytic continuation along γ. There is a number $\epsilon > 0$ such that if $\sigma\colon [0, 1] \to \mathbb{C}$ is any path from a to b with $|\gamma(t) - \sigma(t)| < \epsilon$ for all t, and if $\{(g_t, B_t)\colon 0 \le t \le 1\}$ is any continuation along σ with $[g_0]_a = [f_0]_a$; then $[g_1]_b = [f_1]_b$.*

Proof. For $0 \le t \le 1$ let $R(t)$ be the radius of convergence of the power series expansion of f_t about $z = \gamma(t)$. It is left to the reader to show that if $R(t) \equiv \infty$ then any value of ϵ will suffice. So suppose $R(t) < \infty$ for all t. Since, by the preceding lemma, R is a continuous function and since $R(t) > 0$ for all t, R has a positive minimum value. Let

3.3 $0 < \epsilon < \frac{1}{2} \min \{R(t)\colon 0 \le t \le 1\}$

and suppose that σ and $\{(g_t, B_t)\}$ are as in the statement of this lemma.

Furthermore, suppose that D_t is a disk of radius $R(t)$ about $\gamma(t)$. The truth of the conclusion will not be affected by this assumption (Why?), and the exposition will be greatly simplified by it.

Since $|\sigma(t)-\gamma(t)| < \epsilon < R(t)$, $\sigma(t) \in B_t \cap D_t$ for all t. Thus, it makes sense to ask whether $g_t(z) = f_t(z)$ for all z in $B_t \cap D_t$. Indeed, to complete the proof we must show that this is precisely the case for $t = 1$. Define the set $T = \{t \in [0, 1]: f_t(z) = g_t(z)$ for z in $B_t \cap D_t\}$; and show that $1 \in T$. This is done by showing that T is a non-empty open and closed subset of $[0, 1]$.

From the hypothesis of the lemma, $0 \in T$ so that $T \neq \square$. To show T is open fix t in T and choose $\delta > 0$ such that

3.4
$$\begin{cases} |\gamma(s)-\gamma(t)| < \epsilon, \ [f_s]_{\gamma(s)} = [f_t]_{\gamma(s)}, \\ |\sigma(s)-\sigma(t)| < \epsilon, \ [g_s]_{\sigma(s)} = [g_t]_{\sigma(s)}, \text{ and} \\ \sigma(s) \in B_t \end{cases}$$

whenever $|s-t| < \delta$. We will now show that $B_s \cap B_t \cap D_s \cap D_t \neq \square$ for $|s-t| < \delta$; in fact, we will show that $\sigma(s)$ is in this intersection. If $|s-t| < \delta$ then

$$|\sigma(s)-\gamma(s)| < \epsilon < R(s)$$

so that $\sigma(s) \in D_s$. Also

$$|\sigma(s)-\gamma(t)| \le |\sigma(s)-\gamma(s)|+|\gamma(s)-\gamma(t)| < 2\epsilon < R(t)$$

by (3.3); so $\sigma(s) \in D_t$. Since we already have that $\sigma(s) \in B_s \cap B_t$ by (3.4),

$$\sigma(s) \in B_s \cap B_t \cap D_s \cap D_t = G.$$

Since $t \in T$ it follows that $f_t(z) = g_t(z)$ for all z in G. Also, from (3.4) $f_s(z) = f_t(z)$ and $g_s(z) = g_t(z)$ for all z in G. Thus $f_s(z) = g_s(z)$ for z in G; but since G has a limit point in $B_s \cap D_s$ it must be that $s \in T$. That is, $(t-\delta, t+\delta) \subset T$ and T is open. The proof that T is closed is similar and will be left to the reader. ∎

3.5 Definition. Let (f, D) be a function element and let G be a region which contains D; then (f, D) *admits unrestricted analytic continuation in* G if for any path γ in G with initial point in D there is an analytic continuation of (f, D) along γ.

If $D = \{z: |z-1| < 1\}$ and f is the principal branch of $\sqrt{z}$ or $\log z$ then (f, D) admits unrestricted continuation in the punctured plane but not in the whole plane (see Exercise 2.7).

It has been stated before that an existence theorem for analytic continuations will not be proved. In particular, if (f, D) is a function element and G is a region containing D, no criterion will be given which implies that (f, D) admits unrestricted continuation in G. The Monodromy Theorem assumes that G has this property and states a uniqueness criterion.

3.6 Monodromy Theorem. *Let (f, D) be a function element and let G be a region containing D such that (f, D) admits unrestricted continuation in G.*

Let $a \in D$, $b \in G$ and let γ_0 and γ_1 be paths in G from a to b; let $\{(f_t, D_t):$ $0 \le t \le 1\}$ and $\{(g_t, D_t): 0 \le t \le 1\}$ be analytic continuations of (f, D) along γ_0 and γ_1 respectively. If γ_0 and γ_1 are FEP homotopic in G then

$$[f_1]_b = [g_1]_b.$$

Proof. Since γ_0 and γ_1 are fixed-end-point homotopic in G there is a continuous function $\Gamma: [0, 1] \times [0, 1] \to G$ such that

$$\Gamma(t, 0) = \gamma_0(t) \qquad \Gamma(t, 1) = \gamma_1(t)$$

$$\Gamma(0, u) = a \qquad \Gamma(1, u) = b$$

for all t and u in $[0, 1]$. Fix u, $0 \le u \le 1$ and consider the path γ_u, defined by $\gamma_u(t) = \Gamma(t, u)$, from a to b. By hypothesis there is an analytic continuation

$$\{(h_{t,u}, D_{t,u}): 0 \le t \le 1\}$$

of (f, D) along γ_u. It follows from Proposition 2.4 that $[g_1]_b = [h_{1,1}]_b$ and $[f_1]_b = [h_{1,0}]_b$. So it suffices to show that

$$[h_{1,0}]_b = [h_{1,1}]_b.$$

To do this introduce the set

$$U = \{u \in [0, 1]: [h_{1,u}]_b = [h_{1,0}]_b\},$$

and show that U is a non-empty open and closed subset of $[0, 1]$. Since $0 \in U$, $U \ne \square$. To show that U is both open and closed we will establish the following.

3.7 Claim. For u in $[0, 1]$ there is a $\delta > 0$ such that if $|u - v| < \delta$ then $[h_{1,u}]_b = [h_{1,v}]_b$. For a fixed u in $[0, 1]$ apply Lemma 3.2 to find an $\epsilon > 0$ such that if σ is any path from a to b with $|\gamma_u(t) - \sigma(t)| < \epsilon$ for all t, and if $\{(k_t, E_t)\}$ is any continuation of (f, D) along σ, then

3.8 $$[h_{1,u}]_b = [k_1]_b.$$

Now Γ is a uniformly continuous function, so there is a $\delta > 0$ such that if $|u - v| < \delta$ then $|\gamma_u(t) - \gamma_v(t)| = |\Gamma(t, u) - \Gamma(t, v)| < \epsilon$ for all t. Claim 3.7 now follows by applying (3.8).

Suppose $u \in U$ and let $\delta > 0$ be the number given by Claim 3.7. By the definition of U, $(u - \delta, u + \delta) \subset U$; so U is open. If $u \in U^-$ and δ is again chosen as in (3.7) then there is a v in U such that $|u - v| < \delta$. But by (3.7) $[h_{1,u}]_b = [h_{1,v}]_b$; and since $v \in U$, $[h_{1,v}]_b = [h_{1,0}]_b$. Therefore $[h_{1,u}]b = [h_{1,0}]_b$ so that $u \in U$, that is, U is closed. ∎

The following corollary is the most important consequence of the Monodromy Theorem.

3.9 Corollary. Let (f, D) be a function element which admits unrestricted continuation in the simply connected region G. Then there is an analytic function $F: G \to \mathbb{C}$ such that $F(z) = f(z)$ for all z in D.

Proof. Fix a in D and let z be any point in G. If γ is a path in G from a to z

and $\{(f_t, D_t): 0 \le t \le 1\}$ is an analytic continuation of (f, D) along γ then let $F(z, \gamma) = f_1(z)$. Since G is simply connected $F(z, \gamma) = F(z, \sigma)$ for any two paths γ and σ in G from a to z. Thus, $F(z) = F(z, \gamma)$ gives a well defined function $F: G \to \mathbb{C}$. To show that F is analytic let $z \in G$ and let γ and $\{(f_t, D_t)\}$ be as above. A simple argument gives that $F(w) = f_1(w)$ for w in a neighborhood of z (Verify!); so F must be analytic. ∎

Exercises

1. Prove that the set T defined in the proof of Lemma 3.2 is closed.
2. Let (f, D) be a function element and let $a \in D$. If $\gamma: [0, 1] \to \mathbb{C}$ is a path with $\gamma(0) = a$ and $\gamma(1) = b$ and $\{(f_t, D_t): 0 \le t \le 1\}$ is an analytic continuation of (f, D) along γ, let $R(t)$ be the radius of convergence of the power series expansion of f_t at $z = \gamma(t)$.

   (a) Show that $R(t)$ is independent of the choice of continuation. That is, if a second continuation $\{(g_t, B_t)\}$ along γ is given with $[g_0]_a = [f]_a$ and $r(t)$ is the radius of convergence of the power series expansion of g_t about $z = \gamma(t)$ then $r(t) = R(t)$ for all t.

   (b) Suppose that $D = B(1; 1)$, f is the restriction of the principal branch of the logarithm to D, and $\gamma(t) = 1 + at$ for $0 \le t \le 1$ and $a > 0$. Find $R(t)$.

   (c) Let (f, D) be as in part (b), let $0 < a < 1$ and let $\gamma(t) = (1 - at) \exp(2\pi it)$ for $0 \le t \le 1$. Find $R(t)$.

   (d) For each of the functions $R(t)$ obtained in parts (b) and (c), find $\min \{R(t): 0 \le t \le 1\}$ as a function of a and examine the behavior of this function as $a \to \infty$ or $a \to 0$.
3. Let $\Gamma: [0, 1] \times [0, 1] \to G$ be a continuous function such that $\Gamma(0, u) = a$, $\Gamma(1, u) = b$ for all u. Let $\gamma_u(t) = \Gamma(t, u)$ and suppose that $\{(f_{t,u}, D_{t,u}): 0 \le t \le 1\}$ is an analytic continuation along γ_u such that $[f_{0,u}]_a = [f_{0,v}]_a$ for all u and v in $[0, 1]$. Let $R(t, u)$ be the radius of convergence of the power series expansion of $f_{t,u}$ about $z = \Gamma(t, u)$. Show that either $R(t, u) \equiv \infty$ or $R: [0, 1] \times [0, 1] \to (0, \infty)$ is a continuous function.
4. Use Exercise 3 to give a second proof of the Monodromy Theorem.

§4. Topological Spaces and Neighborhood Systems

The notion of a topological space arises by abstracting one of the most important concepts in the theory of metric spaces—that of an open set. Recall that in Chapter II we were given a metric or distance function on a set X and this metric was used to define what is meant by an open set. In a topological space we are given a collection of subsets of a set X which are called open sets, but there is no metric available. After axiomatizing the properties of open sets, it will be our purpose to recreate as much of the theory of metric spaces as is possible.

4.1 Definition. A *topological space* is a pair $(X, \mathcal{T})$ where X is a set and $\mathcal{T}$ is a collection of subsets of X having the following properties:

(a) $\square \in \mathcal{T}$ and $X \in \mathcal{T}$;

(b) if $U_1, \ldots, U_n$ are in $\mathcal{T}$ then $\bigcap_{j=1}^{n} U_j \in \mathcal{T}$;

(c) if $\{U_i : i \in I\}$ is any collection of sets in $\mathcal{T}$ then $\bigcup_{i \in I} U_i$ is in $\mathcal{T}$.

The collection of sets $\mathcal{T}$ is called a *topology* on X, and each member of $\mathcal{T}$ is called an *open set*.

Notice that properties (a), (b), and (c) of this definition are the properties of open subsets of a metric space that were proved in Proposition II. 1.9. So if (X,d) is a metric space and $\mathcal{T}$ is the collection of all open subsets of X then $(X,\mathcal{T})$ is a topological space.

When it is said that a topological space is an abstraction of a metric space, the reader should not get the impression that he is merely playing a game by discarding the metric. That is, no one should believe that there is a distance function in the background, but the reader is now required to prove theorems without resorting to it. This is quite false. There are topological spaces $(X, \mathcal{T})$ such that for no metric d on X is $\mathcal{T}$ the collection of open sets obtained via d. We will see such an example shortly, but it is first necessary to further explore this concept of a topology.

The statement "Let X be a topological space" is, of course, meaningless; a topological space consists of a topology $\mathcal{T}$ as well as a set X. However, this phase will be used when there is no possibility of confusion.

4.2 Definition. A subset F of a topological space X is *closed* if $X - F$ is open. A point a in X is a *limit point* of a set A if for every open set U that contains a there is a point x in $A \cap U$ such that $x \neq a$.

Many of the proofs of propositions in this section follow along the same lines as corresponding propositions in Chapter II. When this is the case the proof will be left to the reader. Such is the case with the following two propositions.

4.3 Proposition. *Let $(X, \mathcal{T})$ be a topological space. Then:*

(a) *$\square$ and X are closed sets;*
(b) *if $F_1, \ldots, F_n$ are closed sets then $F_1 \cup \ldots \cup F_n$ is closed;*
(c) *if $\{F_i : i \in I\}$ is a collection of closed sets then $\bigcap_{i \in I} F_i$ is a closed set.*

4.4 Proposition. *A subset of a topological space is closed iff it contains all its limit points.*

Now for an example of a topological space that is not a metric space. Let $X = [0,1] = \{t : 0 \leq t \leq 1\}$ and let $\mathcal{T}$ consist of all sets U such that:

(i) if $0 \in U$ then $X - U$ is either empty or a sequence of points in X;
(ii) if $0 \notin U$ then U can be any set.

It is left to the reader to prove that $(X, \mathcal{T})$ is a topological space. Some of the examples of open sets in this topology are: the set of all irrational numbers in X; the set of all irrational numbers together with zero. To see that no metric can give the collection of open sets $\mathscr{T}$, suppose that there is such a metric and obtain a contradiction. Suppose that d is a metric on X such that $U \in \mathscr{T}$ iff for each x in U there is an $\epsilon > 0$ such that $B(x; \epsilon) = \{y : d(x, y) < \epsilon\} \subset U$. Now let $A = (0, 1)$; if $U \in \mathscr{T}$ and $0 \in U$ then there is a point a in $U \cap A$, $a \neq 0$ (in fact there is an infinity of such points). Hence, 0 is a limit point of A. It follows that there is a sequence $\{t_n\}$ in A such that $d(t_n, 0) \to 0$. But if $U = \{x \in X : x \neq t_n \text{ for any } n\} = X - \{t_1, t_2, \ldots\}$ then $0 \in U$ and U is open. So it must follow that $t_n \in U$ for n sufficiently large; this is an obvious contradiction. Hence, no metric can be found.

This example illustrates a technique that, although available for metric spaces, is of little use for general topological spaces: the convergence of sequences. It is possible to define "convergent sequence" in a topological space (Do it!), but this concept is not as intimately connected with the structure of a topological space as it is with a metric space. For example, it was shown above that a point can be a limit point of a set A but there is no sequence in A that converges to it.

If a topological space $(X, \mathscr{T})$ is such that a metric d on X can be found with the property that a set is in $\mathscr{T}$ iff it is open in (X, d), then $(X, \mathscr{T})$ is said to be *metrisable*. There are many non-metrisable spaces. In addition to inventing non-metrisable topologies as was done above, it is possible to define processes for obtaining new topological spaces from old ones which will put metrisable spaces together to obtain non-metrisable ones. For example, the arbitrary cartesian product of topological spaces can be defined; in this case the product space is not metrisable unless there are only a countable number of coordinates and each coordinate space is itself metrisable. (See VII. 1.18.)

Another example may be obtained as follows. Consider the unit interval $I = [0, 1]$. Stick one copy of I onto another and we have a topological space which still "looks like" I. For example, $[1, 2]$ is a copy of I and if we stick it onto I we obtain $[0, 2]$. In fact, if we "stick" a finite number of closed intervals together another closed interval is obtained. What happens if a countable number of closed intervals are stuck together? The answer is that we obtain the infinite interval $[0, \infty)$. (If the intervals are stuck together on both sides then $\mathbb{R}$ is obtained.) What happens if we put together an uncountable number of copies of I? The resulting space is called the long line. Locally (i.e., near each point) it looks like the real line. However, the long line is not metrisable. As a general rule of thumb, it may be said that if a process is used to obtain new spaces from old ones, a non-metrisable space will result if the process is used an uncountable number of times.

For another example of a space that is non-metrisable let X be a set consisting of three points—say $X = \{a, b, c\}$. Let $\mathscr{T} = \{\Box, X, \{a\}, \{b\}, \{a, b\}\}$; it is easy to check that $\mathscr{T}$ is a topology for X. To see that $(X, \mathscr{T})$ is not metrisable notice that the only open set containing c is the set X itself. There do not exist disjoint open sets U and V such that $a \in U$ and $c \in V$. On the other hand if there was a metric d on X such that $\mathscr{T}$ is the collection of open sets relative to this metric then it would be possible to find such open sets (e.g., let $U = B(a; \epsilon)$ and $V = B(c; \epsilon)$ where $\epsilon < d(a, c)$). In other words, $(X, \mathcal{T})$ fails to be metrisable because $\mathcal{T}$ does not have enough open sets to separate points. (Does the first example of a non-metrisable space also fail because of this deficiency?) The next definition hypothesizes enough open sets to eliminate this difficulty.

4.5 Definition. A topological space $(X, \mathcal{T})$ is said to be a *Hausdorff space* if for any two distinct points a and b in X there are disjoint open sets U and V such that $a \in U$ and $b \in V$.

Every metric space is a Hausdorff space. As we have already seen there are examples of topological spaces which are not Hausdorff spaces. Many authors include in the definition of a topological space the property of a Hausdorff space. This policy is easily defended since most of the examples of topological spaces which one encounters are, indeed, Hausdorff spaces. However there are also some fairly good arguments against this combining of concepts. The first argument is that mathematical pedagogy dictates that ideas should be separated so that they may be more fully understood. The second, and perhaps more substantial reason for not assuming all spaces to be Hausdorff, is that more examples of non-Hausdorff spaces are arising in a natural context. Even though there will be no non-Hausdorff space which will appear in this book, this separation of the two concepts will be maintained for a while longer.

The next step in this development is the definition, in the setting of topological spaces, of certain concepts encountered in the theory of metric spaces and the stating of analogous propositions.

4.6 Definition. A topological space $(X, \mathcal{T})$ is *connected* if the only non-empty subset of X which is both open and closed is the set X itself.

4.7 Proposition. *Let $(X, \mathcal{T})$ be a topological space; then $X = \bigcup \{C_i : i \in I\}$ where each C_i is a component of X (a maximal connected subset of X). Furthermore, distinct components of X are disjoint and each component is closed.*

4.8 Definition. Let $(X, \mathcal{T})$ and $(\Omega, \mathcal{S})$ be topological spaces. A function $f: X \to \Omega$ is *continuous* if $f^{-1}(\Delta) \in \mathcal{T}$ whenever $\Delta \in \mathcal{S}$.

4.9 Proposition. *Let $(X, \mathcal{T})$ and $(\Omega, \mathcal{S})$ be topological spaces and let $f: X \to \Omega$ be a function. Then the following are equivalent:*

(a) *f is continuous;*
(b) *if Λ is a closed subset of Ω then $f^{-1}(\Lambda)$ is a closed subset of X;*
(c) *if $a \in X$ and if $f(a) \in \Delta \in \mathcal{S}$ then there is a set U in $\mathcal{T}$ such that $a \in U$ and $f(U) \subset \Delta$.*

4.10 Proposition. *Let $(X, \mathcal{T})$ and $(\Omega, \mathcal{S})$ be topological spaces and suppose that X is connected. If $f: X \to \Omega$ is a continuous function such that $f(X) = \Omega$, then Ω is connected.*

4.11 Definition. A set $K \subset X$ is *compact* if for every sub-collection $\mathcal{O}$ of $\mathcal{T}$

such that $K \subset \bigcup \{U: U \in \mathcal{O}\}$ there are a finite number of sets $U_1, \ldots, U_n$ in $\mathcal{O}$ such that $K \subset \bigcup_{k=1}^{n} U_k$.

4.12 Proposition. *Let* $(X, \mathcal{T})$ *and* $(\Omega, \mathcal{S})$ *be topological spaces and suppose* K *is a compact subset of* X. *If* $f: X \to \Omega$ *is a continuous function then* $f(K)$ *is compact in* Ω.

If (X, d) is a metric space and $Y \subset X$ then (Y, d) is also a metric space. Is there a way of making a subset of a topological space into a topological space? We could, of course, declare every subset of Y to be open and this would make Y into a topological space. But what is desired is a topology on Y which has some relationship to the topology on X; a natural topology on a subset of a topological space.

If $(X, \mathcal{T})$ is a topological space and $Y \subset X$ then define

$$\mathcal{T}_Y = \{U \cap Y: U \in \mathcal{T}\}.$$

It is easy to check that $\mathcal{T}_Y$ is a topology on Y.

4.13 Definition. If Y is a subset of a topological space $(X, \mathcal{T})$ then $\mathcal{T}_Y$ is called the *relative topology on* Y. A subset W of Y is *relatively open in* Y if $W \in \mathcal{T}_Y$; W is *relatively closed in* Y if $Y - W \in \mathcal{T}_Y$.

Whenever we speak of a subset of a topological space as a topological space it will be assumed that it has the relative topology unless the contrary is explicitly stated.

4.14 Proposition. *Let* $(X, \mathcal{T})$ *be a topological space and let* Y *be a subset of* X.

(a) *If* X *is compact and* Y *is a closed subset of* X *then* $(Y, \mathcal{T}_Y)$ *is compact.*

(b) Y *is a compact subset of* X *iff* $(Y, \mathcal{T}_Y)$ *is a compact topological space.*

(c) *If* $(X, \mathcal{T})$ *is a Hausdorff space then* $(Y, \mathcal{T}_Y)$ *is a Hausdorff space.*

(d) *If* $(X, \mathcal{T})$ *is a Hausdorff space and* $(Y, \mathcal{T}_Y)$ *is compact then* Y *is a closed subset of* X.

Proof. The proofs of (a), (b), and (c) are left as exercises. To prove part (d) it suffices to show that for each point a in $X - Y$ there is an open set U such that $a \in U$ and $U \cap Y = \square$. So fix a point a in $X - Y$; for each point y in Y there are open sets U_y and V_y in X such that $a \in U_y$, $y \in V_y$, and $U_y \cap V_y = \square$ (because $(X, \mathcal{T})$ is a Hausdorff space). Then $\{V_y \cap Y: y \in Y\}$ is a collection of sets in $\mathcal{T}_Y$ which covers Y. So there are points $y_1, \ldots, y_n$ in Y such that $Y \subset \bigcup_{i=1}^{n} (V_{y_i} \cap Y) \subset \bigcup_{i=1}^{n} V_{y_i} = V$. Since $a \in U_{y_i}$ for each i, $a \in U = \bigcap_{i=1}^{n} U_{y_i}$; also $U \in \mathcal{T}$. It is easily verified that $U \cap V = \bigcup_{i=1}^{n} (U \cap V_{y_i}) = \square$ so that $U \cap Y = \square$. ∎

The proof of this proposition yields a stronger result.

4.15 Corollary. *Let* $(X, \mathcal{T})$ *be a Hausdorff space and let* Y *be a compact subset of* X; *then for each point* a *in* $X - Y$ *there are open sets* U *and* V *in* X *such that* $a \in U$, $Y \subset V$. *and* $U \cap V = \square$.

If we return to the consideration of metric spaces we can discover a new way to define a topology. The sequence of steps by which open sets are obtained in a metric space are as follows: the metric is given, then open balls are defined, then open sets are defined as those sets which contain a ball about each of their points. What we wish to mimic now is the introduction of balls; this is done by defining a neighborhood system on a set.

4.16 Definition. Let X be a set and suppose that for each point x in X there is a collection $\mathcal{N}_x$ of subsets of X having the following properties:

   (a) for each U in $\mathcal{N}_x$, $x \in U$;
   (b) if U and $V \in \mathcal{N}_x$, there is a W in $\mathcal{N}_x$ such that $W \subset U \cap V$;
   (c) if $U \in \mathcal{N}_x$ and $V \in \mathcal{N}_y$ then for each z in $U \cap V$ there is a W in $\mathcal{N}_z$ such that $W \subset U \cap V$.

Then the collection $\{\mathcal{N}_x : x \in X\}$ is called a *neighborhood system* on X.

If (X, d) is a metric space and if $x \in X$ then $\mathcal{N}_x = \{B(x; \epsilon): \epsilon > 0\}$ gives a collection $\{\mathcal{N}_x : x \in X\}$ which is a neighborhood system. In fact, this was the prototype of the above definitions.

Notice that condition (c) relates the neighborhood systems of different points. If only conditions (a) and (b) were satisfied, it would not follow that these neighborhoods would be open sets in the topology to be defined. For example, if X is a metric space and $\mathcal{N}_x$ is the collection of closed balls about x then $\{\mathcal{N}_x : x \in X\}$ satisfies conditions (a) and (b) but not (c). Moreover, it is easy to verify that by letting $x = y = z$ condition (b) can be deduced from (c) (Verify!).

The next proposition relates neighborhood systems and topological spaces.

4.17 Proposition. (a) *If $(X, \mathcal{T})$ is a topological space and $\mathcal{N}_x = \{U \in \mathcal{T}: x \in U\}$ then $\{\mathcal{N}_x : x \in X\}$ is a neighborhood system on X.*

   (b) *If $\{\mathcal{N}_x : x \in X\}$ is a neighborhood system on a set X then let $\mathcal{T} = \{U: x$ in U implies there is a V in $\mathcal{N}_x$ such that $V \subset U\}$. Then $\mathcal{T}$ is a topology on X and $\mathcal{N}_x \subset \mathcal{T}$ for each x.*

   (c) *If $(X, \mathcal{T})$ is a topological space and $\{\mathcal{N}_x : x \in X\}$ is defined as in part (a), then the topology obtained as in part (b) is again $\mathcal{T}$.*

   (d) *If $\{\mathcal{N}_x : x \in X\}$ is a given neighborhood system and $\mathcal{T}$ is the topology defined in part (b), then the neighborhood system obtained from $\mathcal{T}$ contains $\{\mathcal{N}_x : x \in X\}$. That is, if V is one of the neighborhoods of x obtained from $\mathcal{T}$ then there is a U in $\mathcal{N}_x$ such that $U \subset V$.*

Proof. The proof of parts (a), (c), and (d) are left to the reader. To prove part (b), first observe that both X and $\square$ are in $\mathcal{T}$ ($\square \in \mathcal{T}$ since the conditions are vacuously satisfied). Let $U_1, \ldots, U_n \in \mathcal{T}$ and put $U = \bigcap_{j=1}^{n} U_j$. If $x \in U$ then for each j there is a V_j in $\mathcal{N}_x$ such that $V_j \subset U_j$. It follows by induction on part (b) of Definition 4.16 that there is a V in $\mathcal{N}_x$ such that $V \subset \bigcap_{j=1}^{n} V_j$. Thus $V \subset U$ and U must belong to $\mathcal{T}$. Since the union of a collection of sets in $\mathcal{T}$ is clearly in $\mathcal{T}$, $\mathcal{T}$ is a topology. Finally, fix x in and let $U \in \mathcal{N}_x$. If $y \in U$ it follows from part (c) of Definition 4.16 that there is a V in $\mathcal{N}_y$ such that $V \subset U$. Thus $U \in \mathcal{T}$. ∎

4.18 Definition. If $\{\mathcal{N}_x: x \in X\}$ is a neighborhood system on X and $\mathcal{T}$ is the topology defined in part (b) of Proposition 4.17, then $\mathcal{T}$ is called *the topology induced by the neighborhood system.*

4.19 Corollary. *If $\{\mathcal{N}_x: x \in X\}$ is a neighborhood system on X and $\mathcal{T}$ is the induced topology then $(X, \mathcal{T})$ is a Hausdorff space iff for any two distinct points x and y in X there is a set U in $\mathcal{N}_x$ and a set V in $\mathcal{N}_y$ such that $U \cap V = \square$.*

Exercises

1. Prove the propositions which were stated in this section without proof.
2. Let $(X, \mathcal{T})$ and $(\Omega, \mathcal{S})$ be topological spaces and let $Y \subset X$. Show that if $f: X \to \Omega$ is a continuous function then the restriction of f to Y is a continuous function of $(Y, \mathcal{T}_Y)$ into $(\Omega, \mathcal{S})$.
3. Let X and Ω be sets and let $\{\mathcal{N}_x: x \in X\}$ and $\{\mathcal{M}_\omega: \omega \in \Omega\}$ be neighborhood systems and let $\mathcal{T}$ and $\mathcal{S}$ be the induced topologies on X and Ω respectively.
  (a) Show that a function $f: X \to \Omega$ is continuous iff when $x \in X$ and $\omega = f(x)$, for each Δ in $\mathcal{M}_\omega$ there is a U in $\mathcal{N}_x$ such that $f(U) \subset \Delta$.
  (b) Let $X = \Omega = \mathbb{C}$ and let $\mathcal{N}_z = \mathcal{M}_z = \{B(z; \epsilon): \epsilon > 0\}$ for each z in $\mathbb{C}$. Interpret part (a) of this exercise for this particular situation.
4. Adopt the notation of Exercise 3. Show that a function $f: X \to \Omega$ is open iff for each x in X and U in $\mathcal{N}_x$ there is a set Δ in $\mathcal{M}_\omega$ (where $\omega = f(x)$) such that $\Delta \subset f(U)$.
5. Adopt the notation of Exercise 3. Let $Y \subset X$ and define $\mathcal{U}_y = \{Y \cap U: U \in \mathcal{N}_y\}$ for each y in Y. Show that $\{\mathcal{U}_y: y \in Y\}$ is a neighborhood system for Y and the topology it induces on Y is $\mathcal{T}_Y$.
6. Adopt the notation of Exercise 3. For each point (x, ω) in $X \times \Omega$ let

$$\mathcal{U}_{(x,\omega)} = \{U \times \Delta: U \in \mathcal{N}_x, \Delta \in \mathcal{M}_\omega\}$$

  (a) Show that $\{\mathcal{U}_{(x,\omega)}: (x, \omega) \in X \times \Omega\}$ is a neighborhood system on $X \times \Omega$ and let $\mathcal{P}$ be the induced topology on $X \times \Omega$.
  (b) If $U \in \mathcal{T}$ and $\Delta \in \mathcal{S}$, call the set $U \times \Delta$ an *open rectangle*. Prove that a set is in $\mathcal{P}$ iff it is the union of open rectangles.
  (c) Define $p_1: X \times \Omega \to X$ and $p_2: X \times \Omega \to \Omega$ by $p_1(x, \omega) = x$ and $p_2(x, \omega) = \omega$. Show that p_1 and p_2 are open continuous maps. Furthermore if $(Z, \mathcal{R})$ is a topological space show that a function $f: (Z, \mathcal{R}) \to (X \times \Omega, \mathcal{P})$ is continuous iff $p_1 \circ f: Z \to X$ and $p_2 \circ f: Z \to \Omega$ are continuous.

§5. *The Sheaf of Germs of Analytic Functions on an Open Set*

This section introduces a topological space which plays a vital role in complex analysis. In addition to the topological structure, an analytical structure is also imposed on this space (in the next section). This will furnish the setting in which to study the complete analytic function as an analytic function.

If (f, D) is a function element, recall that the germ of (f, D) at a point $z = a$ in D is the collection of all function elements (g, B) such that $a \in B$ and $g(z) = f(z)$ for all z in $B \cap D$. This germ is denoted by $[f]_a$.

5.1 Definition. For an open set G in $\mathbb{C}$ let

$$\mathscr{S}(G) = \{(z, [f]_z) : z \in G, f \text{ is analytic at } z\}.$$

Define a map $\rho \colon \mathscr{S}(G) \to \mathbb{C}$ by $\rho(z, [f]_z) = z$. Then the pair $(\mathscr{S}(G), \rho)$ is called *the sheaf of germs of analytic functions on G*. The map ρ is called the *projection map*; and for each z in G, $\rho^{-1}(z) = \rho^{-1}(\{z\})$ is called the *stalk* or *fiber* over z. The set G is called the *base space* of the sheaf.

Notice that for a point $(z, [f]_z)$ to be in $\mathscr{S}(G)$, it is not necessary that f be defined on all of G; it is only required that f be analytic in a neighborhood of z.

How do we picture this sheaf? (There are, of course, too many dimensions to form an accurate geometrical picture.) One way is to follow the agricultural terminology used in the definition. On top of each stalk there is a collection of germs; the stalks are tied together into a sheaf. A better feeling for $\mathscr{S}(G)$ can be obtained by examining the notation for points in $\mathscr{S}(G)$. When we consider a point $(z, [f]_z)$ in $\mathscr{S}(G)$, think of a function element (f, D) in the germ $[f]_z$ instead of the germ itself. For every point w in D there is a point $(w, [f]_w)$ in $\mathscr{S}(G)$. Thus about $(z, [f]_z)$ there is a sheet or surface $\{(w; [f]_w) : w \in D\}$. In fact, $\mathscr{S}(G)$ is entirely made up of such sheets and they overlap in various ways. Alternately, we can think of $\mathscr{S}(G)$ as the union of graphs; each point $(z, [f]_z)$ in $\mathscr{S}(G)$ corresponding to the point $(z, f(z))$ on the graph of (f, D). (The graph of (f, D) is a subset of $\mathbb{C}^2$ or $\mathbb{R}^4$.) Two function elements are equivalent at a point z if their graphs coincide near z.

A topology will be defined on $\mathscr{S}(G)$ by defining a neighborhood system. For an open set D contained in G and a function $f \colon D \to \mathbb{C}$ analytic on D define

5.2 $N(f, D) = \{(z, [f]_z) : z \in D\}.$

That is, $N(f, D)$ is defined for each function element (f, D). If we think of $\mathscr{S}(G)$ as a collection of sheets lying above G which are indexed by the germs, then $N(f, D)$ is the part of that sheet indexed by f and which lies above D.

5.3 Theorem. *For each point $(a, [f]_a)$ in $\mathscr{S}(G)$ let*

$$\mathscr{N}_{(a, [f]_a)} = \{N(g, B) : a \in B \text{ and } [g]_a = [f]_a\}.$$

then $\{\mathscr{N}_{(a, [f]_a)} : (a, [f]_a) \in \mathscr{S}(G)\}$ is a neighborhood system on $\mathscr{S}(G)$ and the induced topology is Hausdorff. Furthermore. the induced topology makes the map $\rho \colon \mathscr{S}(G) \to G$ continuous.

Proof. Fix $(a, [f]_a)$ in $\mathscr{S}(G)$; since it is clear that condition (a) of Definition

4.16 holds, it remains to verify that condition (c) holds. (Condition (b) is a consequence of (c).) Let $N(g_1, B_1)$ and $N(g_2, B_2) \in \mathcal{N}_{(a,[f]_a)}$ and let

5.4 $(b, [h]_b) \in N(g_1, B_1) \cap N(g_2, B_2).$

It is necessary to find a function element (k, W) such that $N(k, W) \in \mathcal{N}_{(b,[h]_b)}$ and $N(k, W) \subset N(g_1, B_1) \cap N(g_2, B_2)$. It follows from (5.4) that $b \in B_1 \cap B_2$ and $[h]_b = [g_1]_b = [g_2]_b$. If $b \in W \subset B_1 \cap B_2$ and h is defined on W then $N(h, W) \subset N(g_1, B_1) \cap N(g_2, B_2)$.

To show that the induced topology is Hausdorff, use Corollary 4.19. So let $(a, [f]_a)$ and $(b, [g]_b)$ be distinct points of $\mathscr{S}(G)$. We must find a neighborhood $N(f, A)$ of $(a, [f]_a)$ and a neighborhood $N(g, B)$ of $(b, [g]_b)$ such that $N(f, A) \cap N(g, B) = \square$. How can it happen that $(a, [f]_a) \neq (b, [g]_b)$? There are two possibilities. Either $a \neq b$ or $a = b$ and $[f]_a \neq [g]_a$. If $a \neq b$ then let A and B be disjoint disks about a and b respectively; it follows immediately that $N(f, A) \cap N(g, B) = \square$. If $a = b$ but $[f]_a \neq [g]_a$, we must work a little harder (but not much). Since $[f]_a \neq [g]_a$ there is a disk $D = B(a; r)$ such that both f and g are defined on D and $f(z) \neq g(z)$ for $0 < |z-a| < r$. (It may happen that $f(a) = g(a)$ but this is inconsequential.)

Claim. $N(f, D) \cap N(g, D) = \square$.
In fact, if $(z, [h]_z) \in N(f, D) \cap N(g, D)$ then $z \in D$, $[h]_z = [f]_z$, and $[h]_z = [g]_z$. It follows that f and g agree on a neighborhood of z, and this is a contradiction. Hence the induced topology is Hausdorff.

Let U be an open subset of G; begin the proof that $\rho: \mathscr{S}(G) \to G$ is continuous by calculating $\rho^{-1}(U)$. Since $\rho(z, [f]_z) = z$,

$$\rho^{-1}(U) = \{(z, [f]_z): z \in U\}.$$

So if $(z, [f]_z) \in \rho^{-1}(U)$ and D is a disk about z on which f is defined and such that $D \subset U$, $N(f, D) \subset \rho^{-1}(U)$. It follows that ρ must be continuous (Exercise 4.3). ∎

Consider what was done when we showed that the induced topology was Hausdorff. If $a \neq b$ then $(a, [f]_a)$ and $(b, [g]_b)$ were on different stalks $\rho^{-1}(a)$ and $\rho^{-1}(b)$; so these distinct stalks were separated. In fact, if $a \in A$, $b \in B$ and $A \cap B = \square$ then $\rho^{-1}(A) \cap \rho^{-1}(B) = \square$. If $a = b$ then $(a, [f]_a)$ and $(a, [g]_a)$ lie on the same stalk $\rho^{-1}(a)$. Since $[f]_a \neq [g]_a$ we were able to divide the stalk. That is, one germ was "higher up" on the stalk than the other.

In the remainder of this section some of the properties of $\mathscr{S}(G)$ as a topological space are investigated. In particular, it will be of interest to characterize the components of $\mathscr{S}(G)$. However, we must first digress to study some additional topological concepts.

5.5 Definition. Let $(X, \mathscr{T})$ be a topological space. If x_0 and $x_1 \in X$ then an *arc* (or *path*) in X from x_0 to x_1 is a continuous function $\gamma: [0, 1] \to X$ such that $\gamma(0) = x_0$ and $\gamma(1) = x_1$. The point x_0 is called the *initial point* of γ and x_1 is called the *final point* or *terminal point*. The *trace* of γ is the set $\{\gamma\} = \{\gamma(t): 0 \le t \le 1\}$.

A subset A of X is said to be *arcwise* or *pathwise* connected if for any two points x_0 and x_1 in A there is a path from x_0 to x_1 whose trace lies in A. The topological space $(X, \mathcal{T})$ is called *locally arcwise* or *pathwise connected* if for each point x in X and each open set U which contains x there is an open arcwise connected set V such that $x \in V$ and $V \subset U$.

For each x in X let $\mathcal{N}_x$ be the collection of all open arcwise connected subsets of X which contain x. Then X is locally arcwise connected iff $\{\mathcal{N}_x : x \in X\}$ is a neighborhood system which induces the original topology on X. (Verify!)

The proof of the following proposition is left to the reader.

5.6 Proposition. *Let $(X, \mathcal{T})$ be a topological space.*

(a) *If A is an arcwise connected subset of X then A is connected.*

(b) *If X is locally arcwise connected then each component of X is an open set.*

The converse to part (a) of the preceding proposition is not true. For example let

$$X = \left\{ t + i \sin \frac{1}{t} : t > 0 \right\} \cup \{si : -1 \le s \le 1\}.$$

Since X is the closure of a connected set it is itself connected. However, there is no arc from the point $1/\pi$ to i which lies in X. X is also an example of a topological space which is connected but not locally arcwise connected.

Suppose X is connected and locally arcwise connected; does it follow that X is arcwise connected? The answer is yes. In fact, this is an abstract version of a theorem which was proved about open connected subsets of the plane. Since disks in the plane are connected, it follows that open subsets of $\mathbb{C}$ are locally arcwise connected. Recall that in Theorem II. 2.3 it was proved that for an open connected subset G of the plane, any two points in G can be joined by a polygon which lies in G. Hence, a partial generalization of this (the concept of a polygon in an abstract metric space is meaningless) is the following proposition whose proof is virtually identical to the proof of II. 2.3.

5.7 Proposition. *If X is locally arcwise connected then an open connected subset of X is arcwise connected.*

We now return to the sheaf of germs of analytic functions on an open set G.

5.8 Proposition. *Let G be an open subset of the plane and let U be an open connected subset of G such that there is an analytic function f defined on U. Then $N(f, U)$ is arcwise connected in $\mathscr{S}(G)$.*

Proof. Let $(a, [f]_a)$ and $(b, [f]_b)$ be two generic points in $N(f, U)$; then $a, b \in U$. Since U is a region there is a path $\gamma : [0, 1] \to U$ from a to b. Define $\sigma : [0, 1] \to N(f, U)$ by

$$\sigma(t) = (\gamma(t), [f]_{\gamma(t)}).$$

Clearly $\sigma(0) = (a, [f]_a)$ and $\sigma(1) = (b, [f]_b)$; if it can be shown that σ is continuous then σ is the desired arc.

Fix t in $[0, 1]$ and let $N(g, V)$ be a neighborhood of $\sigma(t)$. Then $\gamma(t) \in V$ and $[f]_{\gamma(t)} = [g]_{\gamma(t)}$. So there is a number $r > 0$ such that $B(\gamma(t); r) \subset U \cap V$ and $f(z) = g(z)$ for $|z - \gamma(t)| < r$. Also, since γ is continuous there is a $\delta > 0$ such that $|\gamma(s) - \gamma(t)| < r$ whenever $|s - t| < \delta$. Combining these last two facts gives that

$$(t - \delta, t + \delta) \subset \sigma^{-1}(N(g, V)).$$

It follows from Exercise 4.3 that σ is continuous. ∎

5.9 Corollary. $\mathscr{S}(G)$ *is locally arcwise connected and the components of* $\mathscr{S}(G)$ *are open arcwise connected sets.*

Proof. The first part of this corollary is a direct consequence of the preceding proposition. The second half follows from Proposition 5.6(b) and Proposition 5.7. ∎

In light of this last corollary, it is possible to gain insight into the nature of the components of $\mathscr{S}(G)$ by studying the curves in $\mathscr{S}(G)$.

5.10 Theorem. *There is a path in* $\mathscr{S}(G)$ *from* $(a, [f]_a)$ *to* $(b, [g]_b)$ *iff there is a path* γ *in* G *from* a *to* b *such that* $[g]_b$ *is the analytic continuation of* $[f]_a$ *along* γ.

Proof. Suppose that $\sigma: [0, 1] \to \mathscr{S}(G)$ is a path with $\sigma(0) = (a, [f]_a)$, $\sigma(1) = (b, [g]_b)$. Then $\gamma = \rho \circ \sigma$ is a path in G from a to b. Since $\sigma(t) \in \mathscr{S}(G)$ for each t, there is a germ $[f_t]_{\gamma(t)}$ such that

$$\sigma(t) = (\gamma(t), [f_t]_{\gamma(t)}).$$

We claim that $\{[f_t]_{\gamma(t)}: 0 \le t \le 1\}$ is the required continuation of $[f]_a$ along γ. Since $[f]_a = [f_0]_a$ and $[g]_b = [f_1]_b$, it is only necessary to show that $\{[f_t]_{\gamma(t)}\}$ is a continuation. For each t let D_t be a disk about $z = \gamma(t)$ such that $D_t \subset G$ and f_t is defined on D_t. Fix t in $[0, 1]$; since $N(f_t, D_t)$ is a neighborhood of $\sigma(t)$ and σ is continuous, there is a $\delta > 0$ such that

$$(t - \delta, \ t + \delta) \subset \sigma^{-1}(N(f_t, D_t)).$$

That is, if $|s - t| < \delta$ then $(\gamma(s), [f_s]_{\gamma(s)}) = \sigma(s) \in N(f_t, D_t)$. But, by definition, this gives that $\gamma(s) \in D_t$ and $[f_s]_{\gamma(s)} = [f_t]_{\gamma(s)}$; and this is precisely the condition needed to insure that $\{(f_t, D_t): 0 \le t \le 1\}$ is a continuation along γ (Definition 2.2).

Now suppose that γ is a curve in G from a to b and $\{[f_t]_{\gamma(t)}: 0 \le t \le 1\}$ is a continuation along γ such that $[f_0]_a = [f]_a$ and $[f_1]_b = [g]_b$. Define $\sigma: [0, 1] \to \mathscr{S}(G)$ by $\sigma(t) = (\gamma(t), [f_t]_{\gamma(t)})$; it is claimed that σ is a path from $(a, [f]_a)$ to $(b, [g]_b)$. Since the initial and final points of σ are the correct ones it is only necessary to show that σ is continuous. Because the details of this argument consist in retracing the steps of the first half of this proof, their execution is left to the reader. ∎

5.11 Theorem. *Let* $\mathscr{C} \subset \mathscr{S}(G)$ *and let* $(a, [f]_a) \in \mathscr{C}$; *then* $\mathscr{C}$ *is a component of* $\mathscr{S}(G)$ *iff*

$$\mathscr{C} = \{(b, [g]_b): [g]_b \text{ is the continuation of } [f]_a \text{ along some curve in } G\}.$$

Proof. Suppose $\mathscr{C}$ is a component of $\mathscr{S}(G)$; by Corollary 5.9 $\mathscr{C}$ is an open arcwise connected subset of $\mathscr{S}(G)$. So by the preceding theorem, for each point $(b, [g]_b)$ in $\mathscr{C}$, $[g]_b$ is the continuation of $[f]_a$ along some curve in G. Conversely, if $[g]_b$ is the continuation of $[f]_a$ along some curve in G then $(b, [g]_b)$ belongs to the component of $\mathscr{S}(G)$ which contains $(a, [f]_a)$; that is, $(b, [g]_b) \in \mathscr{C}$.

Now suppose that $\mathscr{C}$ consists of all points $(b, [g]_b)$ such that $[g]_b$ is a continuation of $[f]_a$. Then $\mathscr{C}$ is arcwise connected and hence is connected. If $\mathscr{C}_1$ is the component of $\mathscr{S}(G)$ containing $\mathscr{C}$ then by the first half of this proof it follows that $\mathscr{C} = \mathscr{C}_1$. ∎

Notice that the point $(a, [f]_a)$ in the statement of the preceding theorem has a transitory role; any point in the component will do.

Fix a function element (f, D) and recall that the complete analytic function $\mathscr{F}$ associated with (f, D) is the collection of all germs $[g]_z$ which are analytic continuations of $[f]_a$ for any a in D (see Definition 2.7). Let

5.12 $$G = \{z: \text{there is a germ } [g]_z \text{ in } \mathscr{F}\};$$

it follows that G is open. In fact, if $z \in G$ then there is a germ $[g]_z$ in $\mathscr{F}$ and a disk B about z on which g is defined. Clearly $B \subset G$ so that G is open.

It follows from Theorem 5.11 that

5.13 $$\mathscr{R} = \{(z, [g]_z): [g]_z \in \mathscr{F}\}$$

is a component of $\mathscr{S}(\mathbb{C})$ and that $\rho(\mathscr{R}) = G$. (It is also true that $\mathscr{R} \subset \mathscr{S}(G)$.)

5.14 Definition. Let $\mathscr{F}$ be a complete analytic function. If $\mathscr{R}$ is the set defined in (5.13) and ρ is the projection map of the sheaf $\mathscr{S}(\mathbb{C})$ then the pair $(\mathscr{R}, \rho)$ is called the *Riemann Surface* of $\mathscr{F}$. The open set G defined in (5.12) is called the *base space* of $\mathscr{F}$.

5.15 Theorem. *Let $\mathscr{F}$ be a complete analytic function with base space G and let $(\mathscr{R}, \rho)$ be its Riemann Surface. Then $\rho: \mathscr{R} \to G$ is an open continuous map. Also, if $(a, [f]_a)$ is a point in $\mathscr{R}$ then there is a neighborhood $N(f, D)$ of $(a, [f]_a)$ such that ρ maps $N(f, D)$ homeomorphically onto an open disk in the plane.*

Proof. Consider $\mathscr{R}$ as a component of $\mathscr{S}(\mathbb{C})$. Since $\rho: \mathscr{S}(\mathbb{C}) \to \mathbb{C}$ is continuous it follows that $\rho: \mathscr{R} \to G$ is continuous. To show that ρ is open it is sufficient to show that $\rho(N(f, U))$ is open for each (f, U) (Exercise 4.4). But $\rho(N(f, U)) = U$ which is open.

If $(a, [f]_a) \in \mathscr{R}$ let D be an open disk such that $(f, D) \in [f]_a$. Then $\rho: N(f, D) \to D$ is an open, continuous, onto map. To show that ρ is a homeomorphism it only remains to show that ρ is one-one on $N(f, D)$. But if $(b, [f]_b)$ and $(c, [f]_c)$ are distinct points of $N(f, D)$ then $b \neq c$ which gives that ρ is one-one. ∎

Remarks. In its standard usage the Riemann surface of a complete analytic function consists not only of what is here called a Riemann Surface but also some extra points—the branch points. These branch points correspond to singularities and permit a deeper analysis of the complete analytic function.

Exercises

1. Define $F: \mathscr{S}(\mathbb{C}) \to \mathbb{C}$ by $F(z, [f]_z) = f(z)$ and show that F is continuous.

2. Let $\mathscr{F}$ be the complete analytic function obtained from the principal branch of the logarithm and let $G = \mathbb{C} - \{0\}$. If D is an open subset of G and $f: D \to \mathbb{C}$ is a branch of the logarithm show that $[f]_a \in \mathscr{F}$ for all a in D. Conversely, if (f, D) is a function element such that $[f]_a \in \mathscr{F}$ for some a in D, show that $f: D \to \mathbb{C}$ is a branch of the logarithm. (Hint: Use Exercise 2.8.)

3. Let $G = \mathbb{C} - \{0\}$, let $\mathscr{F}$ be the complete analytic function obtained from the principal branch of the logarithm, and let $(\mathscr{R}, \rho)$ be the Riemann surface of $\mathscr{F}$ (so that G is the base of $\mathscr{R}$). Show that $\mathscr{R}$ is homeomorphic to the graph $\Gamma = \{(z, e^z): z \in G\}$ considered as a subset of $\mathbb{C} \times \mathbb{C}$. (Use the map $h: \mathscr{R} \to \Gamma$ defined by $h(z, [f]_z) = (f(z), z)$ and use Exercise 2.) State and prove an analogous result for branches of $z^{1/n}$.

4. Consider the sheaf $\mathscr{S}(\mathbb{C})$, let $B = \{z: |z-1| < 1\}$, let ℓ be the principal branch of the logarithm defined on B, and let $\ell_1(z) = \ell(z) + 2\pi i$ for all z in B. (a) Let $D = \{z: |z| < 1\}$ and show that $(\frac{1}{2}, [\ell]_{\frac{1}{2}})$ and $(\frac{1}{2}, [\ell_1]_{\frac{1}{2}})$ belong to the same component of $\rho^{-1}(D)$. (b) Find two disjoint open subsets of $\mathscr{S}(\mathbb{C})$ each of which contains one of the points $(\frac{1}{2}, [\ell]_{\frac{1}{2}})$ and $(\frac{1}{2}, [\ell_1]_{\frac{1}{2}})$.

§6. Analytic Manifolds

In this section a structure will be defined on a topological space which, when it exists, enables us to define an analytic function on the space. Before making the necessary definitions it is instructive to consider a previously encountered example of such a structure. The extended plane $\mathbb{C}_\infty$ can be endowed with an analytic structure in a neighborhood of each of its points. If $a \in \mathbb{C}_\infty$ and $a \neq \infty$ then a finite neighborhood U of a is an open subset of the plane. If $\varphi: U \to \mathbb{C}$ is the identity map, $\varphi(z) = z$, then φ gives a "coordinatization" of the neighborhood U. (Do not become confused over the preceding triviality. The introduction of the identity function φ seems an unnecessary nuisance. After all, U is an open subset of $\mathbb{C}$ so we know what it means to have a function analytic on U. Why bring up φ? The answer is that for the general definition it is necessary to consider pairs such as (U, φ); trivialities appear here because this is a trivial example.) If $a = \infty$ then let $U_\infty = \{z: |z| > 1\} \cup \{\infty\}$ and define $\varphi_\infty: U_\infty \to \mathbb{C}$ by $\varphi_\infty(z) = z^{-1}$ for $z \neq \infty$, $\varphi_\infty(\infty) = 0$. So φ_∞ is a homeomorphism of U_∞ onto the open disk $B(0; 1)$. Hence to each point a in $\mathbb{C}_\infty$ a pair (U_a, φ_a) is attached such that U_a is a neighborhood of a and φ_a is a homeomorphism of U_a onto an open subset of the plane. What happens if two of the sets U_a and U_b intersect? Suppose for example that $a \neq \infty$ and $U_a \cap U_\infty \neq \Box$. Let $G_\infty = B(0; 1) = \varphi_\infty(U_\infty)$ and let $G_a = \varphi_a(U_a) \ (= U_a)$. Then $\varphi_\infty^{-1}(z) = z^{-1}$ for all z in G_∞; thus $\varphi_a \circ \varphi_\infty^{-1}(z) = z^{-1}$ for all z in $\varphi_\infty(U_\infty \cap U_a)$. Since $\infty \notin U_a$, $0 \notin \varphi_\infty(U_\infty \cap U_a)$

so that $\varphi_a \circ \varphi_\infty^{-1}$ is analytic on its domain. Similarly, if both a and b are finite then $\varphi_a \circ \varphi_b^{-1}$ is analytic on its domain. This is the crucial property.

6.1 Definition. Let X be a topological space; a *coordinate patch* on X is a pair (U, φ) where U is an open subset of X and φ is a homeomorphism of U onto an open subset of the plane. If $a \in U$ then the coordinate patch (U, φ) is said to *contain a*.

6.2 Definition. An *analytic manifold* is a pair (X, Φ) where X is a Hausdorff connected topological space and Φ is a collection of coordinate patches on X such that: (i) each point of X is contained in at least one member of Φ, and (ii) if (U_a, φ_a), $(U_b, \varphi_b) \in \Phi$ with $U_a \cap U_b \neq \square$ then $\varphi_a \circ \varphi_b^{-1}$ is an analytic function of $\varphi_b(U_a \cap U_b)$ onto $\varphi_a(U_a \cap U_b)$. The set Φ of coordinate patches is called an *analytic structure* on X.

An analytic manifold is also called an *analytic surface*.

Note immediately that $\varphi_a \circ \varphi_b^{-1}$ is one-one since both φ_a and φ_b are. Henceforward, for the sake of brevity, care will not be taken in mentioning the appropriate domain of a function such as $\varphi_a \circ \varphi_b^{-1}$.

Next, it must be emphasized that the definition of an analytic manifold is tied to its analytic structure. It is possible to give the open disk two incompatible analytic structures (one of them the natural one). This is also the case with the torus which can be made into an analytic manifold in an uncountable number of incompatible ways. (Exercise 4.) But this investigation must be postponed until we have the notion of an isomorphism between analytic surfaces.

With the collection of coordinate patches introduced prior to Definition 6.1, $\mathbb{C}_\infty$ becomes an analytic manifold. However, a closed disk is not a manifold since the points on the boundary cannot be surrounded by a coordinate patch. Similarly, the union of two intersecting planes in $\mathbb{R}^3$ is not an analytic manifold.

A number of examples of analytic surfaces will be available after substantiating the following observations which are gathered into a proposition.

6.3 Proposition. (a) *Suppose (X, Φ) is an analytic surface and V is an open connected subset of X. If*

$$\Phi_V = \{(U \cap V, \varphi) : (U, \varphi) \in \Phi\}$$

then (V, Φ_V) is an analytic surface. (b) *If (X, Φ) is an analytic surface and Ω is a topological space such that there is a homeomorphism h of X onto Ω, then with*

$$\Psi = \{(h(U), \varphi \circ h^{-1}) : (U, \varphi) \in \Phi\},$$

(Ω, Ψ) is an analytic surface.

Proof. (a) This is a triviality.

(b) It is clear that Ω is connected, that Ψ is a collection of coordinate patches for Ω, and that each point of Ω is contained in at least one member of Ψ. So let (U,φ) and $(V,\mu) \in \Phi$ such that $h(U) \cap h(V) \neq \square$. But then $U \cap V \neq \square$ since $h(U) \cap h(V) = h(U \cap V)$. So, $(\varphi \circ h^{-1}) \circ (\mu \circ h^{-1}) = (\varphi \circ h^{-1}) \circ (h \circ \mu^{-1}) = \varphi \circ \mu^{-1}$ which is analytic by the condition on Φ. Thus (Ω, Ψ) is an analytic manifold. ∎

In virtue of part (a) of the preceding proposition the following assumption is made on all analytic structures Φ that will be discussed:

If $(U,\varphi) \in \Phi$ and V is an open subset of U then $(V,\varphi) \in \Phi$.

Of course, when an analytic structure is defined one does not bother to give all the coordinate patches, but only those that generate Φ in the above manner.

Proposition 6.3 also implies that any space that is homeomorphic to a region in the plane is an analytic surface. Hence a piece of paper with a crease in it is an analytic surface. Is this a shock to you? If the reader has ever seen an introduction to differentiable manifolds, he may be surprised at this.

If X is a subset of $\mathbb{R}^3$ then X is a differentiable 2-manifold if each point in X is contained in a coordinate patch (U,φ) such that $\varphi^{-1}: \varphi(U) \rightarrow U \subset \mathbb{R}^3$ has coordinate functions with continuous partial derivatives. That is, let $G = \varphi(U)$ and $\varphi^{-1}(s,t) = (\xi(s,t), \eta(s,t), \zeta(s,t))$ for all (s,t) in G. It is required that ξ, η, and ζ be functions from G into $\mathbb{R}$ with continuous partial derivatives. A folded piece of paper is not a differentiable 2-manifold. In fact, if (U,φ) is a patch that contains a point on the crease then φ^{-1} has at least one non-differentiable coordinate function. Since analyticity is a stronger notion than differentiability this seems confusing, but the explanation is simple. If h is a homeomorphism of X onto the region G in the plane then an analytic structure is imposed on X via h. In fact, by considering X with this structure we are only considering G under a different guise. In the definition of a differentiable 2-manifold there is no differentiable structure "imposed" on X; the structure is restricted by conditions that it inherits as a subset of $\mathbb{R}^3$ (where there is already a differentiable structure).

In a similar fashion the surface of a cube in $\mathbb{R}^3$ is an analytic surface since it is homeomorphic to $\mathbb{C}_\infty$.

Suppose now that G is a region and $f: G \rightarrow \mathbb{C}$ is an analytic function such that $f'(z) \neq 0$ for any z in G. We wish to give the graph

$$\Gamma = \{(z, f(z)): z \in G\}$$

an analytic structure. If $p: \Gamma \rightarrow G$ is defined by $p(z, f(z)) = z$ then p is a homeomorphism of Γ onto G; consequently Γ inherits the analytic structure of G as in 6.3(c). But this does not use the analyticity of f, let alone the fact that f' doesn't vanish. This is surely an uninteresting structure. So, to rephrase the problem: put an analytic structure on Γ that is "connected" to the analytic properties of f.

Fix $\alpha = (a, f(a))$ in Γ. Since $f'(a) \neq 0$ there is a disk D_a about a such that $D_a \subset G$ and f is one-one on D_a. Let

6.4 $$U_\alpha = \{(z, f(z)): z \in D_a\}$$

and define $\varphi_\alpha: U_\alpha \to \mathbb{C}$ by

6.5 $$\varphi_\alpha(z, f(z)) = f(z)$$

for each $(z, f(z))$ in U_α.

6.6 Proposition. *Let G be a region in the plane and let f be an analytic function on G with non-vanishing derivative. If Γ is the graph of f and*

$$\Phi = \{(U_\alpha, \varphi_\alpha): \alpha \in \Gamma, \ U_\alpha \ and \ \varphi_\alpha \ as \ in \ (6.4) \ and \ (6.5)\}$$

then (Γ, Φ) is an analytic manifold.

Proof. Since Γ is homeomorphic to G it must be connected. Fix $\alpha = (a, f(a))$ in Γ; it is left to the reader to show that φ_α is a homeomorphism of U_α onto $f(D_a)$. Suppose .that $\beta = (b, f(b)) \in \Gamma$ with $U_\alpha \cap U_\beta \neq \square$ and compute $\varphi_\alpha \circ \varphi_\beta^{-1}$. Since $f: D_b \to \mathbb{C}$ is one-one there is an analytic function $g: \Omega \to D_b$, where $\Omega = f(D_b)$, such that $f(g(\omega)) = \omega$ for all ω in Ω. Since $\varphi_\beta(U_\beta) = \Omega$ it follows that $\varphi_\beta^{-1}(\omega) = (g(\omega), \omega)$; thus $\varphi_\alpha \circ \varphi_\beta^{-1}(\omega) = \varphi_\alpha(g(\omega), \omega) = \omega$ for each ω in $\varphi_\beta(U_\alpha \cap U_\beta)$. In particular $\varphi_\alpha \circ \varphi_\beta^{-1}$ is analytic. ■

Henceforward, whenever the graph of an analytic function with non-vanishing derivative is considered as an analytic manifold, it will be assumed that it has the analytic structure given in the preceding proposition.

6.7 Theorem. *If $(\mathscr{R}, \rho)$ is the Riemann surface of a complete analytic function and $\Phi = \{(U, \rho): U \ is \ open \ in \ \mathscr{R}, \ \rho \ is \ one\text{-}one \ on \ U\}$, then $(\mathscr{R}, \Phi)$ is an analytic manifold.*

Proof. It follows from Theorem 5.15 that each point of $\mathscr{R}$ is contained in an open set on which ρ is one-one and that ρ is a homeomorphism there. Furthermore, if (U, ρ) and $(V, \rho) \in \Phi$ and $U \cap V \neq \square$ then $\rho \circ (\rho|U)^{-1}$ is the identity map. (The notation $\rho|U$ is used to denote the restriction of ρ to U.) Since $\mathscr{R}$ is connected it is an analytic surface. ■

6.8 Definition. Let (X, Φ) and (Ω, Ψ) be analytic manifolds and let $f: X \to \Omega$ be a continuous function; let $a \in X$ and $\alpha = f(a)$. The function f is *analytic at* a if for any patch (Λ, ψ) in Ψ which contains α there is a patch (U, φ) in Φ which contains a such that:

(i) $f(U) \subset \Lambda$;
(ii) $\psi \circ f \circ \varphi^{-1}$ is analytic on $\varphi(U) \subset \mathbb{C}$.

The function is *analytic on X* if it is analytic at each point of X.

The condition that (U, φ) can be found such that $a \in U$ and $f(U) \subset \Lambda$ is a consequence of the continuity of f (Proposition 4.9(c)). The heart of the definition is the requirement that $\psi \circ f \circ \varphi^{-1}$ be an analytic function from $\varphi(U) \subset \mathbb{C}$ into $\mathbb{C}$.

For two given analytic surfaces there may be many analytic functions from one to the other or there may be very few. Clearly every constant function is analytic; but there may be no other analytic functions. For example, if $X = \mathbb{C}$ and Ω is a bounded region in the plane then Liouville's Theorem implies there are no non-constant analytic functions from X into Ω. Also, suppose that $f: \mathbb{C}_\infty \to \mathbb{C}$ is an analytic function; then $f(\mathbb{C}_\infty)$ is compact so that the restriction of f to $\mathbb{C}$ is a bounded analytic function on $\mathbb{C}$. Again, Liouville's Theorem says that each such f is a constant function. On the other hand, if p is a polynomial and $a \in \mathbb{C}$ then both $p(z)$ and $p\left(\dfrac{1}{z-a}\right)$ are analytic functions from $\mathbb{C}_\infty$ to $\mathbb{C}_\infty$. In fact, these are practically the only analytic functions from $\mathbb{C}_\infty$ to $\mathbb{C}_\infty$ (Exercise 7).

If (X, Φ) is an analytic surface then there are many analytic functions defined on open subsets of X. For example, if $(U, \varphi) \in \Phi$ then $\varphi: U \to \mathbb{C}$ is analytic. It follows (Proposition 6.10 below) that $f \circ \varphi: U \to \mathbb{C}$ is analytic for any analytic function $f: \varphi(U) \to \mathbb{C}$.

Before proving some of the basic properties of analytic functions on manifolds, one further example will be given. This example is stated as a theorem and justifies the terminology "complete analytic function".

6.9 Theorem. *Let $\mathscr{F}$ be a complete analytic function with Riemann Surface $(\mathscr{R}, \rho)$. If $\mathscr{F}: \mathscr{R} \to \mathbb{C}$ is defined by*

$$\mathscr{F}(z, [f]_z) = f(z)$$

then $\mathscr{F}$ is an analytic function.

Proof. Fix $\alpha = (a, [f]_a)$ in $\mathscr{R}$ and let D be a disk about a on which f is defined and analytic. Let U be the component of $\rho^{-1}(D)$ which contains α; so (U, ρ) is a coordinate patch. Let ρ^{-1} denote the inverse of $\rho: U \to \rho(U)$. We must show that $\mathscr{F} \circ \rho^{-1}$ is analytic on $\rho(U) \subset \mathbb{C}$. But for z in $\rho(U)$,

$$\mathscr{F} \circ \rho^{-1}(z) = \mathscr{F}(z, f(z)) = f(z);$$

that is, $\mathscr{F} \circ \rho^{-1} = f$ which is analytic. ∎

The next several results are generalizations of theorems about analytic functions defined on regions in the plane.

6.10 Proposition. *Suppose (X, Φ), (Y, Ψ), and (Z, Σ) are analytic manifolds and $f: X \to Y$ and $g: Y \to Z$ are analytic functions; then $g \circ f: X \to Z$ is an analytic function.*

The proof is left to the reader.

6.11 Theorem. *Let (X, Φ) and (Ω, Ψ) be analytic manifolds and let f and g be* *analytic functions from X into* Ω. *If* $\{x \in X : f(x) = g(x)\}$ *has a limit point in* X *then* $f = g$.

Proof. Define the subset A of X by

$$A = \{x : f \text{ and } g \text{ agree on a neighborhood of } x\}.$$

This set A will be shown to be non-empty and the reader will be required to prove that A is both open and closed in X. By hypothesis there is a point a in X such that for every neighborhood U of a there is a point x in U with $x \neq a$ and $f(x) = g(x)$. It is easy to conclude that $f(a) = g(a) = \alpha$. If $(\Lambda, \psi) \in \Psi$ and $\alpha \in \Lambda$ then there is a patch (U, φ) in Φ such that $f(U)$ and $g(U)$ are contained in Λ with both $\psi \circ f \circ \varphi^{-1}$ and $\psi \circ g \circ \varphi^{-1}$ analytic in a disk D about $z_0 = \varphi(\alpha)$. But the hypothesis gives that z_0 is a limit point of $\{z \in D : \psi \circ f \circ \varphi^{-1}(z) = \psi \circ g \circ \varphi^{-1}(z)\} = F$. In fact, if $f(x) = g(x)$ then $\varphi(f(x)) \in F$. Thus $\psi \circ f \circ \varphi^{-1}(z) = \psi \circ g \circ \varphi^{-1}(z)$ for all z in D; or $f(x) = g(x)$ for all x in $U \cap \varphi^{-1}(D)$. Hence $a \in A$ and $A \neq \square$. ∎

6.12 Maximum Modulus Theorem. *Let* (X, Φ) *be an analytic manifold and let* $f : X \to \mathbb{C}$ *be an analytic function. If there is a point* $a \in X$ *and a neighborhood* U *of* a *such that* $|f(a)| \geq |f(x)|$ *for all* x *in* U *then* f *is a constant function.*

The proof is left to the reader.

The Maximum Modulus Theorem allows us to generalize Liouville's Theorem in the following way.

6.13 Liouville's Theorem. *If* (X, Φ) *is a compact analytic manifold then there is no non-constant analytic function from* X *into* $\mathbb{C}$.

6.14 Open Mapping Theorem. *Let* (X, Φ) *and* (Ω, Ψ) *be analytic manifolds and let* $f : X \to \Omega$ *be a non-constant analytic function. If* U *is an open subset of* X *then* $f(U)$ *is open in* Ω.

Proof. Let U be an open subset of X and let $\alpha \in f(U)$. If $a \in U$ such that $\alpha = f(a)$ and $(\Lambda, \psi) \in \Psi$ which contains α, then let $(V, \varphi) \in \Phi$ such that $f(V) \subset \Lambda$ and $\psi \circ f \circ \varphi^{-1}$ is analytic. Let $W = U \cap V$; then W is open and so $\varphi(W)$ is an open subset of the plane. Since f is not a constant function it follows from Theorem 6.11 that $\psi \circ f \circ \varphi^{-1}$ is not constant. Hence, the Open Mapping Theorem for functions of a complex variable implies that $\psi(f(W)) = \psi \circ f \circ \varphi^{-1}(\varphi(W))$ is open in $\mathbb{C}$. But then $f(W) = \psi^{-1}(\psi(f(W)))$ is open and $\alpha \in f(W) \subset f(U)$. So $f(U)$ must be open. ∎

6.15 Definition. If (X, Φ) and (Ω, Ψ) are analytic manifolds, an *isomorphism* of X onto Ω is an analytic function $f : X \to \Omega$ which is one-one and onto. If an isomorphism exists then (X, Φ) is said to be *isomorphic* to (Ω, Ψ).

If $f : X \to \Omega$ is an isomorphism then f is an open mapping by (6.14). It follows that $f^{-1} : \Omega \to X$ is a homeomorphism. Is f^{-1} also analytic? The answer is yes and the reader is asked to prove the next proposition.

6.16 Proposition. *If* $f : X \to \Omega$ *is an isomorphism then* $f^{-1} : \Omega \to X$ *is also an isomorphism.*

Recall that the Riemann Mapping Theorem states that if G is a simply connected region in the plane and $G \neq \mathbb{C}$, then G is isomorphic to the open unit disk if both are considered as analytic surfaces. Also recall that Proposition 6.3(c) states that if (X, Φ) is an analytic manifold and $h\colon X \to \Omega$ is a homeomorphism of X onto the topological space Ω, then h induces an analytic structure on Ω; denote this structure by $\Phi \circ h^{-1}$. Suppose Ω already has an analytic structure Ψ. It is easy to see that $\Phi \circ h^{-1} = \Psi$ iff h is analytic; that is, iff h is an isomorphism from (X, Φ) onto (Ω, Ψ).

As an example let $X = \mathbb{C}$ and $\Omega = \{z\colon |z| < 1\}$, and define $h\colon X \to \Omega$ by

$$h(z) = \frac{z}{1+|z|} \, .$$

Then h is a homeomorphism of $\mathbb{C}$ onto Ω, but it is clearly not analytic. So using h and the analytic structure on $\mathbb{C}$, a structure can be induced on Ω which is completely unrelated to its natural structure. For example, with this induced structure Ω will have no bounded non-constant analytic functions.

Consider the following situation: let G be a region in the plane and let $f\colon G \to \mathbb{C}$ be an analytic function with non-vanishing derivative. Let (g, D) be a function element such that $g(D) \subset G$ and $f(g(z)) = z$ for all z in D. (That is, g is a "local" inverse of f.) If $\mathscr{F}$ is the complete analytic function obtained from (g, D) and $(\mathscr{R}, \rho)$ is its Riemann surface, let us examine whether $\mathscr{R}$ and the graph of f are isomorphic analytic manifolds. (The analytic structure for the graph of f was introduced in Proposition 6.6.) The answer to this question is yes provided that the domain of f is not restricted. To illustrate what can go wrong let $G = \{z\colon |z| < 1\}$ and let f be the exponential function e^z. There $\mathscr{F}$ consists of all germs of branches of the logarithm (Exercise 5.3); so $\mathscr{R}$ is rather large and complicated. However, $\{(z, e^z)\colon z \in G\}$ is a simple copy of the disk G. The difficulty arises because the domain of e^z has been artificially restricted. If instead G is the whole complex plane then $\mathscr{F}$ and the graph of G are indeed isomorphic analytic surfaces (see Exercise 5.3).

6.17 Definition. Let $f\colon G \to \mathbb{C}$ be an analytic function with non-vanishing derivative. If $a \in G$ and $\alpha = f(a)$, let (g, D) be a function element such that $\alpha \in D$ and $f(g(z)) = z$ for all z in D. If $\mathscr{F}$ is the complete analytic function obtained from (g, D) then $\mathscr{F}$ is called *the complete analytic function of local inverses for f.*

There are two questions that arise in connection with this definition. First, does the definition of the complete analytic function of local inverses for f depend on the choice of the function element? Could we have started with another local inverse and still have obtained the same complete analytic function? Second, does $\mathscr{F}$ contain the germ of every local inverse of f? The answer to each of these questions is given in the following proposition; it is "yes".

6.18 Proposition. *Let f be an analytic function with non-vanishing derivative on a region G. Let a and $b \in G$, $\alpha = f(a)$, $\beta = f(b)$; and let Δ_0 and Δ_1 be disks*

*about α and β respectively such that there are analytic functions $g_0 \colon \Delta_0 \to \mathbb{C}$
and $g_1 \colon \Delta_1 \to \mathbb{C}$ with $g_0(\alpha) = a$, $g_1(\beta) = b$, $f(g_0(\zeta)) = \zeta$ for all ζ in Δ_0,
$f(g_1(\zeta)) = \zeta$ for all ζ in Δ_1. Then there is a path σ in $f(G)$ from α to β such
that (g_1, Δ_1) is the continuation of (g_0, Δ_0) along σ.*

Proof. Since G is connected there is a path γ in G from a to b. For $0 \le t \le 1$
let D_t be a disk about $\gamma(t)$ such that $D_t \subset G$ and on which f is one-one. Let
$\sigma = f \circ \gamma$ and let Δ_t be a disk about $\sigma(t)$ such that $\Delta_t \subset f(D_t)$. Finally, let
$g_t \colon \Delta_t \to \mathbb{C}$ be an analytic function such that

$$f(g_t(\zeta)) = \zeta \text{ for } \zeta \text{ in } \Delta_t$$

$$g_t(\sigma(t)) = \gamma(t)$$

Claim. $\{(g_t, \Delta_t)\}$ is an analytic continuation along σ. To show this fix t and
let δ be chosen so that $\gamma(s) \in f^{-1}(\Delta_t) \cap D_t$ whenever $|s-t| < \delta$. Now fix s

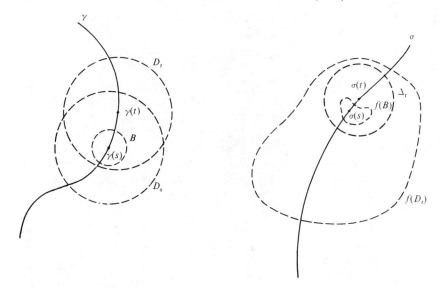

with $|s-t| < \delta$ and let B be a disk about $\gamma(s)$ such that $B \subset f^{-1}(\Delta_t) \cap D_s$
$\cap D_t$. So $f(B)$ is an open set containing $\sigma(s) = f(\gamma(s))$ and $f(B) \subset f(D_t)$.
By definition $g_s(f(z)) = z$ for z in B; thus $f(g_s(\zeta)) = \zeta$ for all ζ in $f(B)$. But
$f(B) \subset \Delta_t$ which gives that $f(g_t(\zeta)) = \zeta$ for all ζ in $f(B)$. But for ζ in $f(B)$
both $g_s(\zeta)$ and $g_t(\zeta)$ are in $f^{-1}(\Delta_t) \cap D_s \cap D_t$ and f is one-one here. Hence
$g_s(\zeta) = g_t(\zeta)$ for all ζ in $f(B)$; alternately, $[g_s]_{\sigma(s)} = [g_t]_{\sigma(s)}$ whenever $|s-t| < \delta$.
This substantiates the claim. ∎

Recall that if $(\mathcal{R}, \rho)$ is the Riemann surface of a complete analytic function $\mathcal{F}$, the symbol $\mathcal{F}$ is also used to denote the analytic function $\mathcal{F} \colon$
$\mathcal{R} \to \mathbb{C}$ defined by $\mathcal{F}(z, [f]_z) = f(z)$ (Theorem 6.9).

6.19 Proposition. *Let f be an analytic function on a region G which has a non-
vanishing derivative, let $\mathcal{F}$ be the complete analytic function of local inverses*

of f, and let $(\mathcal{R}, \rho)$ *be the Riemann surface of* $\mathcal{F}$. *If* $\mathcal{F}(\mathcal{R}) = G$ *and* $\rho(\mathcal{R}) = f(G)$ *then*

$$\tau(z, [g]_z) = (g(z), z)$$

defines an isomorphism between the analytic manifold $\mathcal{R}$ *and graph* (f).

Before proving this proposition we must show that under the same hypothesis each member of $\mathcal{F}$ is a local inverse of f. This provides a partial converse to Proposition 6.18.

6.20 Lemma. *Let* f, G, $\mathcal{F}$ *and* $(\mathcal{R}, \rho)$ *be as in the preceding proposition. If* $[g]_a \in \mathcal{F}$ *then there is a disk* D *about* a *on which* g *is defined and such that* $f(g(z)) = z$ *for all* z *in* D.

Proof. Fix $[g]_a$ in $\mathcal{F}$ (so $(a, [g]_a) \in \mathcal{R}$); then, by hypothesis, $g(a) = \mathcal{F}(a, [g]_a) \in G$. If $b = f(g(a))$ then there is a disk B about b and an analytic function $h : B \to \mathbb{C}$ such that $h(b) = g(a)$ and $f(h(z)) = z$ for all z in B. From Proposition 6.18, $[h]_b \in \mathcal{F}$. Also $a = \rho(a, [g]_a) \in \rho(\mathcal{R}) = f(G)$. According to Theorem 5.11 there is a path γ in $f(G)$ from a to b such that $[h]_b$ is the continuation along γ of $[g]_a$. Let $\{(g_t, D_t)\}$ be a continuation along γ such that $[g_0]_a = [g]_a$ and $[g_1]_b = [h]_b$. Define a subset T of $[0, 1]$ by

$$T = \{t : f(g_t(z)) = z \text{ for all } z \text{ in } D_t\}.$$

We want to show that $T = [0, 1]$. In fact, once this is proved it follows that $0 \in T$ so that (g, D) must be a local inverse of f.

Since T contains 1 it is non-empty; it must be shown that T is both open and closed in $[0, 1]$. For any number t in $[0, 1]$ let $\delta > 0$ such that $[g_s]_{\gamma(s)} = [g_t]_{\gamma(s)}$ whenever $|s-t| < \delta$. If $t \in T$ then $f(g_s(z)) = f(g_t(z)) = z$ for all z in $D_s \cap D_t$ and $|s-t| < \delta$. It follows that $(t-\delta, t+\delta) \subset T$ and so T is open. If $t \in T^-$ then there is an s in T such that $|s-t| < \delta$. For z in $D_s \cap D_t$ we have that $f(g_t(z)) = f(g_s(z)) = t$ so that $t \in T$. That is, T is closed. ∎

Proof of Proposition 6.19. It follows from the preceding lemma that $(g(a), a) \in$ graph (f) if $(a, [g]_a) \in \mathcal{R}$. Thus, τ does indeed map $\mathcal{R}$ into graph (f).

Suppose $(\alpha, f(\alpha)) \in$ graph (f) and $a = f(\alpha)$. If (g, D) is a function element such that $a \in D$, $g(a) = \alpha$, and $f(g(z)) = z$ for all z in D, then $[g]_a \in \mathcal{F}$ and $\tau(a, [g]_a) = (\alpha, f(\alpha))$. That is $\tau(\mathcal{R}) =$ graph (f). To show that τ is one-one let $(a, [g]_a)$ and $(b, [h]_b) \in \mathcal{R}$ such that $\tau(a, [g]_a) = \tau(b, [h]_b)$; that is, $(g(a), a) = (h(b), b)$. Thus, $a = b$ and $g(a) = h(b) = \alpha$. Moreover, Lemma 6.20 implies that

6.21 $$f(g(z)) = f(h(z))$$

for z in a neighborhood of a. But $f'(\alpha) \neq 0$, so that f is one-one in a neighborhood of $\zeta = \alpha$. It follows from (6.21) that $[g]_a = [h]_\alpha$ and therefore, τ is one-one.

It remains to show that τ is analytic. This is actually an easy argument once the question to be answered is made explicit. So fix $(a, [g]_a)$ in $\mathcal{R}$ and put $\alpha = g(a)$. Let Δ be a disk about α on which f is one-one and let D be a disk about a on which g is defined and such that $D \subset f(\Delta)$. Let $U = \{(\zeta,$

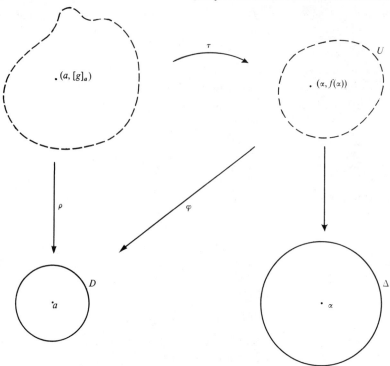

$f(\zeta))$: $\zeta \in \Delta\}$ and define φ: $U \to \mathbb{C}$ by $\varphi(\zeta, f(\zeta)) = f(\zeta)$; so (U, φ) is a coordinate patch on graph (f) (Proposition 6.6). Also $N(f, D) = \{(z, [g]_z)$: $z \in D\}$ gives that $(N(g, D), \rho)$ is a coordinate patch on $\mathscr{R}$ (Theorem 6.7) containing $(a, [g]_a)$, and satisfying $\tau(N(g, D)) \subset U$. Hence to complete the proof it must be shown that $\varphi \circ \tau \circ \rho^{-1}$ is an analytic function on D. This is trivial. In fact, if $z \in D$ then

$$\varphi \circ \tau \circ \rho^{-1}(z) = \varphi \circ \tau(z, [g]_z)$$

$$= \varphi(g(z), z)$$

$$= z;$$

that is, $\varphi \circ \tau \circ \rho^{-1}$ is the identity function on D. ∎

If G is the punctured plane and $f(z) = z^n$ for some n or if $G = \mathbb{C}$ and $f(z) = e^z$ then the hypothesis of Proposition 6.19 is satisfied. Let us examine this a little more closely for the case where $f(z) = z^2$, $z \in G = \mathbb{C} - \{0\}$. So $\mathscr{R}$ is isomorphic to the graph of z^2; let $\Gamma = \{(z, z^2): z \neq 0\}$. Now $\tau: \mathscr{R} \to \Gamma$ is defined by $\tau(a, [g]_a) = (g(a), a)$. Recall that $\mathscr{F}: \mathscr{R} \to \mathbb{C}$ is defined by $\mathscr{F}(a, [g]_a) = g(a)$. For this case $\mathscr{F}$ acts like the square root function. In other words, we have found a natural domain of definition of $z^{\frac{1}{2}}$. The corresponding function on Γ would project (z, z^2) to its first coordinate.

Even though we have shown that $\mathscr{R}$ (a very abstract object) is equivalent to a less abstract object (the graph Γ), this is still somewhat dissatisfying. After all, the graph is a subset of $\mathbb{C}^2$ which is beyond our geometric intuition.

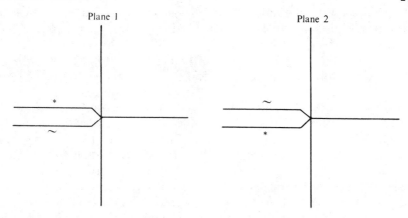

Therefore, we would like to have a more geometric picture of the Riemann surface. Consider two copies of the plane that have been slit along the negative real axis. So imagine the planes as having two negative real axes and label them * and $\sim$ as shown in the figure. The space X we will describe will be the union of the two planes but where the two *-axes are identified and the two $\sim$-axes are identified. So if a curve in Plane 1 approaches the *-axis and hits it at $-x$ then it exits in Plane 2 at $-x$ on the *-axis. For the point -1 on the *-axis a typical neighborhood would consist of a half disk about -1 above the *-axis in Plane 1 and half a disk about -1 below the *-axis in Plane 2. This is a representation of the Riemann surface of local inverses for z^2 (the Riemann surface of $\sqrt{z}$ for short). To see this define a map $k: X \to \mathscr{R}$ as follows. If z is in Plane 1, z not on the negative real axis, let $h(z) = (z, [g]_z)$ where g is the principal branch of the square root. If z is in Plane 2 but not on the negative real axis let $h(z) = (z, [g]_z)$ where $-g$ is the principal branch of the square root. It remains to define $h(z)$ for z on the *-axis and the $\sim$-axis. This we leave to the reader along with the proof that the resulting function h is an isomorphism (the space X has a natural analytic structure).

In the case $f(z) = z^n$ we can carry out the same construction, but here n copies of the plane are required. If $f(z) = e^z$ the same ideas are again employed but now it is necessary to use an infinite number of planes indexed by all the integers. In the case of the surface for $z^{1/n}$, a curve which passes through the negative real axis of one plane exits through the negative real axis of the next one. If it is in the n-th plane, then it exits through the negative axis of the first plane. For the surface of $\log z$, a curve can continue hopping from one plane to the next and will never return to the plane where it started unless it "retraces its steps".

Exercises

1. Show that an analytic manifold is locally compact. That is, prove that if $a \in X$ and U is an open neighborhood of a then there is an open neighborhood V of a such that $V^- \subset U$ and V^- is compact.

2. Which of the following are analytic manifolds? What is its analytic structure if it is a manifold? (a) A cone in $\mathbb{R}^3$. (b) $\{(x_1, x_2, x_3) \in \mathbb{R}^3 : x_1^2 + x_2^2 + x_3^2 = 1$ or $x_1^2 + x_2^2 > 1$ and $x_3 = 0\}$

3. The following is a generalization of Proposition 6.3(b). Let (X, Φ) be an analytic manifold, let Ω be a topological space, and suppose there is a continuous function h of X onto Ω that is locally one-one (that is, if $x \in X$ there is an open set U such that $x \in U$ and h is one-one on U). If $(U, \varphi) \in \Phi$ and h is one-one on U let $\Delta = h(U)$ and let $\psi : \Delta \to \mathbb{C}$ be defined by $\psi(\omega) = \varphi \circ (h/U)^{-1}(\omega)$. Let Ψ be the collection of all such pairs (Δ, ψ). Prove that (Ω, Ψ) is an analytic manifold and h is an analytic function from X to Ω.

4. Let $T = \{z : |z| = 1\} \times \{z : |z| = 1\}$; then T is a torus. (This torus is homeomorphic to the usual hollow doughnut in $\mathbb{R}^3$.) If ω and ω' are complex numbers such that Im $(\omega/\omega') \neq 0$ then ω and ω', considered as elements of the vector space $\mathbb{C}$ over $\mathbb{R}$, are linearly independent. So each z in $\mathbb{C}$ can be uniquely represented as $z = t\omega + t'\omega'$; t, t' in $\mathbb{R}$. Define $h : \mathbb{C} \to T$ by $h(t\omega + t\omega') = (e^{2\pi i t}, e^{2\pi i t'})$. Show that h induces an analytic structure on T. (Use Exercise 3.) (b) If ω, ω' and ζ, ζ' are two pairs of complex numbers such that Im $(\omega/\omega') \neq 0$ and Im $(\zeta/\zeta') \neq 0$, define $\sigma(s\zeta + s'\zeta') = (e^{2\pi i s}, e^{2\pi i s'})$ and $\tau(t\omega + t'\omega') = (e^{2\pi i t}, e^{2\pi i t'})$. Let $G = \{t\omega + t'\omega' : 0 < t < 1, 0 < t' < 1\}$ and $\Omega = \{s\zeta + s'\zeta' : 0 < s < 1, 0 < s' < 1\}$; show that both σ and τ are one-one on G and Ω respectively. (Both G and Ω are the interiors of parallelograms.) If Φ_τ and Φ_σ are the analytic structures induced on T by τ and σ respectively, and if the identity map of (T, Φ_τ) into (T, Φ_σ) is analytic then show that the function $f : G \to \Omega$ defined by $f = \sigma^{-1} \circ \tau$ is analytic. (To say that the identity map of (T, Φ_τ) into (T, Φ_σ) is analytic is to say that Φ_τ and Φ_σ are equivalent structures.) (c) Let $\omega = 1, \omega' = i, \zeta = 1, \zeta' = \alpha$ where Im $\alpha \neq 0$; define σ, τ, G, Ω and f as in part (b). Show that Φ_τ and Φ_σ are equivalent analytic structures if and only if $\alpha = i$. (Hint: Use the Cauchy-Riemann equations.) (d) Can you generalize part (c)? Conjecture a generalization?

5. (a) Let f be a meromorphic function defined on $\mathbb{C}$ and suppose f has two independent periods ω and ω'. That is, $f(z) = f(z + n\omega + n'\omega')$ for all z in $\mathbb{C}$ and all integers n and n', and Im $(\omega/\omega') \neq 0$. Using the notation of Exercise 4(a) show that there is an analytic function $F : T \to \mathbb{C}_\infty$ such that $f = F \circ h$. (For an example of a meromorphic function with two independent periods see Exercise VIII. 3.2(g).)

(b) Prove that there is no non-constant entire function with two independent periods.

6. Show that an analytic surface is arcwise connected.

7. Suppose that $f : \mathbb{C}_\infty \to \mathbb{C}_\infty$ is an analytic function.

(a) Show that either $f \equiv \infty$ or $f^{-1}(\infty)$ is a finite set.

(b) If $f \not\equiv \infty$, let $a_1, \ldots, a_n$ be the points in $\mathbb{C}$ where f takes on the value ∞. Show that there are polynomials $p_0, p_1, \ldots, p_n$ such that

$$f(z) = p_0(z) + \sum_{k=1}^{n} p_k \left(\frac{1}{z - a_k} \right)$$

for z in $\mathbb{C}$.

(c) If f is one-one, show that either $f(z) = az+b$ (some a, b in $\mathbb{C}$) or $f(z) = \dfrac{a}{z-c} + b$ (some a, b, c in $\mathbb{C}$).

8. Furnish the details of the discussion of the surface for $\sqrt{z}$ at the end of this section.

9. Let $G = \left\{ z: -\dfrac{\pi}{2} < \operatorname{Re} z < \dfrac{\pi}{2} \right\}$ and define $f: G \to \mathbb{C}$ by $f(z) = \sin z$. Give a discussion for f similar to the discussion of $\sqrt{z}$ at the end of this section.

§7. Covering spaces

In this section the concept of a covering space will be introduced and some of its elementary properties will be deduced. One byproduct of this study is the fact that two closed curves in the punctured plane are homotopic iff they have the same winding number about the origin.

Intuitively, a topological space X is a covering space for the topological space Ω if X can be wrapped around Ω in such a way that it can be easily unwrapped. What is meant by "wrapping" one space around another? This seems to indicate that we want a function from X onto Ω. To say that it must be easily unwrapped must mean that we can find an inverse for the function.

7.1 Definition. If Ω is a topological space then a *covering space* of Ω is a pair (X, ρ) where X is a connected topological space and ρ is a continuous function of X onto Ω such that: for each ω in Ω there is a neighborhood Δ of ω such that each component of $\rho^{-1}(\Delta)$ is open, and ρ maps each of these components homeomorphically onto Δ. Such an open set Δ is called *fundamental* and Δ is *properly covered*.

Both $(\mathbb{C}, \exp)$ and $(\mathbb{C} - \{0\}, z^n)$ are covering spaces of the punctured plane; also, each open disk in the punctured plane is fundamental. If $\Gamma = \{z: |z| = 1\}$ and $\rho: \mathbb{R} \to \Gamma$ is defined by $\rho(t) = \exp(2\pi it)$, then $(\mathbb{R}, \rho)$ is a covering space of Γ. Every proper arc in Γ is fundamental.

The following is a list of properties of covering spaces. Their proof is left to the reader.

7.2 Proposition. *Let (X, ρ) be a covering space of Ω.*

(a) *ρ is an open mapping of X onto Ω.*

(b) *If $x \in X$ then there is an open neighborhood U of x on which ρ is a homeomorphism.*

(c) *Every fundamental open set is connected.*

(d) *If Ω is locally arcwise connected then so is X.*

In light of part (b) of the preceding proposition it is natural to ask if every locally one-one function ρ of X onto Ω makes (X, ρ) a covering space of Ω. The answer is no. For example, let $X = \{z \in \mathbb{C}: z \neq 0, 0 < \arg z < 5\pi/4\}$, $\rho(z) = z^2$, and let $\Omega = \rho(X) = \mathbb{C} - \{0\}$; then ρ is locally one-one. Let $\Delta_r = \{\xi: |\xi - i| < r\}$ for any r, $0 < r < 1$. It is easy to see that $\rho^{-1}(\Delta_r)$ consists of two components. One of these (the one in the first quadrant) is mapped homeomorphically onto Δ_r, while the other is not. In fact, $\xi = i$ is not in the image of this second component.

One of the most important properties of covering spaces is the fact that a curve in Ω can be lifted to a curve in X.

7.3 Definition. Let (X, ρ) be a covering space of Ω and let γ be a path in Ω. A path $\tilde{\gamma}$ in X is called a *lifting* of γ if $\rho \circ \tilde{\gamma} = \gamma$.

A useful way of understanding what a lifting is is to consider the following diagrams. If $I = [0, 1]$ and the path γ is given then we have the following mapping diagram:

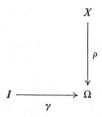

To say that γ can be lifted is to say that the diagram can be completed in such a way that it is a commutative diagram:

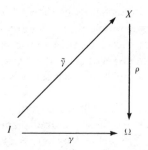

that is, one can go from one place in the diagram to another without being concerned about which path is taken.

It is an important property of covering spaces that every path may be lifted when the base space is locally arcwise connected. Actually a stronger result which will be of use later can be proved.

7.4 Theorem. *Let (X, ρ) be a covering space of the space Ω. If $F: [0, 1] \times [0, 1] \to \Omega$ is a continuous function with $F(0, 0) = \omega_0$ and if x_0 is any point in X with $\rho(x_0) = \omega_0$, then there is a unique continuous function $\tilde{F}: [0, 1] \times [0, 1] \to X$ such that $\tilde{F}(0, 0) = x_0$ and $\rho \circ \tilde{F} = F$.*

Before giving the proof of this theorem let us state an important corollary.

7.5 Corollary. *Let (X, ρ) be a covering space of the space Ω. If γ is a path in Ω with initial point ω_0 and $\rho(x_0) = \omega_0$, then there is a unique lifting $\tilde{\gamma}$ of γ with initial point x_0.*

Proof. Define $F: [0, 1] \times [0, 1] \to \Omega$ by $F(s, t) = \gamma(s)$; so F is continuous and $F(0, 0) = \omega_0$. According to Theorem 7.4 there is a unique function $\tilde{F}$: $[0, 1] \times [0, 1] \to X$ such that $\tilde{F}(0, 0) = x_0$ and $\rho \circ \tilde{F} = F$. Let $\tilde{\gamma}(s) = \tilde{F}(s, 0)$; then $\tilde{\gamma}$ has initial point x_0 and is a lifting of γ. To prove the uniqueness of $\tilde{\gamma}$, suppose $\tilde{\sigma}$ is also a path in X with initial point x_0 and which lifts γ. Define $\tilde{K}: [0, 1] \times [0, 1] \to X$ by $\tilde{K}(s, t) = \tilde{\sigma}(s)$. Then $\tilde{K}: (0, 0) = x_0$ and $\rho \circ \tilde{K}(s, t) = \rho \circ \tilde{\sigma}(s) = \gamma(s) = F(s, t)$. By the uniqueness part of Theorem 7.4, $\tilde{F} = \tilde{K}$. Thus $\tilde{\gamma}(s) = \tilde{F}(s, 0) = \tilde{K}(s, 0) = \tilde{\sigma}(s)$. ∎

Proof of Theorem 7.4. Let $\{0 = s_0 < s_1 < \ldots < s_n = 1\}$ and $\{0 = t_0 < t_1 < \ldots < t_n = 1\}$ be partitions off $[0, 1]$ such that for $0 \le i, j \le n-1$,

$$F([s_i, s_{i+1}] \times [t_j, t_{j+1}])$$

is contained in a fundamental open set Δ_{ij} in Ω. (Verify that this can be done.) Now $\omega_0 = F(0, 0) \in \Delta_{00}$. Let U_{00} be the component $\rho^{-1}(\Delta_{00})$ which contains x_0. Since $\rho|U_{00}$ is a homeomorphism of U_{00} onto Δ_{00}, it is possible to define $\tilde{F}: [0, s_1] \times [0, t_1] \to X$ by

$$\tilde{F}(s, t) = (\rho|U_{00})^{-1} \circ F(s, t).$$

Now extend $\tilde{F}$ to $[0, s_2] \times [0, t_1]$ as follows. $\tilde{F}(\{s_1\} \times [0, t_1])$ is connected (Why?) and, since $F(\{s_1\} \times [0, t_1]) \subset \Delta_{10}$, it is contained in $\rho^{-1}(\Delta_{10})$. Let U_{10} be the component of $\rho^{-1}(\Delta_{10})$ which contains $F(\{s_1\} \times [0, t_1])$. Then $\rho|U_{10}$ is a homeomorphism; define

$$\tilde{F}(s, t) = (\rho|U_{10})^{-1} \circ F(s, t)$$

for (s, t) in $[s_1, s_2] \times [0, t_1]$. This gives a continuous function $\tilde{F}$ on $[0, s_2] \times [0, t_1]$. (The domain can be written as the union of two closed sets, on each of these sets $\tilde{F}$ is continuous, and $\tilde{F}$ agrees on their intersection; hence, $\tilde{F}$ is continuous on their union.) Continuing this process leads to a continuous function $\tilde{F}: [0, 1] \times [0, 1] \to X$ such that $\rho \circ \tilde{F} = F$ and $\tilde{F}(0, 0) = x_0$. Since at each stage of this construction the definition of $\tilde{F}$ is unique (because ρ is a homeomorphism on each U_{ij}), it follows that $\tilde{F}$ is unique. ∎

The next result is called the Monodromy Theorem. To distinguish this from the theorem of the same name which was obtained in §3, the present version is referred to as the "abstract" theorem. Later it will be shown how the original theorem can be deduced from this abstract one.

7.6 Abstract Monodromy Theorem. *Let (X, ρ) be a covering space of Ω and let γ and σ be two paths in Ω with the same initial and final points. Let $\tilde{\gamma}$ and $\tilde{\sigma}$ be paths in X with the same initial points such that $\tilde{\gamma}$ and $\tilde{\sigma}$ are liftings of γ and σ respectively. If $\gamma \sim \sigma$ (FEP) in Ω then $\tilde{\gamma}$ and $\tilde{\sigma}$ have the same final points and $\tilde{\gamma} \sim \tilde{\sigma}$ (FEP) in X.*

Note. Although we have not defined the concept of FEP homotopy between two curves in an arbitrary topological space, the definition is similar to that given for curves in a region of the plane (Definition IV. 6.11).

Proof. Let ω_0 and ω_1 be the initial and final points, respectively, of γ and σ.

Let $x_0 \in \rho^{-1}(\omega_0)$ such that $\tilde{\gamma}(0) = \tilde{\sigma}(0) = x_0$. By hypothesis there is a continuous function $F: [0, 1] \times [0, 1] \to \Omega$ such that $F(0, t) = \omega_0$, $F(1, t) = \omega_1$, $F(s, 0) = \gamma(s)$, and $F(s, 1) = \sigma(s)$ for all s and t in $[0, 1]$. According to Theorem 7.4 there is a unique continuous function $\tilde{F}: [0, 1] \times [0, 1] \to X$ such that $\tilde{F}(0, 0) = x_0$ and $\rho \circ \tilde{F} = F$. Now $F(\{0\} \times [0, 1]) = \{\omega_0\}$ and $\rho \circ \tilde{F} = F$ implies that $\tilde{F}(\{0\} \times [0, 1]) \subset \rho^{-1}(\omega_0)$. But each component of $\rho^{-1}(\omega_0)$ consists of a single point (Exercise 4) and $\tilde{F}(\{0\} \times [0, 1])$ is connected. Therefore $\tilde{F}(0, t) = x_0$ for all t. Similarly, there is a point x_1 such that $\tilde{F}(1, t) = x_1$ for all t and $\rho(x_1) = \omega_1$.

By the uniqueness of $\tilde{\gamma}$ and the fact that $s \to \tilde{F}(s, 0)$ is a path with initial point x_0 which lifts γ, it must be that $\tilde{\gamma}(s) = \tilde{F}(s, 0)$. Similarly, $\tilde{\sigma}(s) = \tilde{F}(s, 1)$. Therefore $\tilde{\gamma}(1) = \tilde{\sigma}(1) = x_1 = F(1, t)$ for all t, and $\tilde{F}$ demonstrates that $\tilde{\gamma} \sim \tilde{\sigma}(\text{FEP})$ in X. ■

In order to show that the Monodromy Theorem can be deduced from the preceding version, it is necessary to first prove a lemma. This preliminary result is actually the Monodromy Theorem for a disk and the reader is asked to supply an elementary proof.

7.7 Lemma. *Let (g, A) be a function element and let B be a disk such that $A \subset B$ and (g, A) admits unrestricted analytic continuation in B. If γ is a closed curve in B with $\gamma(0) = \gamma(1) = a$ in A and $\{(g_t, A_t)\}$ is an analytic continuation of (g, A) along γ then $[g_0]_a = [g_1]_a$.*

The next theorem will facilitate the deduction of the Monodromy Theorem from the Abstract Monodromy Theorem. Additionally, it has some interest by itself.

7.8 Theorem. *Let (f, D) be a function element which admits unrestricted continuation in the region G, and let $\mathscr{C}$ be the component of the sheaf $(\mathscr{S}(G), \rho)$ that contains $(z_0, [f]_{z_0})$ for some z_0 in D. Then $(\mathscr{C}, \rho)$ is a covering space of G.*

Proof. Let B be any disk such that $B \subset G$ and let $\mathscr{U}$ be the component of $\rho^{-1}(B)$ which is contained in $\mathscr{C}$. The proof will be completed by showing that ρ maps $\mathscr{U}$ homeomorphically onto B.

Fix $(a, [g]_a)$ in $\mathscr{U}$; then

7.9 Claim. $(z, [h]_z) \in \mathscr{U}$ iff $z \in B$ and $[h]_z$ is the continuation of $[g]_a$ along some curve in B.

In fact, if $(z, [h]_z)$ is such a point then there is a curve γ in $\rho^{-1}(B) (\subset \mathscr{S}(B))$ from $(a, [g]_a)$ to $(z, [h]_z)$ (Theorem 5.10). Thus $(z, [h]_z)$ must belong to the same component of $\rho^{-1}(B)$ as does $(a, [g]_a)$; that is, $(z, [h]_z) \in \mathscr{U}$. For the converse, let $(z, [h]_z) \in \mathscr{U}$; since $\mathscr{U}$ is pathwise connected (Proposition 5.7) there is a path in $\mathscr{U}$ from $(a, [g]_a)$ to $(z, [h]_z)$. But this implies that $[h]_z$ is the continuation of $[g]_a$ along a path in $\rho(\mathscr{U}) \subset B$ (Theorem 5.10). So claim 7.9 has been shown.

Since (f, D) admits unrestricted continuation in G and $[g]_a$ is a continuation of $[f]_{z_0}$ (z_0 in D), it is a trivial matter to see that $[g]_a$ admits unrestricted continuation in B. In view of (7.9) this gives that $\rho(\mathscr{U}) = B$.

It only remains to prove that $\rho | \mathscr{U}$ is one-one. This amounts to showing that if $(b, [h]_b)$ and $(b, [k]_b) \in \mathscr{U}$ then $[h]_b = [k]_b$. But since $\mathscr{U}$ is arcwise connected, this is exactly the conclusion of Lemma 7.7. ∎

Let us retain the notation of the preceding theorem. Fix a in D and let γ and σ be paths in G from a to a point $z = b$. Suppose that $\{(f_t, D_t)\}$ and $\{(g_t, B_t)\}$ are continuations of (f, D) along γ and σ respectively; so $[f_0]_a = [g_0]_a = [f]_a$. Now $\tilde{\gamma}(t) = (\gamma(t), [f_t]_{\gamma(t)})$ and $\tilde{\sigma}(t) = (\sigma(t), [g_t]_{\sigma(t)})$ are paths in $\mathscr{C}$ (see the proof of Theorem 5.10) with the same initial point $(a, [f]_a)$. Moreover $\rho \circ \tilde{\gamma} = \gamma$ and $\rho \circ \tilde{\sigma} = \sigma$; so $\tilde{\gamma}$ and $\tilde{\sigma}$ are the unique liftings of γ and σ to $\mathscr{C}$. According to the Abstract Monodromy Theorem, if $\gamma \sim \sigma$(FEP) in G then $\tilde{\gamma}$ and $\tilde{\sigma}$ have the same final point. That is, $(b, [f_1]_b) = \tilde{\gamma}(1) = \tilde{\sigma}(1) = (b, [g_1]_b)$ so that $[f_1]_b = [g_1]_b$. This is precisely the conclusion of Theorem 3.6.

For another application of the Abstract Monodromy Theorem we wish to prove that closed curves in the punctured plane are homotopic iff they have the same winding number about the origin. To do this let $\Gamma = \{z : |z| = 1\}$; as we observed at the beginning of this section, if $\rho(t) = \exp(2\pi i t)$ then $(\mathbb{R}, \rho)$ is a covering space of Γ. If γ is any rectifiable curve in $\mathbb{C} - \{0\}$ it is easy to see that γ is homotopic (in $\mathbb{C} - \{0\}$) to the curve σ defined by $\sigma(t) = \gamma(t)/|\gamma(t)|$ (Exercise IV. 6.4). So assume that $|\gamma(t)| = 1$ for all t. Similarly, we can assume that $\gamma(0) = 1$.

Let $\tilde{\gamma}$ be the unique curve in $\mathbb{R}$ such that $\tilde{\gamma}(0) = 0$ and $\gamma(t) = \exp(2\pi i \tilde{\gamma}(t))$. Since γ is rectifiable it is easy to see that $\tilde{\gamma}$ is also rectifiable. Also

$$n(\gamma; 0) = \frac{1}{2\pi i} \int_\gamma \frac{dz}{z}$$

$$= \frac{1}{2\pi i} \int_0^1 \frac{d\gamma(t)}{\gamma(t)}$$

$$= \frac{1}{2\pi i} \int_0^1 \frac{d \exp(2\pi i \tilde{\gamma}(t))}{\gamma(t)}$$

$$= \int_0^1 d\tilde{\gamma}(t);$$

so

7.10 $$n(\gamma, 0) = \tilde{\gamma}(1)$$

since $\tilde{\gamma}(0) = 0$.

So if σ is also a closed rectifiable curve with $|\sigma(t)| \equiv 1$, $\sigma(0) = 0 = \sigma(1)$ and $n(\gamma; 0) = n(\sigma, 0) = n$ then $\tilde{\sigma}(1) = \tilde{\gamma}(1) = n$, where $\tilde{\sigma}$ is the unique lifting of σ to $\mathbb{R}$ such that $\tilde{\sigma}(0) = 0$. Let $F: [0, 1] \times [0, 1] \to \Gamma$ be defined by

$$F(s, t) = \exp\{2\pi i[(1-t)\tilde{\sigma}(s) + t\tilde{\gamma}(s)]\}$$

then $F(0, t) = 1 = F(1, t)$ (Why?) and F demonstrates that $\gamma \sim \sigma$.

Notice that the rectifiability of γ and σ was only used to define the winding number of γ and σ about the origin. It is possible to extend the definition of winding number to non-rectifiable curves.

7.11 Definition. If γ is any closed curve in $\mathbb{C} - \{0\}$ with $\gamma(0) = \gamma(1) = 1$, let $\gamma_1(t) = \gamma(t)/|\gamma(t)|$ and let $\tilde{\gamma}$ be the unique curve in $\mathbb{R}$ such that $\tilde{\gamma}(0) = 0$ and $\gamma_1(t) = \exp(2\pi i \tilde{\gamma}(t))$; *the winding number of γ about the origin* is

$$n(\gamma; 0) = \tilde{\gamma}(1).$$

In view of (7.10) this definition agrees with the former definition for rectifiable curves.

7.12 Theorem. *Let γ and σ be two closed curves in $\mathbb{C} - \{0\}$ such that $\sigma(0) = \gamma(0) = 1$; then $\gamma \sim \sigma$ in $\mathbb{C} - \{0\}$ iff $n(\gamma; 0) = n(\sigma; 0)$.*

Proof. It was shown above that if $n(\gamma; 0) = n(\sigma; 0)$ then $\gamma \sim \sigma$. Conversely, if $\gamma \sim \sigma$ then the Abstract Monodromy Theorem implies that the liftings $\tilde{\gamma}$ and $\tilde{\sigma}$ such that $\tilde{\gamma}(0) = \tilde{\sigma}(0) = 0$ have the same end point. That is, $n(\gamma; 0) = n(\sigma; 0)$. ∎

Exercises

1. Suppose that (X, ρ) is a covering space of Ω and (Ω, π) is a covering space of Y; prove that $(X, \pi \circ \rho)$ is also a covering space of Y.

2. Let (X, ρ) and (Y, σ) be covering spaces of Ω and Λ respectively. Define $\rho \times \sigma \colon X \times Y \to \Omega \times \Lambda$ by $(\rho \times \sigma)(x, y) = (\rho(x), \sigma(y))$ and show that $(X \times Y, \rho \times \sigma)$ is a covering space of $\Omega \times \Lambda$.

3. Let (Ω, ψ) be an analytic manifold and let (X, ρ) be a covering space of Ω. Show that there is an analytic structure Φ on X such that ρ is an analytic function from (X, Φ) to (Ω, ψ).

4. Let (X, ρ) be a covering space of Ω and let $\omega \in \Omega$. Show that each component of $\rho^{-1}(\omega)$ consists of a single point and $\rho^{-1}(\omega)$ has no limit points in X.

5. Let Ω be a pathwise connected space and let (X, ρ) be a covering space of Ω. If ω_1 and ω_2 are points in Ω, show that $\rho^{-1}(\omega_1)$ and $\rho^{-1}(\omega_2)$ have the same cardinality. (Hint: Let γ be a path in Ω from ω_1 to ω_2; if $x_1 \in \rho^{-1}(\omega_1)$ and $\tilde{\gamma}$ is the lifting of γ with initial point $x_1 = \tilde{\gamma}(0)$, let $f(x_1) = \tilde{\gamma}(1)$. Show that f is a one-one map of $\rho^{-1}(\omega_1)$ onto $\rho^{-1}(\omega_2)$.)

6. In this exercise all spaces are regions in the plane.

   (a) Let (G, f) be a covering of Ω and suppose that f is analytic; show that if Ω is simply connected then f is one-one. (Hint: If $f(z_1) = f(z_2)$ let γ be a path in G from z_1 to z_2 and consider a certain analytic continuation along $f \circ \gamma$; apply the Monodromy Theorem.)

   (b) Suppose that (G_1, f_1) and (G_2, f_2) are coverings of the region Ω such that both f_1 and f_2 are analytic. Show that if G_1 is simply connected then there is an analytic function $f \colon G_1 \to G_2$ such that (G_1, f) is a covering of G_2 and $f_1 = f_2 \circ f$. That is the diagram is commutative.

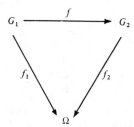

(c) Let (G_1, f_1), (G_2, f_2) and Ω be as in part (b) and, in addition, assume that both G_1 and G_2 are simply connected. Show that there is a one-one analytic function mapping G_1 onto G_2.

7. Let G and Ω be regions in the plane and suppose that $f: G \to \Omega$ is an analytic function such that (G, f) is a covering space of Ω. Show that for every region Ω_1 contained in Ω which is simply connected there is an analytic function $g: \Omega_1 \to G$ such that $f(g_1(\omega)) = \omega$ for all ω in Ω_1.

8. What is a simply connected covering space of the figure eight?

9. Give two nonhomeomorphic covering spaces of the figure eight that are not simply connected.

10. Prove that the closed curve in Exercise IV.6.8 is not homotopic to zero in the doubly punctured plane.

Chapter X

Harmonic Functions

In this chapter harmonic functions will be studied and the Dirichlet Problem will be solved. The Dirichlet Problem consists in determining all regions G such that for any continuous function $f: \partial G \to \mathbb{R}$ there is a continuous function $u: G^- \to \mathbb{R}$ such that $u(z) = f(z)$ for z in ∂G and u is harmonic in G. Alternately, we are asked to determine all regions G such that Laplace's Equation is solvable with arbitrary boundary values.

§1. Basic properties of harmonic functions

We begin by recalling the following definition and giving some examples of harmonic functions.

1.1 Definition. If G is an open subset of $\mathbb{C}$ then a function $u: G \to \mathbb{R}$ is *harmonic* if u has continuous second partial derivatives and

$$\frac{\partial^2 u}{\partial x^2} + \frac{\partial^2 u}{\partial y^2} = 0.$$

This equation is called LAPLACE'S EQUATION.

We also review the following facts about harmonic functions.

(1) (Theorem III.2.29) A function f on a region G is analytic iff $\operatorname{Re} f = u$ and $\operatorname{Im} f = v$ are harmonic functions which satisfy the Cauchy-Riemann equations.

(2) (Theorem VIII.2.2(j)) A region G is simply connected iff for each harmonic function u on G there is a harmonic function v on G such that $f = u + iv$ is analytic on G.

1.2 Definition. If $f: G \to \mathbb{C}$ is an analytic function then $u = \operatorname{Re} f$ and $v = \operatorname{Im} f$ are called *harmonic conjugates*.

With this terminology, Theorem VIII.2.2(j) becomes the statement that every harmonic function on a simply connected region has a harmonic conjugate. If u is a harmonic function on G and D is a disk that is contained in G then there is a harmonic function v on D such that $u + iv$ is analytic on D. In other words, each harmonic function has a harmonic conjugate locally. Finally note that if v_1 and v_2 are both harmonic conjugates of u then $i(v_1 - v_2) = (u + iv_1) - (u + iv_2)$ is an analytic function whose range is contained in the imaginary axis; hence $v_1 = v_2 + c$, for some constant c.

1.3 Proposition. *If $u: G \to \mathbb{R}$ is harmonic then u is infinitely differentiable.*

Proof. Fix $z_0 = x_0 + iy_0$ in G and let δ be chosen such that $B(z_0; \delta) \subset G$.

Then u has a harmonic conjugate v on $B(z_0;\delta)$. That is, $f = u + iv$ is analytic and hence infinitely differentiable on $B(z_0;\delta)$. It now follows that u is infinitely differentiable. ∎

The preceding proposition gives a property that harmonic functions share with analytic functions. The next result is the analogue of the Cauchy Integral Formula.

1.4 Mean Value Theorem. *If* $u: G \to \mathbb{R}$ *is a harmonic function and* $\bar{B}(a; r)$ *is a closed disk contained in* G, *then*

$$u(a) = \frac{1}{2\pi} \int\limits_0^{2\pi} u(a + re^{i\theta})\, d\theta$$

Proof. Let D be a disk such that $\bar{B}(a; r) \subset D \subset G$ and let f be an analytic function on D such that $u = \operatorname{Re} f$. It is easy to deduce from Cauchy's Integral Formula that

$$f(a) = \frac{1}{2\pi} \int\limits_0^{2\pi} f(a + re^{i\theta})\, d\theta.$$

By taking the real part of each side of this equation we complete the proof. ∎

In order to study this property of harmonic functions we isolate it.

1.5 Definition. A continuous function $u: G \to \mathbb{R}$ has the *Mean Value Property* (MVP) if whenever $\bar{B}(a;r) \subset G$

$$u(a) = \frac{1}{2\pi} \int\limits_0^{2\pi} u(a + re^{i\theta})\, d\theta.$$

In the following section it will be shown that any continuous function defined on a region that has the MVP must be a harmonic function. One of the main tools used in showing this is the following analogue of the Maximum Modulus Theorem for harmonic functions.

1.6 Maximum Principle (First Version). *Let* G *be a region and suppose that* u *is a continuous real valued function on* G *with the* MVP. *If there is a point* a *in* G *such that* $u(a) \geq u(z)$ *for all* z *in* G *then* u *is a constant function.*

Proof. Let the set A be defined by

$$A = \{z \in G : u(z) = u(a)\}.$$

Since u is continuous the set A is closed in G. If $z_0 \in A$ let r be chosen such that $\bar{B}(z_0;r) \subset G$. Suppose there is a point b in $B(z_0;r)$ such that $u(b) \neq u(a)$; then, $u(b) < u(a)$. By continuity, $u(z) < u(a) = u(z_0)$ for all z in a neighborhood of b. In particular, if $\rho = |z_0 - b|$ and $b = z_0 + \rho e^{i\beta}$, $0 \leq \beta < 2\pi$ then there is a proper interval I of $[0, 2\pi]$ such that $\beta \in I$ and $u(z_0 + \rho e^{i\theta}) < u(z_0)$ for all θ in I. Hence, by the MVP

$$u(z_0) = \frac{1}{2\pi} \int_0^{2\pi} u(z_0 + \rho e^{i\theta}) \, d\theta < u(z_0),$$

a contradiction. So $B(z_0; r) \subset A$ and A is also open. By the connectedness of G, $A = G$. ∎

1.7 Maximum Principle (Second Version). *Let G be a region and let u and v be two bounded continuous real valued functions on G that have the MVP. If for each point a in the extended boundary $\partial_\infty G$,*

$$\limsup_{z \to a} u(z) \le \liminf_{z \to a} v(z)$$

then either $u(z) < v(z)$ for all z in G or $u = v$.

Proof. Fix a in $\partial_\infty G$ and for each $\delta > 0$ let $G_\delta = G \cap B(a; \delta)$. Then according to the hypothesis,

$$0 \ge \lim_{\delta \to 0} [\sup\{u(z): z \in G_\delta\} - \inf\{v(z): z \in G_\delta\}]$$

$$= \lim_{\delta \to 0} [\sup\{u(z): z \in G_\delta\} + \sup\{-v(z): z \in G_\delta\}]$$

$$\ge \lim_{\delta \to 0} \sup \{u(z) - v(z): z \in G_\delta\}.$$

So $\limsup_{z \to a} [u(z) - v(z)] \le 0$ for each a in $\partial_\infty G$. So it is sufficient to prove the theorem under the assumption that $v(z) = 0$ for all z in G. That is, assume

1.8 $$\limsup_{z \to a} u(z) \le 0$$

for all a in $\partial_\infty G$ and show that either $u(z) < 0$ for all z in G or $u \equiv 0$. By virtue of the first version of the Maximum Principle, it suffices to show that $u(z) \le 0$ for all z in G.

Suppose that u satisfies (1.8) and there is a point b in G with $u(b) > 0$. Let $\epsilon > 0$ be chosen so that $u(b) > \epsilon$ and let $B = \{z \in G : u(z) \ge \epsilon\}$. If $a \in \partial_\infty G$ then (1.8) implies there is a $\delta = \delta(a)$ such that $u(z) < \epsilon$ for all z in $G \cap B(a; \delta)$. Using the Lebesgue Covering Lemma, a δ can be found that is independent of a. That is, there is a $\delta > 0$ such that if $z \in G$ and $d(z, \partial_\infty G) < \delta$ then $u(z) < \epsilon$. Thus,

$$B \subset \{z \in G : d(z, \partial_\infty G) \ge \delta\}.$$

This gives that B is bounded in the plane; since B is clearly closed, it is compact. So if $B \ne \square$, there is a point z_0 in B such that $u(z_0) \ge u(z)$ for all z in B. Since $u(z) < \epsilon$ for z in $G - B$, this gives that u assumes a maximum value at a point in G. So u must be constant. But this constant must be $u(z_0)$ which is positive and this contradicts (1.8). ∎

The following corollary is a useful special case of the Maximum Principle.

1.9 Corollary. *Let G be a bounded region and suppose that* $w : G^- \to \mathbb{R}$ *is a continuous function that satisfies the* MVP *on G. If* $w(z) = 0$ *for all z in* ∂G *then* $w(z) = 0$ *for all z in G.*

Proof. First take $w = u$ and $v = 0$ in Theorem 1.7. So $w(z) < 0$ for all z or $w(z) \equiv 0$. Now take $w = v$ and $u = 0$ in (1.7); so either $w(z) > 0$ for all z or $w(z) \equiv 0$. Since both of these hold, $w \equiv 0$. ∎

Even though Theorem 1.7 is called the Maximum Principle, it is also a Minimum Principle. For the sake of completeness, a Minimum Principle corresponding to Theorem 1.6 is stated below. It can be proved either by appealing to (1.7) or by considering the function $-u$ and appealing to (1.6).

1.10 Minimum Principle. *Let G be a region and suppose that u is a continuous real valued function on G with the* MVP. *If there is a point a in G such that* $u(a) \le u(z)$ *for all z in G then u is a constant function.*

Exercises

1. Show that if u is harmonic then so are $u_x = \dfrac{\partial u}{\partial x}$ and $u_y = \dfrac{\partial u}{\partial y}$.

2. If u is harmonic, show that $f = u_x - iu_y$ is analytic.

3. Let $p(x, y) = \displaystyle\sum_{k,l=0}^{n} a_{kl} x^k y^l$ for all x, y in $\mathbb{R}$.
Show that p is harmonic iff:

   (a) $k(k-1)a_{k,l-2} + l(l-1)a_{k-2,l} = 0$ for $2 \le k, l \le n$;
   (b) $a_{n-1,l} = a_{n,l} = 0$ for $2 \le l \le n$;
   (c) $a_{k,n-1} = a_{k,n} = 0$ for $2 \le k \le n$.

4. Prove that a nonconstant harmonic function on a region is an open map. (Hint: Use the fact that the connected subsets of $\mathbb{R}$ are intervals.)

5. If f is analytic on G and $f(z) \ne 0$ for any z show that $u = \log|f|$ is harmonic on G.

6. Let u be harmonic in G and suppose $\bar{B}(a;R) \subset G$. Show that

$$u(a) = \frac{1}{\pi R^2} \iint\limits_{\bar{B}(a;R)} u(x, y) \, dx \, dy.$$

7. For $|z| < 1$ let

$$u(z) = \mathrm{Im}\left[\left(\frac{1+z}{1-z} \right)^2 \right].$$

Show that u is harmonic and $\lim\limits_{r \to 1-} u(re^{i\theta}) = 0$ for all θ. Does this violate Theorem 1.7? Why?

8. Let $u : G \to \mathbb{R}$ be a function with continuous second partial derivatives and define $U(r, \theta) = u(r \cos \theta, r \sin \theta)$.

   (a) Show that

$$r^2 \left[\frac{\partial^2 u}{\partial x^2} + \frac{\partial^2 u}{\partial y^2} \right] = r^2 \frac{\partial^2 U}{\partial r^2} + r \frac{\partial U}{\partial r} + \frac{\partial^2 U}{\partial \theta^2}$$

$$= r \frac{\partial}{\partial r} \left(r \frac{\partial U}{\partial r} \right) + \frac{\partial^2 U}{\partial \theta^2} .$$

So if $0 \notin G$ then u is harmonic iff

$$r \frac{\partial}{\partial r} \left(r \frac{\partial U}{\partial r} \right) + \frac{\partial^2 U}{\partial \theta^2} = 0.$$

(b) Let u have the property that it depends only on $|z|$ and not $\arg z$. That is, $u(z) = \varphi(|z|)$. Show that u is harmonic iff $u(z) = a \log |z| + b$ for some constants a and b.

9. Let $u: G \to \mathbb{R}$ be harmonic and let $A = \{z \in G : u_x(z) = u_y(z) = 0\}$; that is, A is the set of zeros of the gradient of u. Can A have a limit point in G?

10. State and prove a Schwarz Reflection Principle for harmonic functions.

11. Deduce the Maximum Principle for analytic functions from Theorem 1.6.

§2. Harmonic functions on a disk

Before studying harmonic functions in the large it is necessary to study them locally. That is, we must study these functions on disks. The plan is to study harmonic functions on the open unit disk $\{z : |z| < 1\}$ and then interpret the results for arbitrary disks. Of basic importance is the Poisson kernel.

2.1 Definition. The function

$$P_r(\theta) = \sum_{n=-\infty}^{\infty} r^{|n|} e^{in\theta},$$

for $0 \leq r < 1$ and $-\infty < \theta < \infty$, is called the *Poisson kernel*.

Let $z = re^{i\theta}$, $0 \leq r < 1$; then

$$\frac{1+re^{i\theta}}{1-re^{i\theta}} = (1+z)(1+z+z^2+ \ldots)$$

$$= 1 + 2 \sum_{n=1}^{\infty} z^n$$

$$= 1 + 2 \sum_{n=1}^{\infty} r^n e^{in\theta}$$

Hence,

$$\text{Re} \left(\frac{1+re^{i\theta}}{1-re^{i\theta}} \right) = 1 + 2 \sum_{n=1}^{\infty} r^n \cos n\theta$$

$$= 1 + \sum_{n=1}^{\infty} r^n (e^{in\theta} + e^{-in\theta})$$

$$= P_r(\theta)$$

Also $\dfrac{1+re^{i\theta}}{1-re^{i\theta}} = \dfrac{1+re^{i\theta}-re^{-i\theta}-r^2}{|1-re^{i\theta}|^2}$ so that

2.2
$$P_r(\theta) = \frac{1-r^2}{1-2r\cos\theta+r^2} = \mathrm{Re}\left(\frac{1+re^{i\theta}}{1-re^{i\theta}}\right)$$

2.3 Proposition. *The Poisson kernel satisfies the following:*

(a) $\dfrac{1}{2\pi} \displaystyle\int_{-\pi}^{\pi} P_r(\theta)\,d\theta = 1;$

(b) $P_r(\theta) > 0$ *for all* θ, $P_r(-\theta) = P_r(\theta)$, *and* P_r *is periodic in* θ *with period* 2π;

(c) $P_r(\theta) < P_r(\delta)$ *if* $0 < \delta < |\theta| \leq \pi;$

(d) *for each* $\delta > 0$, $\lim_{r\to1-} P_r(\theta) = 0$ *uniformly in* θ *for* $\pi \geq |\theta| \geq \delta$.

Proof. (a) For a fixed value of r, $0 \leq r < 1$, the series (2.1) converges uniformly in θ. So

$$\frac{1}{2\pi}\int_{-\pi}^{\pi} P_r(\theta)\,d\theta = \sum_{n=-\infty}^{\infty} r^{|n|}\frac{1}{2\pi}\int_{-\pi}^{\pi} e^{in\theta}\,d\theta = 1$$

(b) From equation (2.2), $P_r(\theta) = (1-r^2)|1-re^{i\theta}|^{-2} > 0$ since $r < 1$. The rest of (b) is an equally trivial consequence of (2.2).

(c) Let $0 < \delta < \theta \leq \pi$ and define $f:[\delta,\ \theta] \to \mathbb{R}$ by $f(t) = P_r(t)$. Using (2.2), a routine calculation shows that $f'(t) < 0$ so that $f(\delta) > f(\theta)$.

(d) We must show that

$$\lim_{r\to1-}[\sup\{P_r(\theta): \delta \leq |\theta| \leq \pi\}] = 0$$

But according to part (c), $P_r(\theta) \leq P_r(\delta)$ if $\delta \leq |\theta| \leq \pi$; so it suffices to show that $\lim_{r\to1-} P_r(\delta) = 0$. But, again, this is a trivial consequence of equation (2.2).∎

Before going to the applications of the Poisson kernel, the reader should take time to consider the significance of Proposition 2.3. Think of $P_r(\theta)$ not as a function of r and θ but as a family of functions of θ, indexed by r. As r approaches 1 these functions converge to zero uniformly on any closed subinterval of $[-\pi, \pi]$ which does not contain $\theta = 0$ (part d). Nevertheless, part (a) is still valid. So as r approaches 1, the graph of P_r becomes closer to the θ axis for θ away from zero but rises sharply near zero so that 2.3(a) is maintained.

The next theorem states that the Dirichlet Problem can be solved for the unit disk.

2.4 Theorem. *Let* $D = \{z:|z| < 1\}$ *and suppose that* $f:\partial D \to \mathbb{R}$ *is a continuous function. Then there is a continuous function* $u:D^- \to \mathbb{R}$ *such that*

(a) $u(z) = f(z)$ *for* z *in* ∂D;

(b) u *is harmonic in* D.

Moreover u is unique and is defined by the formula

2.5
$$u(re^{i\theta}) = \frac{1}{2\pi} \int_{-\pi}^{\pi} P_r(\theta - t) f(e^{it})\, dt$$

for $0 \le r < 1, 0 \le \theta \le 2\pi$.

Proof. Define $u: \bar{D} \to \mathbb{R}$ by letting $u(re^{i\theta})$ be as in (2.5) if $0 \le r < 1$ and letting $u(e^{i\theta}) = f(e^{i\theta})$. Clearly u satisfies part (a); it remains to show that u is continuous on D^- and harmonic in D.

(i) u *is harmonic in* D. If $0 \le r < 1$ then

$$u(re^{i\theta}) = \frac{1}{2\pi} \int_{-\pi}^{\pi} \mathrm{Re}\left[\frac{1 + re^{i(\theta - t)}}{1 - re^{i(\theta - t)}}\right] f(e^{it})\, dt$$

$$= \mathrm{Re}\left\{\frac{1}{2\pi} \int_{-\pi}^{\pi} f(e^{it})\left[\frac{1 + re^{i(\theta - t)}}{1 - re^{i(\theta - t)}}\right] dt\right\}.$$

$$= \mathrm{Re}\left\{\frac{1}{2\pi} \int_{-\pi}^{\pi} f(e^{it})\left[\frac{e^{it} + re^{i\theta}}{e^{it} - re^{i\theta}}\right] dt\right\}.$$

So define $g: D \to \mathbb{C}$ by

$$g(z) = \frac{1}{2\pi} \int_{-\pi}^{\pi} f(e^{it})\left[\frac{e^{it} + z}{e^{it} - z}\right] dt.$$

Since $u = \mathrm{Re}\, g$ we need only show that g is analytic. But this is an easy consequence of Exercise IV.2.2.

(ii) u *is continuous on* D^-. Since u is harmonic on D it only remains to show that u is continuous at each point of the boundary of D. To accomplish this we make the following

2.6 *Claim.* Given α in $[-\pi, \pi]$ and $\epsilon > 0$ there is a ρ, $0 < \rho < 1$, and an arc A of ∂D about $e^{i\alpha}$ such that for $\rho < r < 1$ and $e^{i\theta}$ in A,

$$|u(re^{i\theta}) - f(e^{i\alpha})| < \epsilon$$

Once claim 2.6 is proved the continuity of u at $e^{i\alpha}$ is immediate since f is a continuous function.

To avoid certain notational difficulties, the claim will only be proved for $\alpha = 0$. (The general case can be obtained from this one by an argument which involves a rotation of the variables.) Since f is continuous at $z = 1$ there is a $\delta > 0$ such that

2.7
$$|f(e^{i\theta}) - f(1)| < \frac{1}{3}\epsilon$$

if $|\theta| < \delta$. Let $M = \max \{|f(e^{i\theta})|:|\theta| \leq \pi\}$; from Proposition 2.3 (d) there is a number ρ, $0 < \rho < 1$, such that

2.8 $$P_r(\theta) < \frac{\epsilon}{3M}$$

for $\rho < r < 1$ and $|\theta| \geq \frac{1}{2}\delta$. Let A be the arc $\{e^{i\theta}:|\theta| < \frac{1}{2}\delta\}$. Then if $e^{i\theta} \in A$ and $\rho < r < 1$,

$$u(re^{i\theta})-f(1) = \frac{1}{2\pi}\int_{-\pi}^{\pi} P_r(\theta-t)f(e^{it})\,dt -f(1)$$

$$= \frac{1}{2\pi}\int_{|t|<\delta} P_r(\theta-t)[f(e^{it})-f(1)]dt + \frac{1}{2\pi}\int_{|t|\geq\delta} P_r(\theta-t)[f(e^{it})-f(1)]dt.$$

If $|t| \geq \delta$ and $|\theta| \leq \frac{1}{2}\delta$ then $|t-\theta| \geq \frac{1}{2}\delta$; so from (2.7) and (2.8) it follows that

$$|u(re^{i\theta})-f(1)| \leq \frac{1}{3}\epsilon+2M\left(\frac{\epsilon}{3M}\right) = \epsilon.$$

This proves Claim 2.6.

Finally, to show that u is unique, suppose that v is a continuous function on D^- which is harmonic on D and $v(e^{i\theta}) = f(e^{i\theta})$ for all θ. Then $u-v$ is harmonic in D and $(u-v)(z) = 0$ for all z in ∂D. It follows from Corollary 1.9 that $u-v \equiv 0$. ∎

2.9 Corollary. *If $u: D^- \to \mathbb{R}$ is a continuous function that is harmonic in D then*

$$u(re^{i\theta}) = \frac{1}{2\pi}\int_{-\pi}^{\pi} P_r(\theta-t)u(e^{it})\,dt$$

for $0 \leq r < 1$ and all θ. Moreover, u is the real part of the analytic function

$$f(z) = \frac{1}{2\pi}\int_{-\pi}^{\pi} \frac{e^{it}+z}{e^{it}-z} u(e^{it})\,dt.$$

Proof. The first part of the corollary is a direct consequence of the theorem. The second part follows from the fact that f is an analytic function (Exercise IV.2.2) and formula (2.2). ∎

2.10 Corollary. *Let $a \in \mathbb{C}$, $\rho > 0$, and suppose h is a continuous real valued function on $\{z:|z-a| = \rho\}$; then there is a unique continuous function $w:\bar{B}(a;\rho) \to \mathbb{R}$ such that w is harmonic on $B(a;\rho)$ and $w(z) = h(z)$ for $|z-a| = \rho$.*

Proof. Consider $f(e^{i\theta}) = h(a+\rho e^{i\theta})$; then f is continuous on ∂D. If $u:D^- \to \mathbb{R}$ is a continuous function such that u is harmonic in D and $u(e^{i\theta}) = f(e^{i\theta})$ then it is an easy matter to show that $w(z) = u\left(\dfrac{z-a}{\rho}\right)$ is the desired function on $\bar{B}(a;\rho)$. ∎

It is now possible to give the promised converse to the Mean Value Theorem.

2.11 Theorem. *If $u: G \to \mathbb{R}$ is a continuous function which has the MVP then u is harmonic.*

Proof. Let $a \in G$ and choose ρ such that $\bar{B}(a;\rho) \subset G$; it is sufficient to show that u is harmonic on $B(a;\rho)$. But according to Corollary 2.10 there is a continuous function $w: \bar{B}(a;\rho) \to \mathbb{R}$ which is harmonic in $B(a;\rho)$ and $w(a+\rho e^{i\theta}) = u(a+\rho e^{i\theta})$ for all θ. Since $u-w$ satisfies the MVP and $(u-w)(z) = 0$ for $|z-a| = \rho$, it follows from Corollary 1.9 that $u \equiv w$ in $B(a;\rho)$; in particular, u must be harmonic. ∎

To prove the above theorem we used Corollary 2.10, which concerns functions harmonic in an arbitrary disk. It is desirable to derive a formula for the Poisson kernel of an arbitrary disk; to do this one need only make a change of variables in the formula (2.2).

If $R > 0$ then substituting r/R for r in the middle of (2.2) gives

2.12
$$\frac{R^2 - r^2}{R^2 - 2rR\cos\theta + r^2}$$

for $0 \le r < R$ and all θ. So if u is continuous on $\bar{B}(a;R)$ and harmonic in $B(a;R)$ then

2.13 $\qquad u(a+re^{i\theta}) = \dfrac{1}{2\pi}\displaystyle\int_{-\pi}^{\pi}\left[\dfrac{R^2 - r^2}{R^2 - 2rR\cos(\theta-t) + r^2}\right]u(a+Re^{it})\,dt$

Now (2.12) can also be written

$$\frac{R^2 - r^2}{|Re^{it} - re^{i\theta}|^2}$$

and $R - r \le |Re^{it} - re^{i\theta}| \le R + r$. Therefore

$$\frac{R-r}{R+r} \le \frac{R^2 - r^2}{R^2 - 2rR\cos(\theta-t) + r^2} \le \frac{R+r}{R-r}.$$

If $u \ge 0$ then equation (2.13) yields the following.

2.14 Harnack's Inequality. *If $u: \bar{B}(a;R) \to \mathbb{R}$ is continuous, harmonic in $B(a;R)$, and $u \ge 0$ then for $0 \le r < R$ and all θ*

$$\frac{R-r}{R+r}u(a) \le u(a+re^{i\theta}) \le \frac{R+r}{R-r}u(a)$$

Before proceeding, the reader is advised to review the relevant definitions and properties of the metric space $C(G, \mathbb{R})$ (Section VII.1).

2.15 Definition. If G is an open subset of $\mathbb{C}$ then Har(G) is the *space of harmonic functions on G*. Since Har(G) $\subset C(G, \mathbb{R})$ it is given the metric that it inherits from $C(G, \mathbb{R})$.

2.16 Harnack's Theorem. *Let G be a region.* (a) *The metric space* Har(G) *is complete.* (b) *If* $\{u_n\}$ *is a sequence in* Har(G) *such that* $u_1 \le u_2 \le \ldots$ *then either* $u_n(z) \to \infty$ *uniformly on compact subsets of G or* $\{u_n\}$ *converges in* Har(G) *to a harmonic function.*

Proof. (a) To show that Har(G) is complete, it is sufficient to show that it is a closed subspace of $C(G, \mathbb{R})$. So let $\{u_n\}$ be a sequence in Har(G) such that $u_n \to u$ in $C(G, \mathbb{R})$. Then (Lemma IV.2.7) it follows that u has the MVP and so, by Theorem 2.11, u must be harmonic.

(b) We may assume that $u_1 \ge 0$ (if not, consider $\{u_n - u_1\}$). Let $u(z) = \sup\{u_n(z) : n \ge 1\}$ for each z in G. So for each z in G one of two possibilities occurs: $u(z) = \infty$ or $u(z) \in \mathbb{R}$ and $u_n(z) \to u(z)$. Define

$$A = \{z \in G : u(z) = \infty\}$$

$$B = \{z \in G : u(z) < \infty\};$$

then $G = A \cup B$ and $A \cap B = \square$. We will show that both A and B are open.

If $a \in G$, let R be chosen such that $\bar{B}(a; R) \subset G$. By Harnack's inequality

2.17
$$\frac{R - |z-a|}{R + |z-a|} u_n(a) \le u_n(z) \le \frac{R + |z-a|}{R - |z-a|} u_n(a)$$

for all z in $B(a; R)$ and all $n \ge 1$. If $a \in A$ then $u_n(a) \to \infty$ so that the left half of (2.17) gives that $u_n(z) \to \infty$ for all z in $B(a; R)$. That is, $B(a; R) \subset A$ and so A is open. In a similar fashion, if $a \in B$ then the right half of (2.17) gives that $u(z) < \infty$ for $|z-a| < R$. That is B is open.

Since G is connected, either $A = G$ or $B = G$. Suppose $A = G$; that is $u \equiv \infty$. Again if $\bar{B}(a; R) \subset G$ and $0 < \rho < R$ then $M = (R-\rho)(R+\rho)^{-1} > 0$ and (2.17) gives that $M\, u_n(a) \le u_n(z)$ for $|z-a| \le \rho$. Hence $u_n(z) \to \infty$ uniformly for z in $\bar{B}(a; \rho)$. In other words, we have shown that for each a in G there is a $\rho > 0$ such that $u_n(z) \to \infty$ uniformly for $|z-a| \le \rho$. From this it is easy to deduce that $u_n(z) \to \infty$ uniformly for z in any compact set.

Now suppose $B = G$, or that $u(z) < \infty$ for all z in G. If $\rho < R$ then, as above, there is a constant N, which depends only on a and ρ such that $M\, u_n(a) \le u_n(z) \le N\, u_n(a)$ for $|z-a| \le \rho$ and all n. So if $m \le n$

$$0 \le u_n(z) - u_m(z) \le N\, u_n(a) - M\, u_m(a)$$

$$\le C[u_n(a) - u_m(a)]$$

for some constant C. Thus, $\{u_n(z)\}$ is a uniformly Cauchy sequence on $\bar{B}(a; \rho)$. It follows that $\{u_n\}$ is a Cauchy sequence in Har(G) and so, by part (a), must converge to a harmonic function. Since $u_n(z) \to u(z)$, u is this harmonic function. ∎

It is possible to give alternate proofs of Harnack's Theorem. One involves applying Dini's Theorem (Exercise VII.1.6). Another involves using the Monotone Convergence Theorem from measure theory to obtain that u has the MVP. However, both these approaches necessitate proving that the function u is continuous. This is rather easy to accomplish by appealing to (2.17) and the fact that $u_n(z) \to u(z)$ for all z; these facts imply that

$$\frac{R-|z-a|}{R+|z-a|}\, u(a) \le u(z) \le \frac{R+|z-a|}{R-|z-a|}\, u(a)$$

Hence

$$\frac{-2\,|z-a|}{R+|z-a|}\, u(a) \le u(z)-u(a) \le \frac{2|z-a|}{R-|z-a|}\, u(a);$$

or

$$|u(z)-u(a)| \le \frac{2\,|z-a|}{R-|z-a|}\, u(a)$$

So as $z \to a$, it is clear that $u(z) \to u(a)$.

Exercises

1. Let $D = \{z : |z| < 1\}$ and suppose that $f : D^- \to \mathbb{C}$ is a continuous function such that both $\operatorname{Re} f$ and $\operatorname{Im} f$ are harmonic. Show that

$$f(re^{i\theta}) = \frac{1}{2\pi} \int\limits_{-\pi}^{\pi} f(e^{it}) P_r(\theta - t)\, dt$$

for all $re^{i\theta}$ in D. Using Definition 2.1 show that f is analytic on D iff

$$\int\limits_{-\pi}^{\pi} f(e^{it}) e^{int}\, dt = 0$$

for all $n \ge 1$.

2. In the statement of Theorem 2.4 suppose that f is piecewise continuous on ∂D. Is the conclusion of the theorem still valid? If not, what parts of the conclusion remain true?

3. Let $D = \{z : |z| < 1\}$, $T = \partial D = \{z : |z| = 1\}$

   (a) Show that if $g : D^- \to \mathbb{C}$ is a continuous function and $g_r : T \to \mathbb{C}$ is defined by $g_r(z) = g(rz)$ then $g_r(z) \to g(z)$ uniformly for z in T as $r \to 1-$.

   (b) If $f : T \to \mathbb{C}$ is a continuous function define $\tilde{f} : D^- \to \mathbb{C}$ by $\tilde{f}(z) = f(z)$ for z in T and

$$\tilde{f}(re^{i\theta}) = \frac{1}{2\pi} \int\limits_{-\pi}^{\pi} f(e^{it}) P_r(\theta - t)\, dt$$

(So $\operatorname{Re}\tilde{f}$ and $\operatorname{Im}\tilde{f}$ are harmonic in D). Define $\tilde{f}_r : T \to \mathbb{C}$ by $\tilde{f}_r(z) = \tilde{f}(rz)$. Show that for each $r < 1$ there is a sequence $\{p_n(z,\bar{z})\}$ of polynomials in z and $\bar{z}$ such that $p_n(z,\bar{z}) \to \tilde{f}_r(z)$ uniformly for z in T. (Hint: Use Definition 2.1.)

(c) **Weierstrass approximation theorem for** T. If $f: T \to \mathbb{C}$ is a continuous function then there is a sequence $\{p_n(z,\bar{z})\}$ of polynomials in z and $\bar{z}$ such that $p_n(z,\bar{z}) \to f(z)$ uniformly for z in T.

(d) Suppose $g:[0,1] \to \mathbb{C}$ is a continuous function such that $g(0) = g(1)$. Use part (c) to show that there is a sequence $\{p_n\}$ of polynomials such that $p_n(t) \to g(t)$ uniformly for t in $[0,1]$.

(e) **Weierstrass approximation theorem for** $[0,1]$. If $g:[0,1] \to \mathbb{C}$ is a continuous function then there is a sequence $\{p_n\}$ of polynomials such that $p_n(t) \to g(t)$ uniformly for t in $[0,1]$. (Hint: Apply part (d) to the function $g(t) + (1-t)g(1) + tg(0)$.)

(f) Show that if the function g in part (e) is real valued then the polynomials can be chosen with real coefficients.

4. Let G be a simply connected region and let Γ be its closure in $\mathbb{C}_\infty$; $\partial_\infty G = \Gamma - G$. Suppose there is a homeomorphism φ of Γ onto D^- ($D = \{z : |z| < 1\}$) such that φ is analytic on G.

(a) Show that $\varphi(G) = D$ and $\varphi(\partial_\infty G) = \partial D$.

(b) Show that if $f: \partial_\infty G \to \mathbb{R}$ is a continuous function then there is a continuous function $u: \Gamma \to \mathbb{R}$ such that $u(z) = f(z)$ for z in $\partial_\infty G$ and u is harmonic in G.

(c) Suppose that the function f in part (b) is not assumed to be continuous at ∞. Show that there is a continuous function $u: G^- \to \mathbb{R}$ such that $u(z) = f(z)$ for z in ∂G and u is harmonic in G (see Exercise 2).

5. Let G be an open set, $a \in G$, and $G_0 = G - \{a\}$. Suppose that u is a harmonic function on G_0 such that $\lim\limits_{z \to a} u(z)$ exists and is equal to A. Show that if $U: G \to \mathbb{R}$ is defined by $U(z) = u(z)$ for $z \neq a$ and $U(a) = A$ then U is harmonic on G.

6. Let $f: \{z : \mathrm{Re}\ z = 0\} \to \mathbb{R}$ be a bounded continuous function and define $u: \{z : \mathrm{Re}\ z > 0\} \to \mathbb{R}$ by

$$u(x+iy) = \frac{1}{\pi} \int_{-\infty}^{\infty} \frac{xf(it)}{x^2 + (y-t)^2} \, dt.$$

Show that u is a bounded harmonic function on the right half plane such that for c in $\mathbb{R}$, $f(ic) = \lim\limits_{z \to ic} u(z)$.

7. Let $D = \{z : |z| < 1\}$ and suppose $f: \partial D \to \mathbb{R}$ is continuous except for a jump discontinuity at $z = 1$. Define $u: D \to \mathbb{R}$ by (2.5). Show that u is harmonic. Let v be a harmonic conjugate of u. What can you say about the behavior of $v(r)$ as $r \to 1-$? What about $v(re^{i\theta})$ as $r \to 1-$ and $\theta \to 0$?

§3. Subharmonic and superharmonic functions

In order to solve the Dirichlet Problem generalizations of harmonic functions are introduced. According to Theorem 2.11, a function is harmonic exactly when it has the MVP. With this in mind, the choice of terminology in the next definition becomes appropriate.

3.1 Definition. Let G be a region and let $\varphi: G \to \mathbb{R}$ be a continuous function. φ is a *subharmonic function* if whenever $\bar{B}(a; r) \subset G$,

$$\varphi(a) \le \frac{1}{2\pi} \int_0^{2\pi} \varphi(a + re^{i\theta})\, d\theta.$$

φ is a *superharmonic function* if whenever $\bar{B}(a; r) \subset G$,

$$\varphi(a) \ge \frac{1}{2\pi} \int_0^{2\pi} \varphi(a + re^{i\theta})\, d\theta.$$

The first comment that should be made is that φ is superharmonic iff $-\varphi$ is subharmonic. Because of this, only the results on subharmonic functions will be given and it will be left to the reader to state the analogous result for superharmonic functions. Nevertheless, we will often quote results on superharmonic functions as though they had been stated in detail.

In the definition of a subharmonic function φ it is possible to assume only that φ is upper semi-continuous. However this would make it necessary to use the Lebesgue Integral in the definition instead of the Riemann Integral. So it is assumed that φ is continuous when φ is subharmonic even though there are certain technical advantages that accrue if only upper semi-continuity is assumed.

Clearly every harmonic function is subharmonic as well as superharmonic. In fact, according to Theorem 2.11, u is harmonic iff u is both subharmonic and superharmonic. If φ_1 and φ_2 are subharmonic then so is $a_1\varphi_1 + a_2\varphi_2$ if $a_1, a_2 \ge 0$.

It is interesting to see which of the results on harmonic functions also hold for subharmonic functions. One of the most important of these is the Maximum Principle.

3.2 Maximum Principle (Third Version). *Let G be a region and let $\varphi: G \to \mathbb{R}$ be a subharmonic function. If there is a point a in G with $\varphi(a) \ge \varphi(z)$ for all z in G then φ is a constant function.*

The proof is the same as the proof of the first version of the Maximum Principle. (Notice that only the Minimum Principle holds for superharmonic functions.)

The second version of the Maximum Principle can also be extended, but here both subharmonic and superharmonic functions must be used.

3.3 Maximum Principle (Fourth Version). *Let G be a region and let φ and ψ be bounded real valued functions defined on G such that φ is subharmonic and ψ is superharmonic. If for each point a in $\partial_\infty G$*

$$\limsup_{z \to a} \varphi(z) \le \liminf_{z \to a} \psi(z),$$

then either $\varphi(z) < \psi(z)$ for all z in G or $\varphi = \psi$ and φ is harmonic.

Again, the proof is identical to that of Theorem 1.7 and will not be repeated here.

Notice that we have not excluded the possibility that a subharmonic function may assume a minimum value. Indeed, this does happen. For example, $\varphi(x, y) = x^2 + y^2$ is a subharmonic function and it assumes a minimum at the origin. This failure of the Minimum Principle is due to the fact that if u has the MVP then so does $-u$; however, if φ is subharmonic then $-\varphi$ is never subharmonic unless it is harmonic.

When we say that *a function satisfies the Maximum Principle*, we refer to the third version. That is, we suppose that it does not assume a maximum value in G unless it is constant.

3.4 Theorem. *Let G be a region and $\varphi: G \to \mathbb{R}$ a continuous function. Then φ is subharmonic iff for every region G_1 contained in G and every harmonic function u_1 on G_1, $\varphi - u_1$ satisfies the Maximum Principle on G_1.*

Proof. Suppose that φ is subharmonic and G_1 and u_1 are as in the statement of the theorem. Then $\varphi - u_1$ is clearly subharmonic and must satisfy the Maximum Principle.

Now suppose φ is continuous and has the stated property; let $\bar{B}(a, r) \subset G$. According to Corollary 2.10 there is a continuous function $u: \bar{B}(a, r) \to \mathbb{R}$ which is harmonic in $B(a; r)$ and $u(z) = \varphi(z)$ for $|z - a| = r$. By hypothesis, $\varphi - u$ satisfies the Maximum Principle. But $(\varphi - u)(z) \equiv 0$ for $|z - a| = r$. So $\varphi \le u$ and

$$\varphi(a) \le u(a) = \frac{1}{2\pi} \int_0^{2\pi} u(a + re^{i\theta}) \, d\theta$$

$$= \frac{1}{2\pi} \int_0^{2\pi} \varphi(a + re^{i\theta}) \, d\theta.$$

Therefore φ is subharmonic. ∎

3.5 Corollary. *Let G be a region and $\varphi: G \to \mathbb{R}$ a continuous function; then φ is subharmonic iff for every bounded region G_1 such that $G_1^- \subset G$ and for every continuous function $u_1: G_1^- \to \mathbb{R}$ that is harmonic in G_1 and satisfies $\varphi(z) \le u_1(z)$ for z on ∂G_1, $\varphi(z) \le u_1(z)$ for z in G_1.*

3.6 Corollary. *Let G be a region and φ_1 and φ_2 subharmonic functions on G; if $\varphi(z) = \max\{\varphi_1(z), \varphi_2(z)\}$ for each z in G then φ is a subharmonic function.*

Proof. Let G_1 be a region such that $G_1^- \subset G$ and let u_1 be a continuous function on G_1^- which is harmonic on G_1 with $\varphi(z) \le u_1(z)$ for all z in ∂G_1. Then both $\varphi_1(z)$ and $\varphi_2(z) \le u_1(z)$ on ∂G_1. From Corollary 3.5 we get that $\varphi_1(z)$ and $\varphi_2(z) \le u_1(z)$ for all z in G_1. So $\varphi(z) \le u_1(z)$ for z in G_1, and, again by Corollary 3.5, φ is subharmonic. ∎

3.7 Corollary. *Let φ be a subharmonic function on a region G and let $\bar{B}(a; r) \subset G$. Let φ' be the function defined on G by:*

(i) *$\varphi'(z) = \varphi(z)$ if $z \in G - B(a; r)$;*
(ii) *φ' is the continuous function on $\bar{B}(a; r)$ which is harmonic in $B(a; r)$ and agrees with $\varphi(z)$ for $|z - a| = r$.*
Then φ' is subharmonic.

The proof is left to the reader.

As was mentioned at the beginning of this section, one of the purposes in studying subharmonic functions is that they enter into the solution of the Dirichlet Problem. Indeed, the fourth version of the Maximum Principle gives an insight into how this occurs. If G is a region and $u: G^- \to \mathbb{R}$ is a continuous function ($G^- = $ the closure in $\mathbb{C}_\infty$) which is harmonic in G, then $\varphi(z) \le u(z)$ for all z in G and for all subharmonic functions φ which satisfy $\limsup_{z \to a} \varphi(z) \le u(a)$ for all a in $\partial_\infty G$. Since u is itself such a subharmonic function we arrive at the trivial result that

3.8 $u(z) = \sup\{\varphi(z) : \varphi \text{ is subharmonic and } \limsup_{z \to a} \varphi(z) \le u(a) \text{ for all } a \text{ in } \partial_\infty G\}$.

Although this is a trivial statement, it is nevertheless a beacon that points the way to a solution of the Dirichlet Problem. Equation (3.8) says that if $f: \partial_\infty G \to \mathbb{R}$ is a continuous function and if f can be extended to a function u that is harmonic on G, then u can be obtained from a set of subharmonic functions which are defined solely in terms of the boundary values f. This leads to the following definition.

3.9 Definition. If G is a region and $f: \partial_\infty G \to \mathbb{R}$ is a continuous function then the *Perron Family*, $\mathscr{P}(f, G)$, consists of all subharmonic functions $\varphi: G \to \mathbb{R}$ such that

$$\limsup_{z \to a} \varphi(z) \le f(a)$$

for all a in $\partial_\infty G$.

Since f is continuous, there is a constant M such that $|f(a)| \le M$ for all a in $\partial_\infty G$. So the constant function $-M$ is in $\mathscr{P}(f, G)$ and the Perron Family is never empty.

If $u: G^- \to \mathbb{R}$ is a continuous function which is harmonic in G and $f = u|\partial_\infty G$ then (3.8) becomes

3.10 $$u(z) = \sup\{\varphi(z) : \varphi \in \mathscr{P}(f, G)\}$$

for each z in G. Conversely, if f is given and u is defined by (3.10) then u must be the solution of the Dirichlet Problem with boundary values f; that is, provided the Dirichlet Problem can be solved. In order to show that (3.10) is a solution two questions must be answered affirmatively.

(a) Is u harmonic in G?
(b) Does $\lim_{z \to a} u(z) = f(a)$ for each a in $\partial_\infty G$?

The first question can always be answered "Yes" and this is shown in the next theorem. The second question sometimes has a negative answer and an example will be given which demonstrates this. However, it is possible to impose geometrical restrictions on G which guarantee that the answer to (b) is always yes for any continuous function f. This will be done in the next section.

3.11 Theorem. *Let G be a region and $f: \partial_\infty G \to \mathbb{R}$ a continuous function; then $u(z) = \sup \{\varphi(z) : \varphi \in \mathscr{P}(f, G)\}$ defines a harmonic function u on G.*

Proof. Let $|f(a)| \le M$ for all $a \in \partial_\infty G$. The proof begins by noting that

3.12 $$\varphi(z) \le M \text{ for all } z \text{ in } G, \varphi \text{ in } \mathscr{P}(f, G)$$

This follows because, by definition, $\limsup_{z \to a} \varphi(z) \le M$ whenever $\varphi \in \mathscr{P}(f, G)$; so (3.12) is a direct consequence of the Maximum Principle.

Fix a in G and let $\bar{B}(a; r) \subset G$. Then $u(a) = \sup \{\varphi(a) : \varphi \in \mathscr{P}(f, G)\}$; so there is a sequence $\{\varphi_n\}$ in $\mathscr{P}(f, G)$ such that $u(a) = \lim \varphi_n(a)$. Let $\Phi_n = \max \{\varphi_1, \ldots, \varphi_n\}$; by Corollary 3.6 Φ_n is subharmonic. Let Φ'_n be the subharmonic function on G such that $\Phi'_n(z) = \Phi_n(z)$ for z in $G - B(a; r)$ and Φ'_n is harmonic on $B(a; r)$ (Corollary 3.7). It is left to the reader to verify the following statements:

3.13 $$\Phi'_n \le \Phi'_{n+1}$$

3.14 $$\varphi_n \le \Phi_n \le \Phi'_n$$

3.15 $$\Phi'_n \in \mathscr{P}(f, G).$$

Because of (3.15), $\Phi'_n(a) \le u(a)$; from (3.14) and the choice of $\{\varphi_n\}$, this gives that

3.16 $$u(a) = \lim \Phi'_n(a).$$

Moreover, statement (3.12) gives that $\Phi'_n \le M$ for all n; so using (3.13), Harnack's Theorem implies that there is a harmonic function U on $B(a; r)$ such that $U(z) = \lim \Phi'_n(z)$ uniformly for z in any proper subdisk of $B(a; r)$. It follows from (3.15) and (3.16) that $U \le u$ and $U(a) = u(a)$, respectively.

Now let $z_0 \in B(a; r)$ and let $\{\psi_n\}$ be a sequence in $\mathscr{P}(f, G)$ such that $u(z_0) = \lim \psi_n(z_0)$.

Let $\chi_n = \max \{\varphi_n, \psi_n\}$, $X_n = \max \{\chi_1, \ldots, \chi_n\}$, and let X'_n be the subharmonic function which agrees with X_n off $B(a; r)$ and is harmonic in $B(a; r)$. As above, this leads to a harmonic function U_0 on $B(a; r)$ such that $U_0 \le u$ and $U_0(z_0) = u(z_0)$. But $\Phi_n \le X_n$ so that $\Phi'_n \le X'_n$. Hence $U \le U_0 \le u$ and $U(a) = U_0(a) = u(a)$. Therefore $U - U_0$ is a negative harmonic function on $B(a; r)$ and $(U - U_0)(a) = 0$. By the Maximum Principle, $U = U_0$; so $U(z_0) = u(z_0)$. Since z_0 was arbitrary, $u = U$ in $B(a; r)$. That is, u is harmonic on every disk contained in G. ∎

3.17 Definition. Let G be a region and let $f: \partial_\infty G \to \mathbb{R}$ be a continuous function. The harmonic function u obtained in the preceding theorem is called the *Perron Function associated with f*.

The next step in solving the Dirichlet Problem is to prove that for each point a in $\partial_\infty G$ $\lim_{z \to a} u(z)$ exists and equals $f(a)$. As was mentioned earlier, this does not always hold. The following example illustrates this phenomenon.

Let $G = \{z : 0 < |z| < 1\}$, $T = \{z : |z| = 1\}$; so $\partial G = T \cup \{0\}$. Define $f : \partial G \to \mathbb{R}$ by $f(z) = 0$ if $z \in T$ and $f(0) = 1$. For $0 < \epsilon < 1$ let $u_\epsilon(z) = (\log |z|) (\log \epsilon)^{-1}$; then u_ϵ is harmonic in G, $u_\epsilon(z) > 0$ for z in G, $u_\epsilon(z) = 0$ for z in T, and $u_\epsilon(z) = 1$ if $|z| = \epsilon$. Suppose that $v \in \mathscr{P}(f, G)$; since $|f| \le 1$, $|v(z)| \le 1$ for all z in G. If $R_\epsilon = \{z : \epsilon < |z| < 1\}$ then $\limsup_{z \to a} v(z) \le u_\epsilon(a)$ for all a in ∂R_ϵ; by the Maximum Principle, $v(z) \le u_\epsilon(z)$ for all z in R_ϵ. Since ϵ was arbitrary this gives that for each z in G, $v(z) \le \lim_{\epsilon \to 0} u_\epsilon(z) = 0$. Hence the Perron function associated with f is the identically zero function, and the Dirichlet Problem cannot be solved for the punctured disk. (Another proof of this is available by using Exercise 2.5 and the Maximum Principle.)

Exercises

1. Which of the following functions are subharmonic? superharmonic? harmonic? neither subharmonic nor superharmonic? (a) $\varphi(x, y) = x^2 + y^2$; (b) $\varphi(x, y) = x^2 - y^2$; (c) $\varphi(x, y) = x^2 + y$; (d) $\varphi(x, y) = x^2 - y$; (e) $\varphi(x, y) = x + y^2$; (f) $\varphi(x, y) = x - y^2$.

2. Let Subhar(G) and Superhar(G) denote, respectively, the sets of subharmonic and superharmonic functions on G.

   (a) Show that Subhar(G) and Superhar(G) are closed subsets of $C(G; \mathbb{R})$.

   (b) Does a version of Harnack's Theorem hold for subharmonic and superharmonic functions?

3. If G is a region and if $f : \partial_\infty G \to \mathbb{R}$ is a continuous function let u_f be the Perron Function associated with f. This defines a map $T : C(\partial_\infty G; \mathbb{R}) \to \text{Har}(G)$ by $T(f) = u_f$. Prove:

   (a) T is linear (i.e., $T(a_1 f_1 + a_2 f_2) = a_1 T(f_1) + a_2 T(f_1)$).

   (b) T is positive (i.e., if $f(a) \ge 0$ for all a in $\partial_\infty G$ then $T(f)(z) \ge 0$ for all z in G).

   (c) T is continuous. Moreover, if $\{f_n\}$ is a sequence in $C(\partial_\infty G; \mathbb{R})$ such that $f_n \to f$ uniformly then $T(f_n) \to T(f)$ uniformly on G.

   (d) If the Dirichlet Problem can be solved for G then T is one-one. Is the converse true?

4. In the hypothesis of Theorem 3.11, suppose only that f is a bounded function on $\partial_\infty G$; prove that the conclusion remains valid. (This is useful if G is an unbounded region and g is a bounded continuous function on ∂G. If we define $f : \partial_\infty G \to \mathbb{R}$ by $f(z) = g(z)$ for z in ∂G and $f(\infty) = 0$ then the conclusion of Theorem 3.11 remains valid. Of course there is no reason to expect that the harmonic function will have predictable behavior near ∞ — we could have assigned any value to $f(\infty)$. However, the behavior near points of ∂G can be studied with hope of success.)

5. Let G be a region and $f: \partial_\infty G \to \mathbb{R}$ a continuous function. Define $U(f, G)$ to be the family of all superharmonic functions ψ on G such that $\liminf\limits_{z \to a} \psi(z) \geq f(a)$. If $v: G \to \mathbb{R}$ is defined by $v(z) = \inf \{\psi(z) : \psi \in U(f, G)\}$, prove that v is harmonic on G. If u is the Perron Function associated with f, show that $u(z) \leq v(z)$. Prove that $\lim\limits_{z \to a} u(z) = f(a)$ for all a in $\partial_\infty G$ iff $u(z) = v(z)$ for all z. Can you give a condition in terms of u and v which is necessary and sufficient that $\lim\limits_{z \to a} u(z) = f(a)$ for an individual point a in $\partial_\infty G$?

6. Show that the requirement that G_1 is bounded in Corollary 3.5 is necessary.

7. If $f: G \to \Omega$ is analytic and $\varphi: \Omega \to \mathbb{R}$ is subharmonic, show that $\varphi \circ f$ is subharmonic if f is one-one. What happens if $f'(z) \neq 0$ for all z in G?

§4. The Dirichlet Problem

4.1 Definition. A region G is called a *Dirichlet Region* if the Dirichlet Problem can be solved for G. That is, G is a Dirichlet Region if for each continuous function $f: \partial_\infty G \to \mathbb{R}$ there is a continuous function $u: G^- \to \mathbb{R}$ such that u is harmonic in G and $u(z) = f(z)$ for all z in $\partial_\infty G$.

We have already seen that a disk is a Dirichlet Region, but the punctured disk is not. In this section, we will see conditions that are sufficient for a region to be a Dirichlet Region. The first step in this direction is to suppose that there are functions which can be used to restrict the behavior of the Perron Functions near the boundary.

For a set G and a point a in $\partial_\infty G$, let $G(a; r) = G \cap B(a; r)$ for all $r > 0$.

4.2 Definition. Let G be a region and let $a \in \partial_\infty G$. *A barrier for G at a is* a family $\{\psi_r : r > 0\}$ of functions such that:

(a) ψ_r is defined and superharmonic on $G(a; r)$ with $0 \leq \psi_r(z) \leq 1$;
(b) $\lim\limits_{z \to a} \psi_r(z) = 0$;
(c) $\lim\limits_{z \to w} \psi_r(z) = 1$ for w in $G \cap \{w : |w - a| = r\}$.

The following observation is useful: if $\hat{\psi}_r$ is defined by letting $\hat{\psi}_r = \psi_r$ on $G(a; r)$ and $\hat{\psi}_r(z) = 1$ for z in $G - B(a; r)$, then $\hat{\psi}_r$ is superharmonic. (Verify!) So the functions $\hat{\psi}$ "approach" the function which is one everywhere but $z = a$, where it is zero. The second observation which must be made is that if G is a Dirichlet Region then there is a barrier for G at each point of $\partial_\infty G$. In fact, if $a \in \partial_\infty G$ ($a \neq \infty$) and $f(z) = |z - a|(1 + |z - a|)^{-1}$ for $z \neq \infty$ with $f(\infty) = 1$, then f is continuous on $\partial_\infty G$; so there is a continuous function $u: G^- \to \mathbb{R}$ such that u is harmonic on G and $u(z) = f(z)$ for z in $\partial_\infty G$. In particular, $u(a) = 0$ and a is the only zero of u in G^- (Why?) Let $c_r = \inf \{u(z) : |z - a| = r, z \in G\} = \min \{u(z) : |z - a| = r, z \in G^-\} > 0$.

Define $\psi_r : G(a; r) \to \mathbb{R}$ by $\psi_r(z) = \dfrac{1}{c_r} \min \{u(z), c_r\}$. It is left to the reader to check that $\{\psi_r\}$ is a barrier.

The next result provides a converse to the above facts.

4.3 Theorem. *Let G be a region and let* $a \in \partial_\infty G$ *such that there is a barrier for G at a. If* $f: \partial_\infty G \to \mathbb{R}$ *is continuous and u is the Perron Function associated with f then*

$$\lim_{z \to a} u(z) = f(a)$$

Proof. Let $\{\psi_r : r > 0\}$ be a barrier for G at a and for convenience assume $a \neq \infty$; also assume that $f(a) = 0$ (otherwise consider the function $f - f(a)$). Let $\epsilon > 0$ and choose $\delta > 0$ such that $|f(w)| < \epsilon$ whenever $w \in \partial_\infty G$ and $|w - a| < 2\delta$; let $\psi = \psi_\delta$. Let $\hat{\psi}: G \to \mathbb{R}$ be defined by $\hat{\psi}(z) = \psi(z)$ for z in $G(a; \delta)$ and $\hat{\psi}(z) = 1$ for z in $G - B(a; \delta)$. Then $\hat{\psi}$ is superharmonic. If $|f(w)| \leq M$ for all w in $\partial_\infty G$, then $-M\hat{\psi} - \epsilon$ is subharmonic.

4.4 Claim. $-M\hat{\psi} - \epsilon$ is in $\mathscr{P}(f, G)$.

If $w \in \partial_\infty G - B(a; \delta)$ then $\lim\sup\limits_{z \to w} [-M\hat{\psi}(z) - \epsilon] = -M - \epsilon < f(w)$. Because $\hat{\psi}(z) \geq 0$, it follows that $\lim\sup\limits_{z \to w} [-M\hat{\psi}(z) - \epsilon] \leq -\epsilon$ for all w in $\partial_\infty G$. In particular, if $w \in \partial_\infty G \cap B(a; \delta)$ then $\lim\sup\limits_{z \to w} [-M\hat{\psi}(z) - \epsilon] \leq -\epsilon < f(w)$ by the choice of δ. This substantiates Claim 4.4. Hence

4.5 $$-M\hat{\psi}(z) - \epsilon \leq u(z)$$

for all z in G.

A similar analysis yields

$$\lim_{z \to w} \inf [M\hat{\psi}(z) + \epsilon] \geq \lim_{z \to w} \sup \varphi(z)$$

for all φ in $\mathscr{P}(f, G)$ and w in $\partial_\infty G$. By the fourth version of the Maximum Principle, $\varphi(z) \leq M\hat{\psi}(z) + \epsilon$ for φ in $\mathscr{P}(f, G)$ and z in G. Hence

$$u(z) \leq M\hat{\psi}(z) + \epsilon;$$

or, combining this with (4.5),

4.6 $$-M\hat{\psi}(z) - \epsilon \leq u(z) \leq M\hat{\psi}(z) + \epsilon$$

for all z in G. But $\lim\limits_{z \to a} \hat{\psi}(z) = \lim\limits_{z \to a} \psi(z) = 0$; since ϵ was arbitrary, (4.6) gives that

$$\lim_{z \to a} u(z) = 0 = f(a).$$

This completes the proof. ∎

Notice that the purpose of the barrier was to construct the function $\hat{\psi}$ which "squeezed" u down to zero.

4.7 Corollary. *A region G is a Dirichlet Region iff there is a barrier for G at each point of* $\partial_\infty G$.

The above corollary is not the solution to the problem of characterizing Dirichlet Regions. True, it gives a necessary and sufficient condition that a region be a Dirichlet Region and this condition is formally weaker than the definition. However, there are aesthetic and practical difficulties with Corollary 4.7. One difficulty is that the condition in (4.7) is not easily verified.

Another difficulty is that it is essentially the same type of condition as the definition; both hypothesize the existence of functions with prescribed boundary behavior.

What is desired? Both tradition and aesthetics dictate that we strive for a topological–geometric condition on G which is necessary and sufficient that G be a Dirichlet Region. Such conditions are usually easy to verify for a given region, and the equivalence of a geometric property and an analytic one is the type of beauty after which most mathematicians strive. At the present time no such equivalence is known and we must be content with sufficient conditions.

4.8 Lemma. *Let G be a region in $\mathbb{C}$ and let S be a closed connected subset of $\mathbb{C}_\infty$ such that $\infty \in S$ and $S \cap \partial_\infty G = \{a\}$. If G_0 is the component of $\mathbb{C}_\infty - S$ which contains G then G_0 is a simply connected region in the plane.*

Proof. Let $G_0, G_1, \ldots$ be the components of $\mathbb{C}_\infty - S$ with $G \subset G_0$; note that each G_n is a region in $\mathbb{C}$. If $z \in \partial_\infty G_n$ then $G_n \cup \{z\}$ is connected (Exercise II.2.1). Since G_n is a component it follows that $\partial_\infty G_n \subset S$. By Lemma II.2.6 $G_n \cup S \, (= G_n^- \cup S)$ is connected and, consequently, so is $\bigcup_{n=1}^{\infty} (G_n \cup S)$ $= \mathbb{C}_\infty - G_0$. In virtue of Theorem VIII 3.2(c), G_0 is simply connected. ∎

4.9 Theorem. *Let G be a region in $\mathbb{C}$ and suppose that $a \in \partial_\infty G$ such that the component of $\mathbb{C}_\infty - G$ which contains a does not reduce to a point. Then G has a barrier at a.*

Proof. Let S be the component of $\mathbb{C}_\infty - G$ such that $a \in S$. By considering an appropriate Mobius transformation if necessary, we may assume that $a = 0$ and $\infty \in S$. Let G_0 be the component of $\mathbb{C}_\infty - S$ which contains G. The preceding lemma gives that G_0 is simply connected; since $0 \notin G_0$ there is a branch ℓ of $\log z$ defined on G_0. In particular ℓ is defined on G. For $r > 0$, let $\ell_r(z) = \ell(z/r) = \ell(z) - \log r$ for z in $G(0; r)$. So $-\ell_r(G(0; r))$ is a subset of the right half plane. Now let $C_r = G \cap \{z : |z| = r\}$; then C_r is the union of at most a countable number of pairwise disjoint open arcs γ_k in $\{z : |z| = r\}$. But $-\ell_r(\gamma_k) = (i\alpha_k, i\beta_k) = \{it : \alpha_k < t < \beta_k\}$ for $k \geq 1$. Hence

$$-\ell_r(C_r) = \bigcup_{k=1}^{\infty} (i\alpha_k, i\beta_k)$$

and these intervals are pairwise disjoint. Furthermore, the length of γ_k is $r(\beta_k - \alpha_k)$; so

4.10
$$\sum_{k=1}^{\infty} (\beta_k - \alpha_k) \leq 2\pi.$$

Now if log is the principal branch of the logarithm then

$$h_k(z) = \text{Im} \log \left(\frac{z - i\alpha_k}{z - i\beta_k} \right)$$

is harmonic in the right half plane and $0 < h_k(z) < \pi$ for Re $z > 0$ (see Exercise III.3.19). Moreover

4.11
$$h_k(x+iy) = \int_{\alpha_k}^{\beta_k} \frac{x}{x^2+(y-t)^2}\, dt$$

if $x > 0$. From (4.11) it follows that

$$\sum_{k=1}^{n} h_k(x+iy) \le \int_{-\infty}^{\infty} \frac{x}{x^2+(y-t)^2}\, dt = \pi.$$

Since each $h_k \ge 0$, Harnack's Theorem gives that $h = \sum_{k=1}^{\infty} h_k$ is harmonic in the right half plane. Hence

$$\psi_r(z) = \frac{1}{\pi} h(-\ell_r(z))$$

is harmonic in $G(0; r)$. It will be shown that $\{\psi_r\}$ is a barrier at a.

Fix $r > 0$; then $\lim_{z \to 0} \text{Re } [-\ell_r(z)] = +\infty$. So it suffices to show that $h(z) \to 0$ as Re $z \to +\infty$. Using (4.11) and (4.10) it follows that for $x > 0$,

$$h(x+iy) = \sum_{k=1}^{\infty} h_k(x+iy)$$

$$= \sum_{k=1}^{\infty} \int_{\alpha_k}^{\beta_k} \frac{1/x}{1+(y-t/x)^2}\, dt$$

$$\le \frac{1}{x} \sum_{k=1}^{\infty} (\beta_k - \alpha_k)$$

$$\le \frac{2\pi}{x}.$$

So, indeed, $\lim_{x \to +\infty} h(x+iy) = 0$ uniformly in y; this gives that $\lim_{z \to 0} \psi_r(z) = 0$.

To prove that $\lim_{z \to w} \psi_r(z) = 1$ for w in G with $|w| = r$, it is sufficient to prove that

4.12
$$\lim_{z \to ic} h(z) = \pi \text{ if } \alpha_k < c < \beta_k \text{ for some } k$$

So fix $k \ge 1$ and fix c in (α_k, β_k). The following will be proved.

4.13 Claim. There are numbers α and β such that $\alpha < \alpha_k < \beta_k < \beta$ and if

$$u(z) = \text{Im} \log \left(\frac{z-i\alpha}{z-i\alpha_k} \right),$$

$$v(z) = \text{Im} \log \left(\frac{z-i\beta_k}{z-i\beta} \right)$$

then $x > 0$ and $\alpha_k < y < \beta_k$ implies $0 \le h(x+iy) - h_k(x+iy) \le u(x+iy) + v(x+iy)$.

Once 4.13 is established, Equation 4.12 is proved as follows. From Exercise III.3.19,

$$v(x+iy) = \arctan\left(\frac{y-\beta_k}{x}\right) - \arctan\left(\frac{y-\beta}{x}\right).$$

so if $x+iy \to ic$, $c < \beta_k < \beta$, then $v(x+iy) \to 0$.
Similarly $u(x+iy) \to 0$ as $x+iy \to ic$, with $\alpha < \alpha_k < c$.
Hence, Claim 4.13 yields

4.14 $\lim [h(z) - h_k(z)] = 0.$

But

$$h_k(x+iy) = \arctan\left(\frac{y-\alpha_k}{x}\right) - \arctan\left(\frac{y-\beta_k}{x}\right),$$

so $\lim_{z \to ic} h_k(z) = \pi$; this combined with (4.14) implies Equation (4.12).

It remains to substantiate Claim 4.13; we argue geometrically. Recall (Exercise III.3.19) that $h_j(z)$ is the angle in the figure. Consider all the intervals $(i\alpha_j, i\beta_j)$ lying above $(i\alpha_k, i\beta_k)$ and translate them downward along the imaginary axis, keeping them above $(i\alpha_k, i\beta_k)$ until their endpoints coincide and such that one of the endpoints coincides with $i\beta_k$. Since $\sum(\beta_j - \alpha_j) \le 2\pi$ there is a number $\beta < (\beta_k + 2\pi)$ such that each of the translated intervals lies in $(i\beta_k, i\beta)$. Now if $z = x+iy$, $x > 0$ and $\alpha_k < y < \beta_k$,

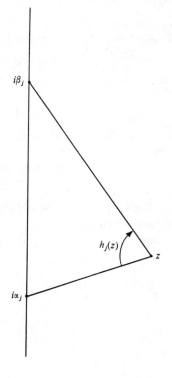

then the angle $h_j(z)$ increases as the interval $(i\alpha_j, i\beta_j)$ is translated downward. Hence $\alpha_k < \text{Im } z < \beta_k$ implies

4.15
$$\sum_j h_j(z) \le v(z),$$

where v is as in the statement of the claim and the sum is over all j such that $\alpha_j \ge \beta_k$. By performing a similar upward translation of the intervals $(i\alpha_j, i\beta_j)$ with $\beta_j \le \alpha_k$, there is a number $\alpha < (\alpha_k - 2\pi)$ such that the translates lie in the interval $(i\alpha, i\alpha_k)$. So if u is as in the claim and $\alpha_k < \text{Im } z < \beta_k$,

4.16
$$\sum_j h_j(z) \le u(z)$$

where the sum is over all j with $\beta_j \le \alpha_k$. By combining (4.15) and (4.16) the claim is established. ∎

4.17 Corollary. *Let G be a region such that no component of $\mathbb{C}_\infty - G$ reduces to a point; then G is a Dirichlet Region.*

4.18 Corollary. *A simply connected region is a Dirichlet Region.*

Proof. If $G \ne \mathbb{C}$ the result is clear since $\mathbb{C}_\infty - G$ has only one component. If $G = \mathbb{C}$ then the result is trivial. ∎

Theorem 4.9 has no converse as the following example illustrates. Let $1 > r_1 > r_2 > \ldots$ with $r_n \to 0$; for each n let γ_n be a proper closed arc of the circle $|z| = r_n$ with length $V(\gamma_n)$. Put $G = B(0; 1) - [\bigcup_{n=1}^{\infty} \{\gamma_m\} \cup \{0\}]$ and suppose that lim $V(\gamma_n)/r_n = 2\pi$. So $\mathbb{C}_\infty - G = \{0\} \cup \bigcup_{n=1}^{\infty} \{\gamma_n\} \cup \{z : |z| \ge 1\}$. According to Theorem 4.9 there is a barrier at each point of $\partial_\infty G = \partial G$ with the possible exception of zero. We will show that there is also a barrier at zero.

If $r_{n-1} > r > r_n$ and if $m > n$, let $B_m = B(0; r) - \bigcup_{j=n}^{m} \{\gamma_j\}$. Let h_m be the continuous function on $\bar{B}_m$ which is harmonic on B_m with $h_m(z) = 1$ for $|z| = r$ and $h_m(z) = 0$ for z in $\bigcup_{j=n}^{m} \{\gamma_j\}$. Then $\{h_m\}$ is a decreasing sequence of positive harmonic functions on $G(0; r)$; by Harnack's Theorem $\{h_m\}$ converges to a harmonic function on $G(0; r)$ which is also positive (Why?) Since $\lim_{z \to w} h(z) = 1$ for $|w| = r$, we need only show that $\lim_{z \to 0} h(z) = 0$. Let k_m be the harmonic function on $B(0; r_m)$ which is 0 on $\{\gamma_m\}$ and 1 on $\{z : |z| = r_m\} - \{\gamma_m\}$ (this does not have continuous boundary values, only piecewise continuous boundary values which are sufficient—see Exercise 2.2). Then $0 \le h \le k_m$ on $B(0; r_m)$ and

$$k_m(0) = \frac{1}{2\pi} \int_0^{2\pi} k_m(r_m e^{i\theta}) \, d\theta$$

$$= \frac{1}{2\pi} \left(2\pi - \frac{V(\gamma_m)}{r_m} \right)$$

Since $\dfrac{V(\gamma_m)}{r_m} \to 2\pi$, $k_m(0) \to 0$; it follows that $h(z) \to 0$ as $z \to 0$. Thus, G has a barrier at zero.

Exercises

1. Let $G = B(0; 1)$ and find a barrier for G at each point of the boundary.
2. Let $G = \mathbb{C} - (\infty, 0]$ and construct a barrier for each point of $\partial_\infty G$.
3. Let G be a region and a a point in $\partial_\infty G$ such that there is a harmonic function $u : G \to \mathbb{R}$ with $\lim_{z \to a} u(z) = 0$ and $\liminf_{z \to w} u(z) > 0$ for all w in $\partial_\infty G$, $w \neq a$. Show that there is a barrier for G at a.
4. This exercise asks for an easier proof of a special case of Theorem 4.9. Let G be a bounded region and let $a \in \partial G$ such that there is a point b with $[a, b] \cap G^- = \{a\}$. Show that G has a barrier at a. (Hint: Consider the transformation $(z-a)(z-b)^{-1}$.)

§5. Green's Function

In this section Green's Function is introduced and its existence is discussed. Green's Function plays a vital role in differential equations and other fields of analysis.

5.1 Definition. Let G be a region in the plane and let $a \in G$. A *Green's Function of G with singularity at a* is a function $g_a : G \to \mathbb{R}$ with the properties:

  (a) g_a is harmonic in $G - \{a\}$;
  (b) $G(z) = g_a(z) + \log|z-a|$ is harmonic in a disk about a;
  (c) $\lim_{z \to w} g_a(z) = 0$ for each w in $\partial_\infty G$.

For a given region G and a point a in G, g_a need not exist. However, if it exists, it is unique. In fact if h_a has the same properties, then, from (b), $h_a - g_a$ is harmonic in G. But (c) implies that $\lim_{z \to w} [h_a(z) - g_a(z)] = 0$ for every w in $\partial_\infty G$; so $h_a = g_a$ by virtue of the Maximum Principle.

A second observation is that a Green's Function is positive. In fact, g_a is harmonic in $G - \{a\}$ and $\lim_{z \to a} g(z) = +\infty$ since $g_a(z) + \log|z-a|$ is harmonic at $z = a$. By the Maximum Principle, $g_a(z) > 0$ for all z in $G - \{a\}$.

Given this observation it is easy to see that $\mathbb{C}$ has no Green's Function with a singularity at zero (or a singularity at any point, for that matter). In fact, suppose g_0 is the Green's Function with singularity at zero and put $g = -g_0$; so $g(z) < 0$ for all z. We will show g must be a constant function, which is a contradiction. To do this, it is sufficient to show that if $0 \neq z_1 \neq z_2 \neq 0$ then $g(z_2) \leq g(z_1)$. If $\epsilon > 0$ then there is a $\delta > 0$ such that

$|g(z)-g(z_1)| < \epsilon$ if $|z_1-z| < \delta$; so $g(z) < g(z_1)+\epsilon$ if $|z-z_1| < \delta$. Let $r > |z_1-z_2| > \delta$; then

$$h_r(z) = \frac{g(z_1)+\epsilon}{\log\left(\dfrac{\delta}{r}\right)} \log\left|\frac{z-z_1}{r}\right|$$

is harmonic in $\mathbb{C}-\{z_1\}$. It is left to the reader to check that $g(z) \leq h_r(z)$ for z on the boundary of the annulus $A = \{z:\delta < |z-z_1| < r\}$. By the Maximum Principle, $g(z) \leq h_r(z)$ for z in A; in particular, $h_r(z_2) \geq g(z_2)$. Letting $r \to \infty$ we get

$$g(z_2) \leq \lim_{r\to\infty} h_r(z_2) = g(z_1)+\epsilon;$$

since ϵ was arbitrary, $g(z_2) \leq g(z_1)$ and g must be a constant function.

When do Green's Functions exist?

5.2 Theorem. *If G is a bounded Dirichlet Region then for each a in G there is a Green's Function on G with singularity at a.*

Proof. Define $f:\partial G \to \mathbb{R}$ by $f(z) = \log |z-a|$, and let $u:G^- \to \mathbb{R}$ be the unique continuous function which is harmonic on G and such that $u(z)=f(z)$ for z in ∂G. Then $g_a(z) = u(z)-\log |z-a|$ is easily seen to be the Green's Function. ∎

This section will close with one last result which says that Green's Functions are conformal invariants.

5.3 Theorem. *Let G and Ω be regions such that there is a one-one analytic function f of G onto Ω; let $a \in G$ and $\alpha = f(a)$. If g_a and γ_α are the Green's Functions for G and Ω with singularities a and α respectively, then*

$$g_a(z) = \gamma_\alpha(f(z)).$$

Proof. Let $\varphi:G \to \mathbb{R}$ be defined by $\varphi = \gamma_\alpha \circ f$. To show that $\varphi = g_a$ it is sufficient to show that φ has the properties of the Green's Function with singularity at $z = a$. Clearly φ is harmonic in $G-\{a\}$. If $w \in \partial_\infty G$ then $\lim_{z\to w} \varphi(z) = 0$ will follow if it can be shown that $\lim \varphi(z_n) = 0$ for any sequence $\{z_n\}$ in G with $z_n \to w$. But $\{f(z_n)\}$ is a sequence in Ω and so there is a subsequence $\{z_{n_k}\}$ such that $f(z_{n_k}) \to w$ in Ω^- (closure in $\mathbb{C}_\infty$). So $\gamma_\alpha(f(z_{n_k})) \to 0$. Since this happens for any convergent subsequence of $\{f(z_n)\}$ it follows that $\lim \varphi(z_n) = \lim \gamma_\alpha(f(z_n)) = 0$. Hence $\lim_{z\to w} \varphi(z) = 0$ for every w in $\partial_\infty G$.

Finally, taking the power series expansion of f about $z = a$,

$$f(z) = \alpha+A_1(z-a)+A_2(z-a)^2+\ldots;$$

or

$$f(z)-\alpha = (z-a)[A_1+A_2(z-a)+\ldots].$$

Hence

5.4 $$\log|f(z)-\alpha| = \log|z-a|+h(z),$$

where $h(z) = \log |A_1 + A_2(z-a) + \ldots|$ is harmonic near $z = a$ since $A_1 \neq 0$.

Suppose that $\gamma_\alpha(w) = \Delta(w) - \log|w - \alpha|$ where Δ is a harmonic function on Ω. Using (5.4)

$$\varphi(z) = \Delta(f(z)) - \log|f(z) - \alpha|$$
$$= [\Delta(f(z)) - h(z)] - \log|z - a|.$$

Since $\Delta \circ f - h$ is harmonic near $z = a$, $\varphi(z) + \log |z - a|$ is harmonic near $z = a$. ∎

Exercises

1. (a) Let G be a simply connected region, let $a \in G$, and let $f: G \to D = \{z: |z| < 1\}$ be a one-one analytic function such that $f(G) = D$ and $f(a) = 0$. Show that the Green's Function on G with singularity at a is $g_a(z) = -\log |f(z)|$.

(b) Find the Green's Functions for each of the following regions: (i) $G = \mathbb{C} - (\infty, 0]$; (ii) $G = \{z: \text{Re } z > 0\}$; (iii) $G = \{z: 0 < \text{Im } z < 2\pi\}$.

2. Let g_a be the Green's Function on a region G with singularity at $z = a$. Prove that if ψ is a positive superharmonic function on $G - \{a\}$ with $\liminf_{z \to a} [\psi(z) + \log |z - a|] > -\infty$, then $g_a(z) \leq \psi(z)$ for $z \neq a$.

3. This exercise gives a proof of the Riemann Mapping Theorem where it is assumed that if G is a simply connected region, $G \neq \mathbb{C}$, then: (i) $\mathbb{C}_\infty - G$ is connected, (ii) every harmonic function on G has a harmonic conjugate, (iii) if $a \notin G$ then a branch of $\log(z - a)$ can be defined.

(a) Let G be a bounded simply connected region and let $a \in G$; prove that there is a Green's Function g_a on G with singularity at a. Let $u(z) = g_a(z) + \log|z - a|$ and let v be the harmonic conjugate of u. If $\varphi = u + iv$ let $f(z) = e^{i\alpha}(z - a)e^{-\varphi(z)}$ for a real number α. (So f is analytic in G.) Prove that $|f(z)| = \exp(-g_a(z))$ and that $\lim_{z \to w} |f(z)| = 1$ for each w in ∂G (Compare this with Exercise 1). Prove that for $0 < r < 1$, $C_r = \{z: |f(z)| = r\}$ consists of a finite number of simple closed curves in G (see Exercise VI.1.3). Let G_r be a component of $\{z: |f(z)| < r\}$ and apply Rouché's Theorem to get that $f(z) = 0$ and $f(z) - w_0 = 0$, $|w_0| < r$, have the same number of solutions in G_r. Prove that f is one-one on G_r. From here conclude that $f(G) = D = \{z: |z| < 1\}$ and $f'(a) > 0$, for a suitable choice of α.

(b) Let G be a simply connected region with $G \neq \mathbb{C}$, but assume that G is unbounded and $0, \infty \in \partial_\infty G$. Let ℓ be a branch of $\log z$ on G, $a \in G$, and $\alpha = \ell(a)$. Show that ℓ is one-one on G and $\ell(z) \neq \alpha + 2\pi i$ for any z in G. Prove that $\varphi(z) = [\ell(z) - \alpha - 2\pi i]^{-1}$ is a conformal map of G onto a bounded simply connected region in the plane. (Show that ℓ omits all values in a neighborhood of $\alpha + 2\pi i$.)

(c) Combine parts (a) and (b) to prove the Riemann Mapping Theorem.

4. (a) Let G be a region such that $\partial G = \gamma$ is a simple continuously differen-

tiable closed curve. If $f: \partial G \to \mathbb{R}$ is continuous and $g(z, a) = g_a(z)$ is the
Green's Function on G with singularity at a, show that

5.5
$$h(a) = \int_{\gamma} f(z) \frac{\partial g}{\partial n} (z, a) \, ds$$

is a formula for the solution of the Dirichlet Problem with boundary values f;
where $\dfrac{\partial g}{\partial n}$ is the derivative of g in the direction of the outward normal to γ and
ds indicates that the integral is with respect to arc length. (Note: these concepts are not discussed in this book but the formula is sufficiently interesting
so as to merit presentation.) (Hint: Apply Green's formula

$$\iint_{\Omega} [u\Delta v - v\Delta u] \, dx \, dy = \int_{\partial\Omega} \left[u \frac{\partial v}{\partial n} - v \frac{\partial u}{\partial n} \right] ds$$

with $\Omega = G - \{z: |z-a| \leq \delta\}$, $\delta < d(a, \{\gamma\})$, $u = h$, $v = g_a(z) = g(z, a)$.)
(b) Show that if $G = \{z: |z| < 1\}$ then (5.5) reduces to equation (2.5).

Chapter XI

Entire Functions

To begin this chapter, let us recall the Weierstrass Factorization Theorem for entire functions (VII. 5.14). Let f be an entire function with a zero of multiplicity $m \geq 0$ at $z = 0$; let $\{a_n\}$ the zeros of f, $a_n \neq 0$, arranged so that a zero of multiplicity k is repeated in this sequence k times. Also assume that $|a_1| \leq |a_2| \leq \ldots$. If $\{p_n\}$ is a sequence of integers such that

0.1
$$\sum_{n=1}^{\infty} \left(\frac{r}{|a_n|}\right)^{p_n+1} < \infty$$

for every $r > 0$ then

0.2
$$P(z) = \prod_{n=1}^{\infty} E_{p_n}(z/a_n)$$

converges uniformly on compact subsets of the plane, where

0.3
$$E_p(z) = (1-z)\exp\left(z + \frac{z^2}{2} + \ldots + \frac{z^p}{p}\right)$$

for $p \geq 1$ and

$$E_0(z) = 1 - z$$

Consequently

0.4
$$f(z) = z^m e^{g(z)} P(z)$$

where g is an entire function. An interesting line of investigation begins if we ask the questions: What properties of f can be deduced if g and P are assumed to have certain "nice" properties? Can restrictions be imposed on f which will imply that g and P have particular properties? The plan that will be adopted in answering these questions is to assume that g and P have certain characteristics, deduce the implied properties of f, and then try to prove the converse of this implication.

How to begin? Clearly the first restriction on g in this program is to suppose that it is a polynomial. It is equally clear that such an assumption must impose a growth condition on $e^{g(z)}$. A convenient assumption on P is that all the integers p_n are equal. From equation (0.1), we see that this is to assume that there is an integer $p \geq 1$ such that

$$\sum_{n=1}^{\infty} |a_n|^{-p} < \infty;$$

that is, it is an assumption on the growth rate of the zeros of f. In the first

279

section of this chapter Jensen's Formula is deduced. Jensen's Formula says that there is a relation between the growth rate of the zeros of f and the growth of $M(r) = \sup\{|f(re^{i\theta})|: 0 \le \theta \le 2\pi\}$ as r increases. In succeeding sections we study the growth of the zeros of f and the growth of $M(r)$. Finally, the chapter culminates in the beautiful factorization theorem of Hadamard which shows an intimate relation between the growth rate of $M(r)$ and these assumptions on g and P.

§1. Jensen's Formula

If f is analytic in an open set containing $\bar{B}(0; r)$ and f doesn't vanish in $\bar{B}(0; r)$ then $\log|f|$ is harmonic there. Hence it has the Mean Value Property (X. 1.4); that is

1.1
$$\log|f(0)| = \frac{1}{2\pi}\int_0^{2\pi} \log|f(re^{i\theta})|\, d\theta.$$

Suppose f has exactly one zero $a = re^{i\alpha}$ on the circle $|z| = r$. If $g(z) = f(z)(z-a)^{-1}$ then (1.1) can be applied to g to obtain

$$\log|g(0)| = \frac{1}{2\pi}\int_0^{2\pi} [\log|f(re^{i\theta})| - \log|re^{i\theta} - re^{i\alpha}|]\, d\theta$$

Since $\log|g(0)| = \log|f(0)| - \log r$, it will follow that (1.1) remains valid where f has a single zero on $|z| = r$, if it can be shown that

$$\frac{1}{2\pi}\int_0^{2\pi} \log|re^{i\theta} - re^{i\alpha}|\, d\theta = \log r;$$

alternately, if it can be shown that

$$\int_0^{2\pi} \log|1 - e^{i\theta}|\, d\theta = 0.$$

But this follows from the fact that

$$\int_0^{2\pi} \log(\sin^2 2\theta)\, d\theta = -4\pi \log 2$$

(Exercise V. 2.2(h)). So (1.1) remains valid if f has a single zero on $|z| = r$; by induction, (1.1) is valid as long as f has no zeros in $B(0; r)$.

The next step is to examine what happens if f has zeros inside $B(0; r)$. In this case, $\log|f(z)|$ is no longer harmonic so that the MVP is not present.

1.2 Jensen's Formula. *Let f be an analytic function on a region containing*

$\bar{B}(0; r)$ and suppose that $a_1, \ldots, a_n$ are the zeros of f in $B(0; r)$ repeated according to multiplicity. if $f(0) \neq 0$ then

$$\log |f(0)| = -\sum_{k=1}^{n} \log \left(\frac{r}{|a_k|}\right) + \frac{1}{2\pi} \int_0^{2\pi} \log |f(re^{i\theta})| \, d\theta.$$

Proof. If $|b| < 1$ then the map $(z-b)(1-\bar{b}z)^{-1}$ takes the disk $B(0; 1)$ onto itself and maps the boundary onto itself. Hence

$$\frac{r^2(z-a_k)}{r^2 - \bar{a}_k z}$$

maps $B(0; r)$ onto itself and takes the boundary to the boundary. Therefore

$$F(z) = f(z) \prod_{k=1}^{n} \frac{r^2 - \bar{a}_k z}{r(z-a_k)}$$

is analytic in an open set containing $\bar{B}(0; r)$, has no zeros in $B(0; r)$, and $|F(z)| = |f(z)|$ for $|z| = r$. So (1.1) applies to F to give

$$\log |F(0)| = \frac{1}{2\pi} \int_0^{2\pi} \log |f(re^{i\theta})| \, d\theta.$$

However

$$F(0) = f(0) \prod_{k=1}^{n} \left(-\frac{r}{a_k}\right)$$

so that Jensen's Formula results. ∎

If the same methods are used but the MVP is replaced by Corollary X. 2.9, $\log |f(z)|$ can be found for $z \neq a_k$, $1 \leq k \leq n$.

1.3 Poisson-Jensen Formula. *Let f be analytic in a region which contains $\bar{B}(0; r)$ and let $a_1, \ldots, a_n$ be the zeros of f in $B(0; r)$ repeated according to multiplicity. If $|z| < r$ and $f(z) \neq 0$ then*

$$\log |f(z)| = -\sum_{k=1}^{n} \log \left|\frac{r^2 - \bar{a}_k z}{r(z-a_k)}\right| + \frac{1}{2\pi} \int_0^{2\pi} \mathrm{Re}\left(\frac{re^{i\theta} + z}{re^{i\theta} - z}\right) \log |f(re^{i\theta})| \, d\theta.$$

Exercises

1. In the hypothesis of Jensen's Formula, do not suppose that $f(0) \neq 0$. Show that if f has a zero at $z = 0$ of multiplicity m then

$$\log \left|\frac{f^{(m)}(0)}{m!}\right| + m \log r = -\sum_{k=1}^{n} \log \left(\frac{r}{|a_k|}\right) + \frac{1}{2\pi} \int_0^{2\pi} \log |f(re^{i\theta})| \, d\theta.$$

2. Let f be an entire function, $M(r) = \sup \{|f(re^{i\theta})|: 0 \le \theta \le 2\pi\}$, $n(r) =$ the number of zeros of f in $B(0; r)$ counted according to multiplicity. Suppose that $f(0) = 1$ and show that $n(r) \log 2 \le \log M(2r)$.

3. In Jensen's Formula do not suppose that f is analytic in a region containing $\bar{B}(0; r)$ but only that f is meromorphic with no pole at $z = 0$. Evaluate

$$\frac{1}{2\pi} \int_0^{2\pi} \log |f(re^{i\theta})| \, d\theta.$$

4. (a) Using the notation of Exercise 2, prove that

$$\int_0^r \frac{n(t)}{t} \, dt = \sum_{k=1}^n \log\left(\frac{r}{|a_k|}\right)$$

where $a_1, \ldots, a_n$ are the zeros of f in $B(0; r)$.

(b) Let f be meromorphic without a pole at $z = 0$ and let $n(r)$ be the number of zeros of f in $B(0; r)$ minus the number of poles (each counted according to multiplicity). Evaluate

$$\int_0^r \frac{n(t)}{t} \, dt.$$

5. Let $D = B(0; 1)$ and suppose that $f: D \to \mathbb{C}$ is an analytic function which is bounded.

(a) If $\{a_n\}$ are the non-zero zeros of f in D counted according to multiplicity, prove that $\sum(1 - |a_n|) < \infty$. (Hint: Use Proposition VII. 5.4.)

(b) If f has a zero at $z = 0$ of multiplicity $m \ge 0$, prove that $f(z) = z^m B(z) \exp(-g(z))$ where B is a Blaschke Product (Exercise VII. 5.4) and g is an analytic function on D with $\mathrm{Re}\, g\,(z) \ge -\log M$ ($M = \sup \{|f(z)|: |z| < 1\}$).

§2. The genus and order of an entire function

2.1 Definition. Let f be an entire function with zeros $\{a_1, a_2, \ldots\}$, repeated according to multiplicity and arranged such that $|a_1| \le |a_2| \le \ldots$. Then f is of *finite rank* if there is an integer p such that

2.2 $$\sum_{n=1}^\infty |a_n|^{-(p+1)} < \infty.$$

If p is the smallest integer such that this occurs, then f is said *to be of rank p*; a function with only a finite number of zeros has rank 0. A function is of *infinite rank* if it is not of finite rank.

From equation (0.1) it is seen that if f has finite rank p then the canonical product P in (0.4) can be taken to be

2.3
$$P(z) = \prod_{n=1}^{\infty} E_p(z/a_n).$$

Notice that if f is of finite rank and p is any integer larger than the rank of f, then (2.2) remains valid. So there is a second canonical product (2.3), and this shows that the factorization (0.4) of f is not unique. However, if the product P is defined by (2.3) where p is the rank of f then the factorization (0.4) is unique except that g may be replaced by $g + 2\pi mi$ for any integer m.

2.4 Definition. Let f be an entire function of rank p with zeros $\{a_1, a_2, \ldots\}$. Then the product defined in (2.3) is said to be in *standard form for f*. If f is understood then it will be said to be in *standard form*.

2.5 Definition. An entire function f has *finite genus* if f has finite rank and if

$$f(z) = z^m e^{g(z)} P(z),$$

where P is in standard form, and g is a polynomial. If p is the rank of f and q is the degree of the polynomial g, then $\mu = \max(p, q)$ is called the *genus of f*.

Notice that the genus of f is a well defined integer because once P is in standard form, then g is uniquely determined up to adding a multiple of $2\pi i$. In particular, the degree of g is determined.

2.6 Theorem. *Let f be an entire function of genus μ. For each positive number α there is a number r_0 such that for $|z| > r_0$*

$$|f(z)| < \exp(\alpha|z|^{\mu+1})$$

Proof. Since f is an entire function of genus μ

$$f(z) = z^m e^{g(z)} \prod_{n=1}^{\infty} E_\mu(z/a_n),$$

where g is a polynomial of degree $\leq \mu$. Notice that if $|z| < \frac{1}{2}$ then

2.7
$$\log|E_\mu(z)| = \text{Re}\{\log(1-z) + z + \ldots + z^\mu/\mu\}$$

$$= \text{Re}\left\{-\frac{1}{\mu+1}z^{\mu+1} - \frac{1}{\mu+2}z^{\mu+2} - \ldots\right\}$$

$$\leq |z|^{\mu+1}\left\{\frac{1}{\mu+1} + \frac{|z|}{\mu+2} + \ldots\right\}$$

$$\leq |z|^{\mu+1}(1 + \tfrac{1}{2} + (\tfrac{1}{2})^2 + \ldots)$$

$$= 2|z|^{\mu+1}$$

Also

$$|E_\mu(z)| \leq (1+|z|)\exp(|z| + \ldots + |z|^\mu/\mu),$$

so that

$$\log|E_\mu(z)| \leq \log(1+|z|) + |z| + \ldots + |z|^\mu/\mu.$$

Hence,

$$\lim_{z \to \infty} \frac{\log |E_\mu(z)|}{|z|^{\mu+1}} = 0$$

So if $A > 0$ then there is a number $R > 0$ such that

2.8 $\log |E_\mu(z)| \le A|z|^{\mu+1}, |z| > R.$

But on $\{z: \frac{1}{2} \le |z| \le R\}$ the function $|z|^{-(\mu+1)} \log |E_\mu(z)|$ is continuous except at $z = +1$, where it tends to $-\infty$. Hence there is a constant $B > 0$ such that

2.9 $\log |E_\mu(z)| \le B|z|^{\mu+1}, \frac{1}{2} \le |z| \le R.$

Combining (2.7), (2.8), and (2.9) gives that

2.10 $\log |E_\mu(z)| \le M|z|^{\mu+1}$

for all z in $\mathbb{C}$, where $M = \max \{2, A, B\}$.

Since $\sum |a_n|^{-(\mu+1)} < \infty$, an integer N can be chosen so that

$$\sum_{n=N+1}^{\infty} |a_n|^{-(\mu+1)} < \frac{\alpha}{4M}.$$

But, using (2.10),

2.11 $\displaystyle\sum_{n=N+1}^{\infty} \log |E_\mu(z/a_n)| \le M \sum_{n=N+1}^{\infty} \left|\frac{z}{a_n}\right|^{\mu+1} \le \frac{\alpha}{4}|z|^{\mu+1}.$

Now notice that in the derivation of (2.8), A could be chosen as small as desired by taking R sufficiently large. So choose $r_1 > 0$ such that

$$\log |E_\mu(z)| \le \frac{\alpha}{4N}|z|^{\mu+1}, \text{ for } |z| > r_1.$$

If $r_2 = \max \{|a_1| r_1, |a_2| r_1, \dots, |a_N| r_1\}$ then

$$\sum_{n=1}^{N} \log |E_\mu(z/a_n)| \le \frac{\alpha}{4}|z|^{\mu+1} \text{ for } |z| > r_2.$$

Combining this with (2.11) gives that

2.12 $\displaystyle\log |P(z)| = \sum_{n=1}^{\infty} \log |E_\mu(z/a_n)| < \frac{\alpha}{2}|z|^{\mu+1}$

for $|z| > r_2$. Since g is a polynomial of degree $\le \mu$,

$$\lim_{z \to \infty} \frac{m \log |z| + |g(z)|}{|z|^{\mu+1}} = 0.$$

So there is an $r_3 > 0$ such that $m \log |z| + |g(z)| < \frac{1}{2} \alpha |z|^{\mu+1}$. Together with (2.12) this yields

$$\log |f(z)| < \alpha |z|^{\mu+1}$$

for $|z| > r_0 = \max\ \{r_2, r_3\}$. By taking the exponential of both sides, the desired inequality is obtained. ∎

The preceding theorem says that by restricting the rate of growth of the zeros of the entire function $f(z) = z^m \exp g(z)P(z)$ and by requiring that g be a polynomial, then the growth of $M(r) = \max\ \{|f(re^{i\theta})|: 0 \le \theta \le 2\pi\}$ is dominated by $\exp (\alpha|z|^{\mu+1})$ for some μ and any $\alpha > 0$.

We wish to prove the converse to this result.

2.13 Definition. An entire function f is of *finite order* if there is a positive constant a and an $r_0 > 0$ such that $|f(z)| < \exp (|z|^a)$ for $|z| > r_0$. If f is not of finite order then f is of *infinite order*.

If f is of finite order then the number $\lambda = \inf\ \{a: |f(z)| < \exp (|z|^a)$ for $|z|$ sufficiently large$\}$ is called the *order of* f.

Notice that if $|f(z)| < \exp (|z|^a)$ for $|z| > r_a > 1$ and $b > a$ then $|f(z)| < \exp (|z|^b)$. The next proposition is an immediate consequence of this observation.

2.14 Proposition. *Let f be an entire function of finite order λ. If $\epsilon > 0$ then $|f(z)| < \exp (|z|^{\lambda+\epsilon})$ for all z with $|z|$ sufficiently large; and a z can be found, with $|z|$ as large as desired, such that $|f(z)| \ge \exp (|z|^{\lambda-\epsilon})$.*

Although the definition of order seems *a priori* weaker than the conclusion of Theorem 2.6, they are, in fact, equivalent. The reader is asked to show this for himself in Exercise 3.

So it is desirable to know if every function of finite order has finite genus (a converse of Theorem 2.6). That this is in fact the case is a result of Hadamard's Factorization Theorem, proved in the next section.

The proof of the next proposition is left to the reader.

2.15 Proposition. *Let f be an entire function of order λ and let $M(r) = \max \{|f(z)|: |z| = r\}$; then*

$$\lambda = \limsup_{r \to \infty} \frac{\log \log M(r)}{\log r}$$

Consider the function $f(z) = \exp (e^z)$; then $|f(z)| = \exp (\text{Re } e^z) = \exp (e^r \cos \theta)$ if $z = re^{i\theta}$. Hence $M(r) = \exp (e^r)$ and

$$\frac{\log \log M(r)}{\log r} = \frac{r}{\log r} \ ;$$

thus, f is of infinite order. On the other hand if $g(z) = \exp (z^n)$, $n \ge 1$, then $|g(z)| = \exp (\text{Re } z^n) = \exp (r^n \cos n\theta)$. Hence $M(r) = \exp (r^n)$ and so

$$\frac{\log \log M(r)}{\log r} = n;$$

thus g is of order n. For further examples see Exercise 7.

Using this terminology, Theorem 2.6 can be rephrased as follows

2.16 Corollary. *If f is an entire function of finite genus μ then f is of finite order $\lambda \le \mu + 1$.*

Exercises

1. Let $f(z) = \sum c_n z^n$ be an entire function of finite genus μ; prove that

$$\lim_{n \to \infty} c_n (n!)^{1/(\mu+1)} = 0.$$

(Hint: Use Cauchy's Estimate.)

2. Let f_1 and f_2 be entire functions of finite orders λ_1 and λ_2 respectively. Show that $f = f_1 + f_2$ has finite order λ and $\lambda \leq \max (\lambda_1, \lambda_2)$. Show that $\lambda = \max (\lambda_1, \lambda_2)$ if $\lambda_1 \neq \lambda_2$ and give an example which shows that $\lambda < \max (\lambda_1, \lambda_2)$ with $f \neq 0$.

3. Suppose f is an entire function and A, B, α are positive constants such that there is a r_0 with $|f(z)| \leq \exp (A|z|^a + B)$ for $|z| > r_0$. Show that f is of finite order $\leq a$.

4. Prove that if f is an entire function of order λ then f' also has order λ.

5. Let $f(z) = \sum c_n z^n$ be an entire function and define the number α by

$$\alpha = \liminf_{n \to \infty} \frac{-\log |c_n|}{n \log n}$$

(a) Show that if f has finite order then $\alpha > 0$. (Hint: If the order of f is λ and $\beta > \lambda$ show that $|c_n| \leq r^{-n} \exp (r^\beta)$ for sufficiently large r, and find the maximum value of this expression.)

(b) Suppose that $0 < \alpha < \infty$ and show that for any $\epsilon > 0$, $\epsilon < \alpha$, there is an integer p such that $|c_n|^{1/n} < n^{-(\alpha-\epsilon)}$ for $n > p$. Conclude that for $|z| = r > 1$ there is a constant A such that

$$|f(z)| < Ar^p + \sum_{n=1}^{\infty} \left(\frac{r}{n^{\alpha-\epsilon}} \right)^n$$

(c) Let p be as in part (b) and let N be the largest integer $\leq (2r)^{1/(\alpha-\epsilon)}$. Take r sufficiently large so that $N > p$ and show that

$$\sum_{m=N+1}^{\infty} \left(\frac{r}{n^{\alpha-\epsilon}} \right)^n < 1 \text{ and } \sum_{n=p+1}^{N} \left(\frac{r}{n^{\alpha-\epsilon}} \right)^n < B \exp ((2r)^{1/(\alpha-\epsilon)} \log r)$$

where B is a constant which does not depend on r.

(d) Use parts (b) and (c) to show that if $0 < \alpha < \infty$ then f has finite order λ and $\lambda \leq \alpha^{-1}$.

(e) Prove that f is of finite order iff $\alpha > 0$, and if f has order λ then $\lambda = \alpha^{-1}$.

6. Find the order of each of the following functions: (a) $\sin z$; (b) $\cos z$; (c) $\cosh \sqrt{z}$; (d) $\sum_{n=1}^{\infty} n^{-an} z^n$ where $a > 0$. (Hint: For part (d) use Exercise 6.)

7. Let f_1 and f_2 be entire functions of finite order λ_1, λ_2; show that $f = f_1 f_2$ has finite order $\lambda \leq \max (\lambda_1, \lambda_2)$.

8. Let $\{a_n\}$ be a sequence of non-zero complex numbers. Let $\rho = \inf \{a: \sum |a_n|^{-a} < \infty\}$; the number ρ is called the *exponent of convergence* of $\{a_n\}$.

(a) If f is an entire function of rank p then the exponent of convergence ρ of the non-zero zeros of f satisfies: $p \le \rho \le p+1$.

(b) If $\rho' =$ the exponent of convergence of $\{a_n\}$ then for every $\epsilon > 0$, $\sum |a_n|^{-(\rho+\epsilon)} < \infty$ and $\sum |a_n|^{-(\rho-\epsilon)} = \infty$.

(c) Let f be an entire function of order λ and let $\{a_1, a_2, \ldots\}$ be the non-zero zeros of f counted according to multiplicity. If ρ is the exponent of convergence of $\{a_n\}$ prove that $\rho \le \lambda$. (Hint: See the proof of (3.5) in the next section.)

(d) Let $P(z) = \prod_{n=1}^{\infty} E_p(z/a_n)$ be a canonical product of rank p, and let ρ be the exponent of convergence of $\{a_n\}$. Prove that the order of P is ρ. (Hint: If λ is the order of P, $\rho \le \lambda$; assume that $|a_1| \le |a_2| \le \ldots$ and fix z, $|z| > 1$. Choose N such that $|a_n| \le 2|z|$ if $n \le N$ and $|a_n| > 2|z|$ if $n \ge N + 1$. Treating the cases $\rho < p+1$ and $\rho = p+1$ separately, use (2.7) to show that for some $\epsilon \ge 0$.

$$\sum_{n=N+1}^{\infty} \log \left| E_p\left(\frac{z}{a_n}\right) \right| < A |z|^{\rho+\epsilon}.$$

Prove that for $|z| \ge \frac{1}{2}$, $\log |E_p(z)| < B |z|^p$ where B is a constant independent of z. Use this to prove that

$$\sum_{n=1}^{N} \log \left| E_p\left(\frac{z}{a_n}\right) \right| < C |z|^{\rho+\epsilon}$$

for some constant C independent of z.)

9. Find the order of the following entire functions:

(a) $f(z) = \displaystyle\prod_{n=1}^{\infty} (1 - a^n z), \quad 0 < |a| < 1$

(b) $f(z) = \displaystyle\prod_{n=1}^{\infty} \left(1 - \frac{z}{n!}\right)$

(c) $f(z) = [\Gamma(z)]^{-1}$.

§3. *Hadamard Factorization Theorem*

In this section the converse of Corollary 2.16 is proved; that is each function of finite order has finite genus. Since a function of finite genus can be factored in a particularly pleasing way this gives a factorization theorem.

3.1 Lemma. *Let f be a non-constant entire function of order λ with $f(0) = 1$, and let $\{a_1, a_2, \ldots\}$ be the zeros of f counted according to multiplicity and arranged so that $|a_1| \le |a_2| \le \ldots$. If an integer $p > \lambda - 1$ then*

$$\frac{d^p}{dz^p}\left[\frac{f'(z)}{f(z)}\right] = -p! \sum_{n=1}^{\infty} \frac{1}{(a_n - z)^{p+1}}$$

for $z \ne a_1, a_2, \ldots$.

Proof. Let $n = n(r) = $ the number of zeros of f in $B(0; r)$; according to the Poisson-Jensen formula

$$\log|f(z)| = -\sum_{k=1}^{n} \log\left|\frac{r^2 - \bar{a}_k z}{r(z - a_k)}\right| + \frac{1}{2\pi}\int_0^{2\pi} \mathrm{Re}\left(\frac{re^{i\theta} + z}{re^{i\theta} - z}\right)\log|f(re^{i\theta})|\, d\theta$$

for $|z| < r$. Using Exercise 1 and Leibniz's rule for differentiating under an integral sign this gives

$$\frac{f'(z)}{f(z)} = \sum_{k=1}^{n}(z - a_k)^{-1} + \sum_{k=1}^{n}\bar{a}_k(r^2 - \bar{a}_k z)^{-1}$$

$$+ \frac{1}{2\pi}\int_0^{2\pi} 2re^{i\theta}(re^{i\theta} - z)^{-2}\log|f(re^{i\theta})|\, d\theta$$

for $|z| < r$ and $z \neq a_1, \ldots, a_n$. Differentiating p times yields:

$$\textbf{3.2}\qquad \frac{d^p}{dz^p}\left[\frac{f'(z)}{f(z)}\right] = -p!\sum_{k=1}^{n}(a_k - z)^{-p-1} + p!\sum_{k=1}^{n}\bar{a}_k^{p+1}(r^2 - \bar{a}_k z)^{-p-1}$$

$$+ (p+1)!\,\frac{1}{2\pi}\int_0^{2\pi} 2re^{i\theta}(re^{i\theta} - z)^{-p-2}\log|f(re^{i\theta})|\, d\theta.$$

Now as $r \to \infty$, $n(r) \to \infty$ so that the result will follow if it can be shown that the last two summands in (3.2) tend to zero as $r \to \infty$.

To see that the second sum converges to zero let $r > 2|z|$; then $|a_k| \le r$ gives $|r^2 - \bar{a}_k z| \ge \frac{1}{2}r^2$ so that $(|\bar{a}_k|\, |r^2 - \bar{a}_k z|^{-1})^{p+1} \le (2/r)^{p+1}$. Hence the second summand is dominated by $n(r)\,(2/r)^{p+1}$. But it is an easy consequence of Jensen's Formula (see Exercise 1.2) that $\log 2\,n(r) \le \log M(r)$. Since f is of order λ, for any $\epsilon > 0$ and r sufficiently large

$$\log 2\,n(r)\,r^{-(p+1)} \le \log[M(r)]\,r^{-(p+1)}$$

$$\le r^{(\lambda+\epsilon)-(p+1)}$$

But $p+1 > \lambda$ so that ϵ may be chosen with $(\lambda+\epsilon)-(p+1) < 0$. Hence $n(r)\,(2/r)^{p+1} \to 0$ as $r \to \infty$; that is, the second summand in (3.2) converges to zero.

To show that the integral in (3.2) converges to zero notice that

$$\int_0^{2\pi} re^{i\theta}(re^{i\theta} - z)^{-p-2}\, d\theta = 0$$

since this integral is a multiple of the integral of $(w - z)^{-p-2}$ around the circle $|w| = r$ and this function has a primitive. So the value of the integral in (3.2) remains unchanged if we substitute $\log|f| - \log M(r)$ for $\log|f|$.

So for $2|z| < r$, the absolute value of the integral in (3.2) is dominated by

3.3 $$(p+1)!\, 2^{p+3} r^{-(p+1)} \frac{1}{2\pi} \int_0^{2\pi} [\log M(r) - \log |f(re^{i\theta})|]\, d\theta.$$

But according to Jensen's Formula,

$$\frac{1}{2\pi} \int_0^{2\pi} \log |f(re^{i\theta})|\, d\theta \geq 0$$

since $f(0) = 1$. Also $\log M(r) \leq r^{\lambda+\epsilon}$ for sufficiently large r so that (3.3) is dominated by

$$(p+1)!\, 2^{p+3} r^{\lambda+\epsilon-(p+1)}$$

As before, ϵ can be chosen so that $r^{\lambda+\epsilon-(p+1)} \to 0$ as $r \to \infty$. ∎

Note that the preceding lemma implicitly assumes that f has infinitely many zeros. However, if f has only a finite number of zeros then the sum in Lemma 3.1 becomes a finite sum and the lemma remains valid.

3.4 Hadamard's Factorization Theorem. *If f is an entire function of finite order λ then f has finite genus $\mu \leq \lambda$.*

Proof. Let p be the largest integer less than or equal to λ; so $p \leq \lambda < p+1$. The first step in the proof is to show that f has finite rank and that the rank is not larger than p. So let $\{a_1, a_2, \ldots\}$ be the zeros of f counted according to multiplicity and arranged such that $|a_1| \leq |a_2| \leq \ldots$. It must be shown that

3.5 $$\sum_{n=1}^{\infty} |a_n|^{-(p+1)} < \infty.$$

There is no loss in generality in assuming that $f(0) = 1$. Indeed, if f has a zero at the origin of multiplicity m and $M(r) = \max \{|f(z)|: |z| = r\}$ then for any $\epsilon > 0$ and $|z| = r$

$$\log |f(z)z^{-m}| \leq \log [M(r)r^{-m}]$$
$$\leq r^{\lambda+\epsilon} - m \log r$$
$$\leq r^{\lambda+2\epsilon}$$

if r is sufficiently large. So $f(z)z^{-m}$ is an entire function of order λ with no zero at the origin. Since multiplication by a scalar does not affect the order, the assumption that $f(0) = 1$ is justified.

Let $n(r) =$ the number of zeros of f in $B(0; r)$. It follows (Exercise 1.2) that $[\log 2]\, n(r) \leq \log M(r)$. Since f has order λ, $\log M(r) \leq r^{\lambda+\frac{1}{2}\epsilon}$ for any $\epsilon > 0$ so that $\lim_{r \to \infty} n(r)r^{-(\lambda+\epsilon)} = 0$. Hence $n(r) \leq r^{\lambda+\epsilon}$ for sufficiently large r. Since $|a_1| \leq |a_2| \leq \ldots$, $k \leq n(|a_k|) \leq |a_k|^{\lambda+\epsilon}$ for all k larger than some integer k_0. Hence,

$$|a_k|^{-(p+1)} \leq k^{-(p+1/\lambda+\epsilon)}$$

for $k > k_0$. So if ϵ is chosen with $\lambda+\epsilon < p+1$ (recall that $\lambda < p+1$) then $\sum |a_k|^{-(p+1)}$ is dominated by a convergent series; (3.5) now follows.

Let $f(z) = P(z) \exp (g(z))$ where P is a canonical product in standard form. Hence for $z \neq a_k$

$$\frac{f'(z)}{f(z)} = g'(z) + \frac{P'(z)}{P(z)}$$

Using Lemma 3.1 gives that

$$-p! \sum_{n=1}^{\infty} (a_n-z)^{-(p+1)} = g^{(p+1)}(z) + \frac{d^p}{dz^p}\left[\frac{P'(z)}{P(z)}\right]$$

However it is easy to show that

$$\frac{d^p}{dz^p}\left[\frac{P'(z)}{P(z)}\right] = -p! \sum_{n=1}^{\infty} (a_n-z)^{-(p+1)}$$

for $z \neq a_1, a_2, \ldots$. Hence $g^{(p+1)} \equiv 0$ and g must be a polynomial of degree $\leq p$. So the genus of $f \leq p \leq \lambda$. ∎

As an application of Hadamard's Theorem a special case of Picard's Theorem can be proved. This theorem is proved in full generality in the next chapter.

3.6 Theorem. *Let f be an entire function of finite order, then f assumes each complex number with one possible exception.*

Proof. Suppose there are complex numbers α and β, $\alpha \neq \beta$, such that $f(z) \neq \alpha$ and $f(z) \neq \beta$ for all z in $\mathbb{C}$. So $f-\alpha$ is an entire function that never vanishes; hence there is an entire function g such that $f(z)-\alpha = \exp g(z)$. Since f has finite order, so does $f-\alpha$; by Hadamard's Theorem g must be a polynomial. But $\exp g(z)$ never assumes the value $\beta-\alpha$ and this means that $g(z)$ never assumes the value $\log (\beta-\alpha)$, a contradiction to the Fundamental Theorem of Algebra. ∎

One might ask how many times f assumes a given value α. If g is a polynomial of degree $n \geq 1$, then every α is assumed exactly n times. However $f = e^g$ assumes each value (with the exception of zero) an infinite number times. Since the order of e^g is n (see Exercise 2.5) the next result lends some confusion to this problem; the confusion will be alleviated in the next chapter.

3.7 Theorem. *Let f be an entire function of finite order λ where λ is not an integer; then f has infinitely many zeros.*

Proof. Suppose f has only a finite number of zeros $\{a_1, a_2, \ldots, a_n\}$ counted according to multiplicity. Then $f(z) = e^{g(z)}(z-a_1)\ldots(z-a_n)$ for an entire function g. By Hadamard's Theorem, g is a polynomial of degree $\leq \lambda$. But it is easy to see that f and e^g have the same order. Since the order of e^g is the degree of g, λ must be an integer. This completes the proof. ∎

3.8 Corollary. *If f is an entire function of order λ and λ is not an integer then f assumes each complex value an infinite number of times.*

Proof. If $\alpha \in f(\mathbb{C})$, apply the preceding theorem to $f - \alpha$. ∎

Exercises

1. Let f be analytic in a region G and suppose that f is not identically zero. Let $G_0 = G - \{z: f(z) = 0\}$ and define $h: G_0 \to \mathbb{R}$ by $h(z) = \log |f(z)|$. Show that $\dfrac{\partial h}{\partial x} - i \dfrac{\partial h}{\partial y} = \dfrac{f'}{f}$ on G_0.

2. Refer to Exercise 2.8 and show that if $\lambda_1 \neq \lambda_2$ then $\lambda = \max (\lambda_1, \lambda_2)$.

3. (a) Let f and g be entire functions of finite order λ and suppose that $f(a_n) = g(a_n)$ for a sequence $\{a_n\}$ such that $\sum |a_n|^{-(\lambda+1)} = \infty$. Show that $f = g$.

(b) Use Exercise 2.9 to show that if f, g and $\{a_n\}$ are as in part (a) with $\sum |a_n|^{-(\lambda+\epsilon)} = \infty$ for some $\epsilon > 0$ then $f = g$.

(c) Find all entire functions f of finite order such that $f(\log n) = n$.

(d) Give an example of an entire function with zeros $\{\log 2, \log 3, \ldots\}$

Chapter XII

The Range of an Analytic Function

In this chapter the range of an analytic function is investigated. A generic problem of this type is the following: Let $\mathscr{F}$ be a family of analytic functions on a region G which satisfy some property P. What can be said about $f(G)$ for each f in $\mathscr{F}$? Are the sets $f(G)$ uniformly big in some sense? Does there exist a ball $B(a; r)$ such that $f(G) \supset B(a; r)$ for each f in $\mathscr{F}$? Needless to say, the answers to such questions depend on the property P that is used to define $\mathscr{F}$.

In fact there are a few theorems of this type that have already been encountered. For example, the Casorati-Weierstrass Theorem says that if $G = \{z: 0 < |z-a| < r\}$ and $\mathscr{F}$ is the set of analytic functions on G with an essential singularity at $z = a$, then for each δ, $0 < \delta < r$, and each f in $\mathscr{F}$ $f(\text{ann }(a; 0; \delta))$ is dense in $\mathbb{C}$ (V. 1.21). Recall (Exercise V. 1.13) that if f is entire and $f(1/z)$ has a pole at $z = 0$, then f is a polynomial. So if f is not a polynomial then $f(1/z)$ has an essential singularity at $z = 0$. So as a corollary to the Casorati-Weierstrass Theorem, $f(\mathbb{C})$ is dense in $\mathbb{C}$ for each entire function (if f is a polynomial then $f(\mathbb{C}) = \mathbb{C}$).

This chapter will culminate in the Great Picard Theorem that substantially improves the Casorati-Weierstrass Theorem. Indeed, it states that if f has an essential singularity at $z = a$ then $f(\text{ann }(a; 0; \delta))$ is equal to the entire plane with possibly one point deleted. Moreover, f assumes each of the values in this punctured disk an infinite number of times. (See Exercise V. 1.10.) As above, this yields that $f(\mathbb{C})$ is also the whole plane, with one possible point deleted, whenever f is an entire function. This is known as the Little Picard Theorem. However, this latter result will be obtained independently.

Before these theorems of Picard are proved, it is necessary to obtain further results about the range of an analytic function—which results are of interest in themselves.

§1. Bloch's Theorem

To fit the result referred to in the title of this section into the general questions posed in the introduction, let $D = B(0; 1)$ and let $\mathscr{F}$ be the family of all functions f analytic on a region containing D^- such that $f(0) = 0$ and $f'(0) = 1$. How "big" can $f(D)$ be? Put another way: because $f'(0) = 1 \neq 0$, f is not constant and so $f(D)$ is open. That is, $f(D)$ must contain a disk of positive radius. As a consequence of Bloch's Theorem, there is a positive constant B such that $f(G)$ contains a disk of radius B for each f in $\mathscr{F}$.

1.1 Lemma. *Let f be analytic in $D = \{z\colon |z| < 1\}$ and suppose that $f(0) = 0$, $f'(0) = 1$, and $|f(z)| \le M$ for all z in D. Then $M \ge 1$ and $f(D) \supset B(0; 1/6M)$.*

Proof. Let $0 < r < 1$ and $f(z) = z + a_2 z^2 + \ldots$; according to Cauchy's Estimate $|a_n| \le M/r^n$ for $n \ge 1$. So $1 = |a_1| \le M$. If $|z| = (4M)^{-1}$ then

$$|f(z)| \ge |z| - \sum_{n=2}^{\infty} |a_n z^n|$$

$$\ge (4M)^{-1} - \sum_{n=2}^{\infty} M(4M)^{-n}$$

$$= (4M)^{-1} - (16M - 4)^{-1}$$

$$\ge (6M)^{-1}$$

since $M \ge 1$.

Suppose $|w| < (6M)^{-1}$; it will be shown that $g(z) = f(z) - w$ has a zero. In fact, for $|z| = (4M)^{-1}$, $|f(z) - g(z)| = |w| < (6M)^{-1} \le |f(z)|$. So, by Rouché's Theorem, f and g have the same number of zeros in $B(0; 1/4M)$. Since $f(0) = 0$, $g(z_0) = 0$ for some z_0; hence $f(D) \supset B(0; 1/6M)$. ∎

1.2 Lemma. *Suppose g is analytic on $B(0; R)$, $g(0) = 0$, $|g'(0)| = \mu > 0$, and $|g(z)| \le M$ for all z, then*

$$g(B(0; R)) \supset B\left(0; \frac{R^2 \mu^2}{6M}\right)$$

Proof. Let $f(z) = [Rg'(0)]^{-1} g(Rz)$ for $|z| < 1$; then f is analytic on $D = \{z\colon |z| < 1\}$, $f(0) = 0$, $f'(0) = 1$, and $|f(z)| \le M/\mu R$ for all z in D. According to the preceding lemma, $f(D) \supset B(0; \mu R/6M)$. If this is translated in terms of the original function g, the lemma is proved. ∎

1.3 Lemma. *Let f be an analytic function on the disk $B(a; r)$ such that $|f'(z) - f'(a)| < |f'(a)|$ for all z in $B(a; r)$, $z \ne a$; then f is one-one.*

Proof. Suppose z_1 and z_2 are points in $B(a; r)$ and $z_1 \ne z_2$. If γ is the line segment $[z_1, z_2]$ then an application of the triangle inequality yields

$$|f(z_1) - f(z_2)| = \left| \int_\gamma f'(z)\, dz \right|$$

$$\ge \left| \int_\gamma f'(a)\, dz \right| - \left| \int_\gamma [f'(z) - f'(a)]\, dz \right|$$

$$\ge |f'(a)|\, |z_1 - z_2| - \int_\gamma |f'(z) - f'(a)|\, |dz|.$$

Using the hypothesis, this gives $|f(z_1) - f(z_2)| > 0$ so that $f(z_1) \ne f(z_2)$ and f is one-one. ∎

1.4 Bloch's Theorem. *Let f be an analytic function on a region containing the closure of the disk $D = \{z\colon |z| < 1\}$ and satisfying $f(0) = 0$, $f'(0) = 1$. Then there is a disk $S \subset D$ on which f is one-one and such that $f(S)$ contains a disk of radius $1/72$.*

Proof. Let $K(r) = \max \{|f'(z)|: |z| = r\}$ and let $h(r) = (1-r)K(r)$. It is easy to see that $h: [0, 1] \to \mathbb{R}$ is continuous, $h(0) = 1$, $h(1) = 0$. Let $r_0 = \sup \{r: h(r) = 1\}$; then $h(r_0) = 1$, $r_0 < 1$, and $h(r) < 1$ if $r > r_0$ (Why?). Let a be chosen with $|a| = r_0$ and $|f'(a)| = K(r_0)$; then

1.5
$$|f'(a)| = (1-r_0)^{-1}.$$

Now if $|z-a| < \frac{1}{2}(1-r_0) = \rho_0$, $|z| < \frac{1}{2}(1+r_0)$; since $r_0 < \frac{1}{2}(1+r_0)$, the definition of r_0 gives

1.6
$$\begin{aligned}
|f'(z)| &\le K(\tfrac{1}{2}(1+r_0)) \\
&= h(\tfrac{1}{2}(1+r_0))\,[1-\tfrac{1}{2}(1+r_0)]^{-1} \\
&< [1-\tfrac{1}{2}(1+r_0)]^{-1} \\
&= 1/\rho_0
\end{aligned}$$

for $|z-a| < \rho_0$. Combining (1.5) and (1.6) gives
$$\begin{aligned}
|f'(z)-f'(a)| &\le |f'(z)| + |f'(a)| \\
&< 3/2\rho_0.
\end{aligned}$$

According to Schwarz's Lemma, this implies that
$$|f'(z)-f'(a)| < \frac{3|z-a|}{2\rho_0^2}$$

for z in $B(a; \rho_0)$. Hence if $z \in S = B(a; \tfrac{1}{3}\rho_0)$,
$$|f'(z)-f'(a)| < \frac{1}{2\rho_0} = |f'(a)|$$

By Lemma 1.4, f is one-one on S.

It remains to show that $f(S)$ contains a disk of radius $1/72$. For this define $g: B(0; \tfrac{1}{3}\rho_0) \to \mathbb{C}$ by $g(z) = f(z+a)-f(a)$ then $g(0) = 0$, $|g'(0)| = |f'(a)| = (2\rho_0)^{-1}$. If $z \in B(0; \tfrac{1}{3}\rho_0)$ then the line segment $\gamma = [a, z+a]$ lies in $S \subset B(a; \rho_0)$. So by (1.6)
$$\begin{aligned}
|g(z)| &= \left|\int_\gamma f'(w)\,dw\right| \\
&\le \frac{1}{\rho_0}|z| \\
&< \tfrac{1}{3}.
\end{aligned}$$

Applying Lemma 1.2 gives that
$$g(B(0; \tfrac{1}{3}\rho_0)) \supset B(0; \sigma)$$

where
$$\sigma = \frac{\left(\dfrac{1}{3}\rho_0\right)^2\left(\dfrac{1}{2\rho_0}\right)^2}{6\left(\dfrac{1}{3}\right)} = \frac{1}{72}$$

If this is translated into a statement about f, it yields that

$$f(S) \supset B\left(f(a); \frac{1}{72}\right). \blacksquare$$

1.7 Corollary. *Let f be an analytic function on a region containing $\bar{B}(0; R)$; then $f(B(0; R))$ contains a disk of radius $\frac{1}{72} R|f'(0)|$.*

Proof. Apply Bloch's theorem to the function $g(z) = [f(Rz) - f(0)]/Rf'(0)$ (the result is trivial if $f'(0) = 0$, so it may be assumed that $f'(0) \neq 0$). $\blacksquare$

1.8 Definition. Let $\mathscr{F}$ be the set of all functions f analytic on a region containing the closure of the disk $D = \{z: |z| < 1\}$ and satisfying $f(0) = 0$, $f'(0) = 1$. For each f in $\mathscr{F}$ let $\beta(f)$ be the supremum of all numbers r such that there is a disk S in D on which f is one-one and such that $f(S)$ contains a disk of radius r. $\left(\text{So } \beta(f) \geq \frac{1}{72}\right)$. *Bloch's constant* is the number B defined by

$$B = \inf \{\beta(f): f \in \mathscr{F}\}.$$

According to Bloch's Theorem, $B \geq \frac{1}{72}$. If one considers the function $f(z) = z$ then clearly $B \leq 1$. However, better estimates than these are known. In fact, it is known that $.43 \leq B \leq .47$. Although the exact value of B remains unknown, it has been conjectured that

$$B = \frac{\Gamma\left(\frac{1}{3}\right)\Gamma\left(\frac{11}{12}\right)}{(1 + \sqrt{3})^{\frac{1}{2}}\Gamma\left(\frac{1}{4}\right)}$$

A related constant is defined as follows.

1.9 Definition. Let $\mathscr{F}$ be as in Definition 1.8. For each f in $\mathscr{F}$ define $\lambda(f) = \sup \{r: f(D)$ contains a disk of radius $r\}$. *Landau's constant* L is defined by

$$L = \inf \{\lambda(f): f \in \mathscr{F}\}.$$

Clearly $L \geq B$ and it is easy to see that $L \leq 1$. Again the exact value of L is unknown but it can be proved that $.50 \leq L \leq .56$. In particular, $L > B$.

1.10 Proposition. *If f is analytic on a region containing the closure of the disk $D = \{a: |z| < 1\}$ and $f(0) = 0, f'(0) = 1$; then $f(D)$ contains a disk of radius L.*

Proof. The proof will be accomplished by showing that $f(D)$ contains a disk of radius $\lambda = \lambda(f)$. For each n there is a point α_n in $f(D)$ such that $B\left(\alpha_n; \lambda - \frac{1}{n}\right) \subset f(D)$. Now $\alpha_n \in f(D) \subset f(D^-)$ and this last set is compact. So there is a point α in $f(D^-)$ and a subsequence $\{\alpha_{n_k}\}$ such that $\alpha_{n_k} \to \alpha$. It is easy to see that we may assume that $\alpha = \lim \alpha_n$. If $|w-\alpha| < \lambda$, choose n_0 such that $|w-\alpha| < \lambda - 1/n_0$. There is an integer $n_1 > n_0$ such that

$$|\alpha_n - \alpha| < \lambda - \frac{1}{n_0} - |w-a|$$

for $n \geq n_1$. Hence

$$|w-\alpha_n| \leq |w-\alpha| + |\alpha-\alpha_n|$$

$$< \lambda - \frac{1}{n_0}$$

$$< \lambda - \frac{1}{n}$$

if $n \geq n_1$. That is $w \in B(\alpha_n; \lambda - 1/n) \subset f(D)$. Since w was arbitrary it follows that $B(\alpha; \lambda) \subset f(D)$. ∎

1.11 Corollary. *Let f be analytic on a region that contains $\bar{B}(0; R)$; then $f(B(0; R))$ contains a disk of radius $R|f'(0)|L$.*

Exercises

1. Examine the proof of Bloch's Theorem to prove that $L \geq 1/24$.
2. Suppose that in the statement of Bloch's Theorem it is only assumed that f is analytic on D. What conclusion can be drawn? (Hint: Consider the functions $f_s(z) = s^{-1}f(sz)$, $0 < s < 1$.) Do the same for Proposition 1.10.

§2. *The Little Picard Theorem*

The principal result of this section generalizes Theorem XI. 3.6. However, before proceeding, a lemma is necessary.

2.1 Lemma. *Let G be a simply connected region and suppose that f is an analytic function on G that does not assume the values 0 or 1. Then there is an analytic function g on G such that*

$$f(z) = -\exp(i\pi \cosh[2g(z)])$$

for z in G.

Proof. Since f never vanishes there is a branch ℓ of $\log f(z)$ defined on G; that is $e^\ell = f$. Let $F(z) = (2\pi i)^{-1}\ell(z)$; if $F(a) = n$ for some integer n then $f(a) = \exp(2\pi i n) = 1$, which cannot happen. Hence F does not assume any integer values. Since F cannot assume the values 0 and 1, it is possible to define

$$H(z) = \sqrt{F(z)} - \sqrt{F(z)-1}.$$

Now $H(z) \neq 0$ for any z so that it is possible to define a branch of g of $\log H$ on G. Hence $\cosh(2g)+1 = \frac{1}{2}(e^{2g}+e^{-2g})+1 = \frac{1}{2}(e^g+e^{-g})^2 =$

$\frac{1}{2}(H+1/H)^2 = 2F = \frac{1}{\pi i}\ell.$ But this gives $f = e^\ell = \exp[\pi i + \pi i \cosh(2g)] =$ $-\exp[\pi i \cosh(2g)].$ ∎

Suppose f and g are as in the lemma, n is a positive integer, and m is any integer. If there is a point a in G with $g(a) = \pm \log(\sqrt{n}+\sqrt{n-1})+\frac{1}{2}im\pi$, then $2 \cosh[2g(a)] = e^{2g(a)}+e^{-2g(a)} = e^{im\pi}(\sqrt{n}+\sqrt{n-1})^{\pm 2}+e^{-im\pi}(\sqrt{n}+\sqrt{n-1})^{\pm 2} = (-1)^m[(\sqrt{n}+\sqrt{n-1})^2+(\sqrt{n}-\sqrt{n-1})^2] = (-1)^m[2(2n-1)];$ or $\cosh[2g(a)] = (-1)^m(2n-1)$. Therefore $f(a) = -\exp[(-1)^m(2n-1)\pi i]$ and, since $(2n-1)$ must be odd, $f(a) = 1$. Hence g cannot assume any of the values

$$\{\pm\log(\sqrt{n}+\sqrt{n-1})+\tfrac{1}{2}im\pi: n \geq 1, m = 0, \pm 1, \ldots\}.$$

These points form the vertices of a grid of rectangles in the plane. The height of an arbitrary rectangle is

$$|\tfrac{1}{2}im\pi - \tfrac{1}{2}i(m+1)\pi| = \tfrac{1}{2}\pi < \sqrt{3}.$$

The width is $\log(\sqrt{n+1}+\sqrt{n})-\log(\sqrt{n}+\sqrt{n-1}) > 0$. Now $\varphi(x) = \log(\sqrt{x+1}+\sqrt{x})-\log(\sqrt{x}+\sqrt{x-1})$ is a decreasing function so that the width of any rectangle $\leq\varphi(1) = \log(1+\sqrt{2}) < \log e = 1$. So the diagonal of the rectangle < 2. This gives the following.

2.2 Lemma. *Let G, f, and g be as in Lemma 2.1. Then $g(G)$ contains no disk of radius 1.*

2.3 Little Picard Theorem. *If f is an entire function that omits two values then f is a constant.*

Proof. If $f(z) \neq a$ and $f(z) \neq b$ for all z then $(f-a)(b-a)^{-1}$ omits the values 0 and 1. So assume that $f(z) \neq 0$ and $f(z) \neq 1$ for all z. According to Lemma 2.2, this gives an entire function g such that $g(\mathbb{C})$ contains no disk of radius 1. Moreover, if f is not a constant function then g is not constant so there is a point z_0 with $g'(z_0) \neq 0$. By considering $g(z+z_0)$ if necessary, it may be supposed that $g'(0) \neq 0$. But according to Corollary 1.11, $g(B(0; R))$ contains a disk of radius $LR|g'(0)|$. If R is chosen sufficiently large this gives that $g(\mathbb{C})$ does contain a disk of radius 1—a contradiction. So f must be constant. ∎

Exercises

1. Show that if f is a meromorphic function on $\mathbb{C}$ such that $\mathbb{C}_\infty - f(\mathbb{C})$ has at least three points then f is a constant. (Hint: What if $\infty \notin f(\mathbb{C})$?)
2. For each integer $n \geq 1$ determine all meromorphic functions f and g on $\mathbb{C}$ such that $f^n + g^n = 1$.

§3. Schottky's Theorem

Let f be a function defined on a simply connected region containing the disk $\bar{B}(0; 1)$ and suppose that f never assumes the values 0 and 1. Let us examine the proof of Lemma 2.1. If ℓ is any branch of $\log f$ let

$$F = \frac{1}{2\pi i}\,\ell,$$

$$H = \sqrt{F} - \sqrt{F-1},$$

$$g = \text{a branch of log } H.$$

There are two places in this scheme where we are allowed a certain amount of latitude; namely, in picking the functions ℓ and g which are branches of $\log f$ and $\log H$, respectively. For the proof of Schottky's Theorem below specify these branches by requiring

3.1 $$0 \le \text{Im } \ell(0) < 2\pi,$$

3.2 $$0 \le \text{Im } g(0) < 2\pi.$$

3.3 Schottky's Theorem. *For each α and β, $0 < \alpha < \infty$ and $0 \le \beta \le 1$, there is a constant $C(\alpha,\beta)$ such that if f is an analytic function on some simply connected region containing $\bar{B}(0;1)$ that omits the values 0 and 1, and such that $|f(0)| \le \alpha$; then $|f(z)| \le C(\alpha,\beta)$ for $|z| \le \beta$.*

Proof. It is only necessary to prove this theorem for $2 \le \alpha < \infty$. The proof is accomplished by looking at two cases.

Case 1. Suppose $\frac{1}{2} \le |f(0)| \le \alpha$. Recalling the functions F, H, and g in Lemma 2.1 (and rediscussed at the beginning of this section), (3.1) gives

$$|F(0)| = \frac{1}{2\pi}\left|\log|f(0)| + i\,\text{Im }\ell(0)\right|$$

$$\le \frac{1}{2\pi}\log\alpha + 1;$$

Let $C_0(\alpha) = \frac{1}{2\pi}\log\alpha + 1$. Also

3.4 $$\left|\sqrt{F(0)} \pm \sqrt{F(0)-1}\right| \le \left|\sqrt{F(0)}\right| + \left|\sqrt{F(0)-1}\right|$$

$$= \exp\left(\tfrac{1}{2}\log|f(0)|\right) + \exp\left(\tfrac{1}{2}\log|F(0)-1|\right)$$

$$= |F(0)|^{\frac{1}{2}} + |F(0)-1|^{\frac{1}{2}}$$

$$\le C_0(\alpha)^{\frac{1}{2}} + [C_0(\alpha)+1]^{\frac{1}{2}}$$

Let $C_1(\alpha) = C_0(\alpha)^{\frac{1}{2}} + [C_0(\alpha)+1]^{\frac{1}{2}}$. Now if $|H(0)| \ge 1$ then (3.2) and (3.4) give

$$|g(0)| = \left|\log|H(0)| + i\,\text{Im }g(0)\right|$$

$$\le \log|H(0)| + 2\pi$$

$$\le \log C_1(\alpha) + 2\pi.$$

If $|H(0)| < 1$ then in a similar fashion,

$$|g(0)| \le -\log|H(0)| + 2\pi$$

$$= \log\left(\frac{1}{|H(0)|}\right) + 2\pi$$

$$= \log\left|\sqrt{F(0)} + \sqrt{F(0)-1}\right| + 2\pi$$

$$\le \log C_1(\alpha) + 2\pi.$$

Let $C_2(\alpha) = \log C_1(\alpha) + 2\pi$.

If $|a| < 1$ then Corollary 1.11 implies that $g(B(a; 1 - |a|))$ contains a disk of radius

3.5 $$L(1 - |a|) |g'(a)|.$$

On the other hand, Lemma 2.2 says that $g(B(0; 1))$ contains no disk of radius 1. Hence, the expression (3.5) must be less than 1; that is,

3.6 $$|g'(a)| < [L(1 - |a|)]^{-1} \text{ for } |a| < 1.$$

If $|a| < 1$, let γ be the line segment $[0, a]$; then

$$|g(a)| \leq |g(0)| + |g(a) - g(0)|$$

$$\leq C_2(\alpha) + \left| \int_\gamma g'(z) \, dz \right|$$

$$\leq C_2(\alpha) + |a| \max \{|g'(z)| : z \in [0, a]\}$$

Using (3.6) this gives

$$|g(a)| \leq C_2(\alpha) + \frac{|a|}{L(1 - |a|)}$$

If $C_3(\alpha, \beta) = C_2(\alpha) + \beta[L(1 - \beta)]^{-1}$ then this gives

$$|g(z)| \leq C_3(\alpha, \beta)$$

if $|z| \leq \beta$. Consequently if $|z| \leq \beta$,

$$|f(z)| = |\exp [\pi i \cosh 2g(z)]|$$

$$\leq \exp [\pi |\cosh 2g(z)|]$$

$$\leq \exp [\pi e^{2|g(z)|}]$$

$$\leq \exp [\pi e^{2C_3(\alpha, \beta)}];$$

define $C_4(\alpha, \beta) = \exp \{\pi \exp [2C_3(\alpha, \beta)]\}$.

Case 2. Suppose $0 < |f(0)| < \frac{1}{2}$. In this case $(1 - f)$ satisfies the conditions of Case 1 so that $|1 - f(z)| \leq C_4(2, \beta)$ if $|z| \leq \beta$. Hence $|f(z)| \leq 1 + C_4(2, \beta)$. If we define

$$C(\alpha, \beta) = \max \{C_4(\alpha, \beta), 1 + C_4(2, \beta)\},$$

the theorem is proved. ∎

3.7 Corollary. *Let f be analytic on a simply connected region containing $\bar{B}(0; R)$ and suppose that f omits the values 0 and 1. If $C(\alpha, \beta)$ is the constant obtained in Schottky's Theorem and $|f(0)| \leq \alpha$ then $|f(z)| \leq C(\alpha, \beta)$ for $|z| \leq \beta R$.*

Proof. Consider the function $f(Rz)$, $|z| \leq 1$. ∎

What Schottky's Theorem (and the Corollary that follows it) says is that a certain family of functions is uniformly bounded on proper subdisks of $B(0; 1)$. By Montel's Theorem, it follows that this family is normal. It is this observation which will be of use in proving the Great Picard Theorem in the next section.

§4. The Great Picard Theorem

The main tool used in the proof of Picard's Theorem is the following result.

4.1 Montel-Caratheodory Theorem. *If $\mathscr{F}$ is the family of all analytic functions on a region G that do not assume the values 0 and 1, then $\mathscr{F}$ is normal in $C(G, \mathbb{C}_\infty)$*

Proof. Fix a point z_0 in G and define the families $\mathscr{G}$ and $\mathscr{H}$ by

$$\mathscr{G} = \{f \in \mathscr{F} : |f(z_0)| \leq 1\},$$

$$\mathscr{H} = \{f \in \mathscr{F} : |f(z_0)| \geq 1\};$$

so $\mathscr{F} = \mathscr{G} \cup \mathscr{H}$. It will be shown that $\mathscr{G}$ is normal in $H(G)$ and that $\mathscr{H}$ is normal in $C(G, \mathbb{C}_\infty)$ (that ∞ is a limit of a sequence in $\mathscr{H}$ is easily seen by considering constant functions). To show that $\mathscr{G}$ is normal in $H(G)$, Montel's Theorem is invoked; that is, it is sufficient to show that $\mathscr{G}$ is locally bounded.

If a is any point in G let γ be a curve in G from z_0 to a; let $D_0, D_1, \ldots, D_n$ be disks in G with centers $z_0, z_1, \ldots, z_n = a$ on $\{\gamma\}$ and such that z_{k-1} and z_k are in $D_{k-1} \cap D_k$ for $1 \leq k \leq n$. Also assume that $D_k^- \subset G$ for $0 \leq k \leq n$. We now apply Schottky's Theorem to D_0. It follows that there is a constant C_0 such that $|f(z)| \leq C_0$ for z in D_0 and f in $\mathscr{G}$. (If $D_0 = B(z_0; r)$ and $R > r$ is such that $\bar{B}(z_0; R) \subset G$ then, according to Corollary 3.7, $|f(z)| \leq C(1, \beta)$ for z in D_0 and f in $\mathscr{G}$ whenever β is chosen with $r < \beta R$). In particular $|f(z_1)| \leq C_0$ so that Schottky's Theorem gives that $\mathscr{G}$ is uniformly bounded by a constant C_1 on D_1. Continuing, we have that $\mathscr{G}$ is uniformly bounded on D_n. Since a was arbitrary, this gives that $\mathscr{G}$ is locally bounded. By Montel's Theorem, $\mathscr{G}$ is normal in $H(G)$.

Now consider $\mathscr{H} = \{f \in \mathscr{F} : |f(z_0)| \geq 1\}$. If $f \in \mathscr{H}$ then $1/f$ is analytic on G because f never vanishes. Also $1/f$ never vanishes and never assumes the value 1; moreover $|(1/f)(z_0)| \leq 1$. Hence $\tilde{\mathscr{H}} = \{1/f : f \in \mathscr{H}\} \subset \mathscr{G}$ and $\tilde{\mathscr{H}}$ is normal in $H(G)$. So if $\{f_n\}$ is a sequence in $\mathscr{H}$ there is a subsequence $\{f_{n_k}\}$ and an analytic function h on G such that $\{1/f_{n_k}\}$ converges in $H(G)$ to h. According to Corollary VII. 2.6 (Corollary to Hurwitz's Theorem), either $h \equiv 0$ or h never vanishes. If $h \equiv 0$ it is easy to see that $f_{n_k}(z) \to \infty$ uniformly on compact subsets of G. If h never vanishes then $1/h$ is analytic and it follows that $f_{n_k}(z) \to 1/h(z)$ uniformly on compact subsets of G. ∎

4.2 Great Picard Theorem. *Suppose an analytic function f has an essential singularity at $z = a$. Then in each neighborhood of a f assumes each complex number, with one possible exception, an infinite number of times.*

Proof. For the sake of simplicity suppose that f has an essential singularity at $z = 0$. Suppose that there is an R such that there are two numbers not in $\{f(z): 0 < |z| < R\}$; we will obtain a contradiction. Again, we may suppose that $f(z) \neq 0$ and $f(z) \neq 1$ for $0 < |z| < R$. Let $G = B(0; R) - \{0\}$ and define $f_n : G \to \mathbb{C}$ by $f_n(z) = f(z/n)$. So each f_n is analytic and no f_n assumes the value 0 or 1. According to the preceding theorem, $\{f_n\}$ is a normal family in $C(G, \mathbb{C}_\infty)$.

Let $\{f_{n_k}\}$ be a subsequence of $\{f_n\}$ such that $f_{n_k} \to \varphi$ uniformly on $\{z: |z| = \frac{1}{2}R\}$, where φ is either analytic on G or $\varphi \equiv \infty$. If φ is analytic, let $M = \max \{|\varphi(z)|: |z| = \frac{1}{2}R\}$; then $|f(z/n_k)| = |f_{n_k}(z)| \le |f_{n_k}(z) - \varphi(z)| + |\varphi(z)| \le 2M$ for n_k sufficiently large and $|z| = \frac{1}{2}R$. Thus $|f(z)| \le 2M$ for $|z| = R/2n_k$ and for sufficiently large n_k. According to the Maximum Modulus Principle, f is uniformly bounded on concentric annuli about zero. This gives that f is bounded by $2M$ on a deleted neighborhood of zero and, so, $z = 0$ must be a removable singularity. Therefore φ cannot be analytic and must be identically infinite.

It is left to the reader to show that if $\varphi \equiv \infty$ then f must have a pole at zero.

So at most one complex number is never assumed. If there is a complex number w which is assumed only a finite number of times then by taking a sufficiently small disk, we again arrive at a punctured disk in which f fails to assume two values. ∎

An alternate phrasing of this theorem is the following.

4.3 Corollary. *If f has an isolated singularity at $z = a$ and if there are two complex numbers that are not assumed infinitely often by f then $z = a$ is either a pole or a removable singularity.*

In the preceding chapter it was shown that an entire function of order λ, where λ is not an integer, assumes each value infinitely often (Corollary XI.3.8). Functions of the form e^g, for g a polynomial, assume each value infinitely often, although there is one excepted value—namely, zero. The Great Picard Theorem yields a general result along these lines (although an exceptional value is possible), so that the following result is not comparable with Corollary XI.3.8).

4.4 Corollary. *If f is an entire function that is not a polynomial then f assumes every complex number, with one exception, an infinite number of times.*

Proof. Consider the function $g(z) = f(1/z)$. Since f is not a polynomial, g has an essential singularity at $z = 0$ (Exercise V.1.13). The result now follows from the Great Picard Theorem. ∎

Notice that Corollary 4.4 is an improvement of the Little Picard Theorem.

Exercises

1. Let f be analytic in $G = B(0; R) - \{0\}$ and discuss all possible values of the integral

$$\frac{1}{2\pi i} \int_\gamma \frac{f'(z)}{f(z) - a} \, dz$$

where γ is the circle $|z| = r < R$ and a is any complex number. If it is assumed that this integral takes on certain values for certain numbers a, does this imply anything about the nature of the singularity at $z = 0$?

2. Show that if f is a one-one entire function then $f(z) = az + b$ for some constants a and b, $a \ne 0$.

Appendix A

Calculus for Complex Valued Functions on an Interval

In this Appendix we would like to indicate a few results for functions defined on an interval, but whose values are in $\mathbb{C}$ rather than $\mathbb{R}$. If $f\colon [a, b] \to \mathbb{C}$ is a given function then one can easily study its calculus type properties by considering the real valued functions $\text{Re } f$ and $\text{Im } f$. For example, the fact that for a complex number $z = x + iy$

$$\max(|x|, |y|) \leq |z| = \sqrt{x^2 + y^2} \leq 2 \max(|x|, |y|),$$

easily allows us to show that f is continuous iff $\text{Re } f$ and $\text{Im } f$ are continuous. However we sometimes wish to have a property defined and explored directly in terms of f without resorting to the real and imaginary parts of f. This is the case with the derivative of f.

A.1 Definition. A function $f\colon [a, b] \to \mathbb{C}$ is *differentiable* at a point x in (a, b) if the limit

$$\lim_{h \to 0} \frac{f(x+h) - f(x)}{h}$$

exists and is finite. The value of this limit is denoted by $f'(x)$. For the points $x = a$ or b we modify this definition by taking right or left sided limits. If f is differentiable at each point of $[a, b]$ then we say that f is a *differentiable function* and we obtain a new function $f'\colon [a, b] \to \mathbb{C}$ which is called the *derivative* of f.

The next Proposition has a trivial proof which we leave to the reader.

A.2 Proposition. *A function $f\colon [a, b] \to \mathbb{C}$ is differentiable iff $\text{Re } f$ and $\text{Im } f$ are differentiable. Also, $f'(x) = (\text{Re } f)'(x) + i(\text{Im } f)'(x)$ for all x in $[a, b]$.*

Of course it makes no sense to talk of the derivative of a complex valued function being positive; accordingly, the geometrical significance of the derivative of a real valued function has no analogue for complex valued functions. However the reader is invited to play a game by assuming that $\text{Re } f$ and $\text{Im } f$ have positive or negative derivatives, and then interpret these conditions for f.

One fact remains true for derivatives and this is the consequence of a vanishing derivative.

A.3 Proposition. *If a function $f\colon [a, b] \to \mathbb{C}$ is differentiable and $f'(x) = 0$ for all x then f is a constant.*

Proof. If $f'(x) = 0$ for all x then $(\text{Re } f)'(x) = (\text{Im } f)'(x) = 0$ for all x. It follows that $\text{Re } f$ and $\text{Im } f$ are constant and, hence, so is f.

One important theorem about the derivative of a real valued function which is not true for complex valued functions is the Mean Value Theorem. In fact, if $f(x) = x^2 + ix^3$ it is easy to show that

$$f(b) - f(a) = f'(c)\,(b - a)$$

for some point c in $[a, b]$ only when $a = b$.

One of the principal applications of the Mean Value Theorem for derivatives is the proof of the Chain Rule. In light of the discussion in the preceding paragraph one might well doubt the validity of the Chain Rule for complex valued functions. The Chain Rule tells us how to calculate the derivative of the composition of two differentiable functions; this leads to two different situations. First suppose that $f : [a,b] \to \mathbb{C}$ is differentiable and let $g : [c,d] \to [a,b]$ also be differentiable. Then $f(g(t)) = \text{Re}f(g(t)) + i\,\text{Im}f(g(t))$; from here the Chain Rule follows by applying the Chain Rule from Calculus. In the second case the result still holds. Let G be an open subset of $\mathbb{C}$ such that $f([a,b]) \subset G$ and suppose $h : G \to \mathbb{C}$ is analytic. We wish to show that $h \circ f$ is differentiable and calculate $(h \circ f)'$. Since the proof of this Chain Rule follows the line of argument used to prove the Chain Rule for the composition of two analytic functions (Proposition III.2.4) we will not repeat it here. We summarize this discussion in the following.

A.4 Proposition. *Let* $f : [a,\,b] \to \mathbb{C}$ *be a differentiable function.*

(a) *If* $g : [c,\,d] \to [a,\,b]$ *is differentiable then* $f \circ g$ *is differentiable and* $(f \circ g)'(t) = f'(g(t))g'(t)$.

(b) *If* G *is an open subset of* $\mathbb{C}$ *containing* $f([a,\,b])$ *and* $h : G \to \mathbb{C}$ *is an analytic function then* $h \circ f$ *is differentiable and* $(h \circ f)'(x) = h'(f(x))f'(x)$.

To discuss integral calculus for complex valued functions we adopt a somewhat different approach. We define the integral in terms of the real and imaginary parts of the function.

A.5 Definition. If $f : [a,\,b] \to \mathbb{C}$ is a continuous function, we define the integral of f over $[a,\,b]$ by

$$\int_a^b f(x)\,dx = \int_a^b \text{Re}\,f(x)\,dx + i\int_a^b \text{Im}\,f(x)\,dx.$$

If the reader wishes to see a direct development of the integral he need only work through Section IV. 2 of the text with $\gamma(t) = t$ for all t. However, this hardly seems worthwhile.

Besides the additivity of the integral the only result which interests us is the Fundamental Theorem of Calculus.

Recall that if $F : [a,\,b] \to \mathbb{C}$ is a function and $f = F'$ then F is called a *primitive* of f.

A.6 Fundamental Theorem of Calculus. *A continuous function* $f: [a, b] \to \mathbb{C}$
*has a primitive and any two primitives differ by a constant. If F is any primitive
of f then*

$$\int_a^b f(x)\, dx = F(b) - F(a)$$

Proof. If g and h are primitives of $\operatorname{Re} f$ and $\operatorname{Im} f$ then $F = g + ih$ is a primitive
of f. The result now easily follows.

Appendix B

Suggestions for Further Study
and Bibliographical Notes

GENERAL

The theory of analytic functions of one complex variable is a vast one. The books by Ahlfors [1], Carathéodory [9], Fuchs [17], Heins [24], Hille [26], Rudin [41], Saks and Zygmund [42], Sansone and Gerretsen [43], and Veech [46] treat some topics not covered in this book. In addition they contain material which further develops some of the topics discussed here. We have not touched upon the theory of functions of several complex variables. The book by Narasimhan [37] contains an elementary introduction to functions of several complex variables. Also Cartan [10] contains an introduction to the subject. The book by Whittaker and Watson [47] contains some very classical analysis and several bibliographical comments. Finally the two volume work by Pólya and Szegö [39] should be looked at by every student. These books contain problems on analysis with the solutions in the back.

CHAPTER III

§1. A more thorough treatment of power series and infinite series in general can be seen in the book by Knopp [28].

§2. There are several ways to define an analytic function. Some books (for example, Ahlfors [1]) define a function to be analytic if it has a derivative at every point in an open subset of the plane. Other books (for example, Cartan [10]) define a function to be analytic in an open subset of the plane if at every point of this open set the function has a power series expansion. This latter approach has one advantage in that it is the standard way of defining a function to be analytic in several variables.

In many ways the study of analytic function theory can be considered as the study of the logarithm function. This will become more evident in the remainder of the book.

§3. More information concerning Möbius transformations can be obtained from the book by Carathéodory [9].

CHAPTER IV

§3. See the paper by Burdick and Lesley [8] for more on uniqueness theorems.

§4 and §5. Cauchy's Theorem first appeared in the treatise [11]. However Cauchy's original statement was far different from the one that appears in this book. Cauchy proved his theorem under the assumption that the function f is differentiable and that the derivative is continuous on and inside a simple closed smooth curve. Goursat [20, 21] removed the assumption that f' is continuous but retained the assumption that the curve over which the integral is to be taken is a simple closed smooth curve. Pringsheim [40] introduced the method of proof that is used by many today. He first proved the theorem for triangles (the method used to prove Goursat's Theorem in §8) and then approximated the contours by polygons. The role of the winding number in Cauchy's theorem and the extension to a system of curves such that the sum of the winding numbers with respect to every point outside of the region of analyticity is zero, seems to have first been observed by Artin [4]. The proof of Theorem IV.5.4 is due to Dixon [13].

CHAPTER V

For more examples of the use of residues to calculate integrals see the books by Lindelöf [31] and Mitrinovic [35]. An interesting paper is the one by Boas and Friedman [7].

§3. The reference for Glicksberg's statement of Rouché's Theorem is [18].

CHAPTER VI

The original paper of Phragmén and Lindelöf [38] is still worth reading.

CHAPTER VII

§1. Much of the material in this section can be done in a more general setting. See for example the books by Kelley [27] and Dugundji [15].

§2. Montel's book [36] on normal families is well worth reading. In this treatise he explores several applications of the theory of normal families to problems in complex analysis.

§4. There is a wealth of material on conformed mappings. Carathéodory [9] has some additional information. Also the books by Bergman [5], Goluzin [19], and Sansone and Gerretsen, vol. II [43] have extensive treatments. It is also possible to use Hilbert space techniques to construct conformal maps. See Bergman [5] and the second volume of Hille [26].

§7. For more information on the gamma function look at the book of Artin [3]. Also the paper [30] contains more information about the gamma function. An interesting survey article is one by Davis [12].

§8. The book by Edwards [16] gives a complete exposition of the Reimann zeta hypothesis from an historical point of view. This book examines Riemann's original paper point by point and fully explicates his results. There is a result of Beurling (see the book by Donoghue [14]) that gives an equivalent formulation of the Riemann zeta hypothesis in terms of functional analysis.

CHAPTER VIII

The reference to Grabiner's treatment of Runge's Theorem is [22]. There are several other proofs of Runge's Theorem. One proof is by "pole pushing"; this was used in the earlier edition of this book and also appears in [42]. A proof using functional analysis appears in Rudin's book [41].

CHAPTER IX

§3. An interesting paper on the monodromy theorem is [45].

§6. The book by Springer [44] gives a readable introduction to the theory of Riemann surfaces. A more advanced treatment is [2].

§7. An excellent treatment of covering spaces can be found in Massey [33].

CHAPTER X

Further results on harmonic functions can be found in Helms [25]. This area of harmonic functions has been extended to functions of more than two variables. In addition the Dirichlet problem can be formulated and solved in this more general setting.

CHAPTER XI

The theory of entire functions is one of the largest branches of analytic function theory. The book by Boas [6] is a standard reference.

CHAPTER XII

The book by Hayman [23] contains many generalizations of the theorems presented in this chapter. Also the paper by MacGregor [32] contains many applications and additional results. This paper also contains an interesting bibliography.

There are essentially two ways of proving Picard's Theorem. The elementary approach used in this chapter is based on the treatment found in Landau's book [29]. The other treatment uses the modular function and can be found in [1] or [46].

References

[1] L. V. AHLFORS, *Complex Analysis*, McGraw-Hill Book Co. (1966). QA 331 A45

[2] L. V. AHLFORS and L. SARIO, *Riemann Surfaces*, Princeton University Press (1960).

[3] E. ARTIN, *The Gamma Function*, Holt, Reinhart, and Winston (1964).

[4] E. ARTIN, *Collected Papers*, Addison-Wesley Publishing Co. (1965).

[5] S. BERGMAN, *The Kernel Function and Conformal Mapping*, American Mathematical Society (1970).

[6] R. P. BOAS, *Entire Functions*, Academic Press (1954).

[7] H. P. BOAS and E. FRIEDMAN, *A simplification of certain contour integrals*, Amer. Math. Monthly, 84 (1977), 467–468.

[8] D. BURDICK AND F. D. LESLEY, *Some uniqueness theorems for analytic functions*, Amer. Math. Monthly, 82 (1975), 152–155.

[9] C. CARATHÉODORY, *Theory of Functions* (2 vols.), Chelsea Publishing Co. (1964).

[10] H. CARTAN, *Elementary Theory of Analytic Functions of One or Several Complex Variables*, Addison-Wesley Publishing Co. (1963).

[11] A. CAUCHY, *Mémoire sur les Intégrales Définies entre des Limites Imaginaires*, Bure Freres (1825).

[12] PHILIP J. DAVIS, *Leonhard Euler's Integral: A historical profile of the gamma function*, Amer. Math. Monthly 66 (1959), 849–869.

[13] JOHN D. DIXON, *A brief proof of Cauchy's integral theorem*, Proc. Amer. Math. Soc., 29 (1971), 625–626.

[14] W. F. DONOGHUE, *Distributions and Fourier Transforms*, Academic Press (1969).

[15] J. DUGUNDJI, *Topology*, Allyn and Bacon (1966).

[16] H. M. EDWARDS, *Riemann's Zeta Function*, Academic Press (1974).

[17] W. H. J. FUCHS, *Topics in the Theory of Functions of One Complex Variable*, D. Van Nostrand Co. (1967). QA331 F786

[18] I. GLICKSBERG, *A remark on Rouché's theorem*, Amer. Math. Monthly, 83 (1976), 186.

[19] G. M. GOLUZIN, *Geometric Theory of Functions of a Complex Variable*, American Mathematical Society (1969).

[20] E. GOURSAT, *Démonstration du Théoreme de Cauchy*, Acta Math., 4 (1884), 197–200.

[21] E. GOURSAT, *Sur la Définition Genéral des Fonctions Analytiques d'après Cauchy*, Trans. Amer. Math. Soc., 1 (1900), 14–16.

[22] S. GRABINER, *A short proof of Runge's theorem*, Amer. Math. Monthly, 83 (1976), 807–808.

[23] W. K. HAYMAN, *Meromorphic Functions*, Oxford at the Clarendon Press (1964).

[24] M. HEINS, *Selected Topics in the Classical Theory of Functions of a Complex Variable*, Holt, Reinhart, and Winston (1962). QA331 H662s

[25] L. L. HELMS, *Introduction to Potential Theory*, Wiley-Interscience (1969).

[26] E. HILLE, *Analytic Function Theory* (2 vols.), Ginn and Co. (1959).

[27] J. L. KELLEY, *General Topology*, Springer-Verlag (1975).

[28] K. KNOPP, *Infinite sequences and series*, Dover Publications (1956).

[29] E. LANDAU, *Darstellung und Begründung einiger neuerer Ergebnisse der Funktionentheorie*, Springer-Verlag (1929).

[30] R. LEIPNIK AND R. OBERG, *Subvex functions and Bohr's uniqueness theorem*, Amer. Math. Monthly, 74 (1967) 1093–1094.

[31] E. LINDELÖF, *Le Calcul des Résidues et ses Applications à la théorie des Fonctions*, Chelsea Publishing Co. (1947).

[32] T. H. MACGREGOR, *Geometric Problems in Complex Analysis*, Amer. Math. Monthly 79 (1972), 447–468.

[33] W. MASSEY, *Algebraic Topology: An Introduction*, Springer-Verlag (1977).

[34] CHARLES A. MCCARTHY, *The Cayley-Hamilton Theorem*, Amer. Math. Monthly 82 (1974), 390–391.

[35] D. S. MITRINOVIC, *Calculus of Residues*, P. Noordhoff (1966).

[36] P. MONTEL, *Leçons sur les séries de polynomes*, Gauthier-Villars (1910).

[37] R. NARASIMHAN, *Several Complex Variables*, The University of Chicago Press (1971).

[38] E. PHRAGMÉN AND E. LINDELÖF, *Sur une extension d'un principe classique de l'analyse et sur quelques propriétés des fonctions monogèues dans le voisinage d'un point singulier*, Acta Math. 31 (1908), 381–406.

311

[39] G. Pólya and G. Szegö, *Problems and Theorems in Analysis* (2 vols.), Springer-Verlag (vol. 1, 1972; vol. 2, 1976).

[40] A. Pringsheim, *Über den Goursat'schen Beweis des Cauchy'schen Integralsatzes*, Trans. Amer. Math. Soc., 2 (1901), 413–421.

[41] W. Rudin, *Real and Complex Analysis*, McGraw-Hill Book Co. (1966).

[42] S. Saks and A. Zygmund, *Analytic Functions*, Monografie Matematyczne (1952).

[43] G. Sansone and J. Gerretsen, *Lectures on the Theory of Functions of a Complex Variable* (2 vols.), P. Noordhoff (vol. 1, 1960; vol. 2, 1969).

[44] G. Springer, *Introduction to Riemann Surfaces*, Addison-Wesley Publishing Co. (1957).

[45] David Styer and C. D. Minda, *The use of the monodromy theorem*, Amer. Math. Monthly, 81 (1974), 639–642.

[46] W. A. Veech, *A Second Course in Complex Analysis*, W. A. Benjamin, Inc. (1967).

[47] E. Whittaker and G. N. Watson, *A Course of Modern Analysis*, Cambridge University Press (1962).

INDEX

LIST OF SYMBOLS

Graduate Texts in Mathematics

continued from page ii